AMERICAN PROFESSORS

AMERICAN PROFESSORS

A National Resource Imperiled

Howard R. Bowen

Jack H. Schuster

New York   Oxford
OXFORD UNIVERSITY PRESS
1986

Oxford University Press

Oxford   London   New York   Toronto
Delhi   Bombay   Calcutta   Madras   Karachi
Kuala Lumpur   Singapore   Hong Kong   Tokyo
Nairobi   Dar es Salaam   Cape Town
Melbourne   Auckland

and associated companies in
Beirut   Berlin   Ibadan   Mexico City   Nicosia

Library of Congress Cataloging in Publication Data
Bowen, Howard Rothmann, 1908—
American professors.
Bibliography: p.
Includes index.
1. College teachers—United States.
2. College teaching—Vocational guidance—United States.
I. Schuster, Jack H.   II. Title.
LB2331.72.B67   1986   378'.12'0973   86–7265
ISBN 0–19–503693–X

Printing (last digit): 9 8 7 6 5 4 3 2 1

Printed in the United States of America
on acid-free paper

Preface

The nation's college and university faculties occupy a strategically important place in contemporary society. As educators, they directly influence the personal development and ideals of a large fraction of each successive generation, and they prepare these same people for a wide range of vocations including virtually all the positions of leadership and technical competence in our society. As researchers and scholars, they are at the forefront in the advancement of learning and culture. And in their roles both as teachers and researchers, they contribute to the broad economic growth and military strength of the nation. They are a major resource which the nation cannot afford to neglect or let waste away, as now seems to be a distinct possibility. Rather, the nation should be augmenting and strengthening this resource.

Our decision to write this book derived from our concern about the many reports of a slow deterioration in the condition of the nation's faculties and a slow decline in their morale. We hoped not only to assemble the facts, but also to offer recommendations for any corrective action that might be warranted by the facts.

After the introduction (Chapter 1), the book is divided into four main sections. The first (Chapters 2–4) is an introduction to the 700,000 people who make up the American faculties. It describes personal backgrounds, characteristics, and values — information that is needed to understand the current condition of the faculties and their aspirations and prospects for the future. The second section (Chapters 5–8) traces trends in the compensation and working conditions of the faculties and describes their situation as of 1985. This section is in part a report on our more than 500 interviews conducted at 38 varied colleges and universities located in all parts of the country. The third section (Chapters 9–11) is an analysis of the academic labor market. It analyzes the flow of people into and out of the academic profession, including the exceptionally brilliant people who are important to higher education and to the nation far beyond their numbers. This third section also provides estimates of the number of faculty positions that must be filled by our colleges and universities over the 25 years from 1985 to 2010. The fourth and concluding section (Chapters 12–13) contains our policy recommendations for colleges and universities and for federal and state government.

The book was conceived at two meetings of interested persons convened by TIAA–CREF[1] in Washington on December 11, 1981, and April 1–2, 1982. Following those meetings, TIAA–CREF and AAHE[2] agreed to sponsor the study and since then have been extraordinarily helpful in many ways. A formal proposal was prepared and presented to the Carnegie Corporation of New York as the lead funding agency on June 10, 1982. Meanwhile, the project was presented to The Ford Foundation and an understanding was reached that the two foundations would jointly underwrite the project. Later the Exxon Education Foundation made significant contributions toward the administrative costs of the study and TIAA–CREF added a modest sum for designated purposes. In total, the combined funds amounted to $206,000. We acknowledge with deep appreciation the financial contributions of the several sponsoring and funding organizations and for the advice and encouragement of the officials of these organizations: Peggy Heim and William Slater of TIAA–CREF, Russell Edgerton of AAHE, E. Alden Dunham of the Carnegie Corporation, Sheila Biddle and Gladys Hardy of The Ford Foundation, and Robert L. Payton of the Exxon Education Foundation. We wish to acknowledge also the contribution of the Claremont Graduate School through President John Maguire who provided space, support, and encouragement without charge for overhead.

During the course of the project there were two meetings of an informal advisory group, one on February 8, 1983, and the other on October 11–12, 1984. The frank suggestions and criticisms emanating from those meetings were enormously helpful. We cannot claim to have met all the criticisms but the manuscript was infinitely improved by the suggestions of this group. The persons who attended at least one of the two initial meetings or either of the two later meetings were:

Sheila Biddle, The Ford Foundation
David Breneman, Kalamazoo College
David Brown, University of North Carolina, Ashville
E. Alden Dunham, Carnegie Corporation of New York
Russell Edgerton, American Association for Higher Education
Doris Fitzgerald, Borough of Manhattan Community College
W. Todd Furniss, American Council on Education
W. Lee Hansen, University of Wisconsin
Gladys Chang Hardy, The Ford Foundation
Peggy Heim, Teachers Insurance and Annuity Association
Richard Johnson, Exxon Education Foundation
Barbara Lee, Rutgers University
Marjorie Lightman, Institute for Research in History
Clara Lovett, George Washington University

---

[1] Teachers Insurance and Annuity Association and College Retirement Equities Fund.
[2] American Association for Higher Educaiton.

Preface

The nation's college and university faculties occupy a strategically important place in contemporary society. As educators, they directly influence the personal development and ideals of a large fraction of each successive generation, and they prepare these same people for a wide range of vocations including virtually all the positions of leadership and technical competence in our society. As researchers and scholars, they are at the forefront in the advancement of learning and culture. And in their roles both as teachers and researchers, they contribute to the broad economic growth and military strength of the nation. They are a major resource which the nation cannot afford to neglect or let waste away, as now seems to be a distinct possibility. Rather, the nation should be augmenting and strengthening this resource.

Our decision to write this book derived from our concern about the many reports of a slow deterioration in the condition of the nation's faculties and a slow decline in their morale. We hoped not only to assemble the facts, but also to offer recommendations for any corrective action that might be warranted by the facts.

After the introduction (Chapter 1), the book is divided into four main sections. The first (Chapters 2–4) is an introduction to the 700,000 people who make up the American faculties. It describes personal backgrounds, characteristics, and values — information that is needed to understand the current condition of the faculties and their aspirations and prospects for the future. The second section (Chapters 5–8) traces trends in the compensation and working conditions of the faculties and describes their situation as of 1985. This section is in part a report on our more than 500 interviews conducted at 38 varied colleges and universities located in all parts of the country. The third section (Chapters 9–11) is an analysis of the academic labor market. It analyzes the flow of people into and out of the academic profession, including the exceptionally brilliant people who are important to higher education and to the nation far beyond their numbers. This third section also provides estimates of the number of faculty positions that must be filled by our colleges and universities over the 25 years from 1985 to 2010. The fourth and concluding section (Chapters 12–13) contains our policy recommendations for colleges and universities and for federal and state government.

The book was conceived at two meetings of interested persons convened by TIAA–CREF[1] in Washington on December 11, 1981, and April 1–2, 1982. Following those meetings, TIAA–CREF and AAHE[2] agreed to sponsor the study and since then have been extraordinarily helpful in many ways. A formal proposal was prepared and presented to the Carnegie Corporation of New York as the lead funding agency on June 10, 1982. Meanwhile, the project was presented to The Ford Foundation and an understanding was reached that the two foundations would jointly underwrite the project. Later the Exxon Education Foundation made significant contributions toward the administrative costs of the study and TIAA–CREF added a modest sum for designated purposes. In total, the combined funds amounted to $206,000. We acknowledge with deep appreciation the financial contributions of the several sponsoring and funding organizations and for the advice and encouragement of the officials of these organizations: Peggy Heim and William Slater of TIAA–CREF, Russell Edgerton of AAHE, E. Alden Dunham of the Carnegie Corporation, Sheila Biddle and Gladys Hardy of The Ford Foundation, and Robert L. Payton of the Exxon Education Foundation. We wish to acknowledge also the contribution of the Claremont Graduate School through President John Maguire who provided space, support, and encouragement without charge for overhead.

During the course of the project there were two meetings of an informal advisory group, one on February 8, 1983, and the other on October 11–12, 1984. The frank suggestions and criticisms emanating from those meetings were enormously helpful. We cannot claim to have met all the criticisms but the manuscript was infinitely improved by the suggestions of this group. The persons who attended at least one of the two initial meetings or either of the two later meetings were:

> Sheila Biddle, The Ford Foundation
> David Breneman, Kalamazoo College
> David Brown, University of North Carolina, Ashville
> E. Alden Dunham, Carnegie Corporation of New York
> Russell Edgerton, American Association for Higher Education
> Doris Fitzgerald, Borough of Manhattan Community College
> W. Todd Furniss, American Council on Education
> W. Lee Hansen, University of Wisconsin
> Gladys Chang Hardy, The Ford Foundation
> Peggy Heim, Teachers Insurance and Annuity Association
> Richard Johnson, Exxon Education Foundation
> Barbara Lee, Rutgers University
> Marjorie Lightman, Institute for Research in History
> Clara Lovett, George Washington University

---

[1] Teachers Insurance and Annuity Association and College Retirement Equities Fund.
[2] American Association for Higher Educaiton.

Robert McCabe, Miami-Dade Community College
Wilbert McKeachie, University of Michigan
E. Arnold Shore, Exxon Education Foundation
William T. Slater, TIAA–CREF
Hoke Smith, Towson State University
Wendell I. Smith, Bucknell University
William Toombs, The Pennsylvania State University
Marjorie Wagner, Santa Rosa, California
Willard Wirtz, National Institute for Work and Learning

We acknowledge also many persons who gave us invaluable assistance in a variety of ways. W. Lee Hansen of the University of Wisconsin wrote chapter 6 on Faculty Compensation. David Alexander of Pomona College and Kenneth M. Greene of the United Chapters of Phi Beta Kappa opened up important data sources relating to exceptionally talented persons. Howard Brooks of Claremont McKenna College, Martin Finkelstein of Seton Hall University, Patricia Foster of Loma Linda University, and Wilbert McKeachie of the University of Michigan conducted interviews with faculty and administrators in colleges and universities around the nation. Patricia Foster and Nancy McGlothin of Scripps College carried out detailed analyses of the interviews.

In the course of preparing this book, extensive literature reviews were conducted by four young scholars as follows:

Part-time Faculty, by Eileen Heveron of The Claremont Graduate School
The Female Professoriate, by Barbara Light of Whittier College
Working Conditions of the Faculty, by Sally Loyd of the Chancellor's Office, California State University
Faculty Performance and Social and Political Attitudes, by Roberta Stathis-Ochoa of California State University, San Bernardino.

These reviews in manuscript form are available postpaid on request to the Faculty in Education, The Claremont Graduate School, Claremont, California 91711 for $10.00 each.

In addition to the formally scheduled interviews, we discussed our project at length with literally scores of authorities on higher education. We would like to recognize the following persons who were especially generous with their time and ideas: Helen Astin, University of California (Los Angeles), Nancy Beck of Educational Testing Service, Herman Blake of Tougaloo College, Thomas Bowen of California State University (Fresno), John Brademas of New York University, Joseph Burton of American Physical Society, Patrick Callan of California Postsecondary Education Commission, John A. Centra of Syracuse University, Burton Clark of the University of California (Los Angeles), Cleveland Dennard of Atlanta, Charles Dickens of the National Science Foundation, Norman Francis of Xavier University of Louisiana, Dennis Gooler of

Northern Illinois University, Kenneth C. Green of the University of California (Los Angeles), Katherine Hanson of the Consortium on the Financing of Higher Education, Rodney Hartnett of Rutgers University, Roger W. Heyns of the Hewlett Foundation, William C. Kelly of the National Research Council, Laura Kent of Santa Monica, Clark Kerr of the University of California, Robert E. Klitgaard of Harvard University, Thomas Linney of the Council of Graduate Schools in the U.S., Seymour Martin Lipset of Stanford University, Hans O. Mauksch of the University of Missouri, Michael McPherson of the Brookings Institution, Walter Metzger of Columbia University, John U. Monro of Tougaloo College, William Muir of the University of Regina, Richard Peterson of the Educational Testing Service, Eugene Rice of University of the Pacific, Kenneth P. Ruscio of the University of California (Los Angeles), Rudolph W. Schulz of The University of Iowa, Laure M. Sharp of the Bureau of Social Science Research, Neil Smelser of the University of California (Berkeley), Irving Spitzberg of the Association of American Colleges, Sidney G. Tickton of the Academy for Education Development, John William Ward of the American Council of Learned Societies, and William Zumeta of the University of California (Los Angeles).

During the course of the study, we benefited from materials provided by thirty-eight discipline-linked professional associations. We visited the national office of twelve of them in New York and Washington, D.C.

A group of our colleagues from The Claremont Colleges were especially helpful in the planning phase of the project and some of them in the final preparation of the manuscript. These included Professors Gordon Bjork, Gordon Douglass, David Drew, Langdon Ellsbree, and Charles Kerchner.

Not least we would express special appreciation to co-workers at the Claremont Graduate School, Dorothy Pearson and Ethel Parker, who performed with incredible skill, perseverance, and good will in preparing many versions of the manuscript, and in keeping our offices running. They were assisted by Marjorie Doken and Viola Keen to whom we are also greatly indebted.

And finally we would convey our gratitude to our wives Diane Schuster and Lois Bowen who have been exceedingly understanding and supportive.

Claremont, California   Howard R. Bowen

1985   Jack H. Schuster

Table of Contents

PART THREE / THE ACADEMIC LABOR MARKET 163

AMERICAN PROFESSORS

CHAPTER ONE

Introduction: A National Resource Imperiled

Will America's colleges and universities be able to deliver teaching and learning of acceptable quantity and quality over the next several decades? The answer will depend partly on external factors such as the finances and governance of high education and the preparation of secondary students. But perhaps the most critical influence of all will be the caliber of the faculties. Without going to the extreme of asserting that "the faculty *is* the university," the talent, training, vitality, and social conscience of the faculties are critical ingredients of the power of each college or university to deliver acceptable teaching and learning. The main duty of every institution of higher education is to place a competent faculty, and the scholarly environment they create, at the disposal of students, and to provide the facilities and encouragement needed by that faculty. The excellence of higher education is a function of the kind of people it is able to enlist and retain on its faculties (Feldman & Newcomb, 1976 pp. 251–66; Ayres & Bennett, 1983).

The nation's faculties are entrusted with the education of about a third to a half of every age cohort of young people, and they touch the lives of millions of other persons in less intensive encounters. They train virtually the entire leadership of the society in the professions, government, business, and, to a lesser extent, the arts. Especially, they train the teachers, clergy, journalists, physicians, and others whose main function is the informing, shaping, and guiding of human development. The nation depends upon the faculties also for much of its basic research and scholarship, philosophical and religious inquiry, public policy analysis, social criticism, cultivation of literature and the fine arts, and technical consulting. The faculties through both their teaching and research are enormously influential in the economic progress and cultural development of the nation.[1] In short, the faculties are

1. Incidentally, they supply the administration leadership of higher education. Four-fifths of all presidents and deans were once faculty members (Moore et al., 1983, pp. 509, 512).

3

a major influence upon the destiny of the nation, and the nation has a clear and urgent interest in assembling and maintaining faculties having adequate numbers of talented, well-trained, highly motivated, and socially responsible people.[2]

Since World War II, American higher education has undergone massive growth and profound changes, and these have affected—some favorably and some unfavorably—the work environment of faculties and their motivations, satisfactions, and compensation. The overall growth in enrollments has been prodigious. Many new colleges and universities have been established, including new kinds of institutions, especially community colleges and regional universities. Most institutions have grown in size and some have become impersonal and bureaucratic. The mix of public and private higher education has changed as enrollment growth in the public sector has outstripped that in the private sector. Research, scholarship, public service, and graduate study have all grown apace, especially under the stimulus of federal support.

The growth has been accompanied by other sweeping changes. The makeup of the student population has been altered greatly as relatively more older persons, more women, more minority persons, and more part-time commuters have been attending college. Many of the "new" students have come to college with insufficient secondary school preparation, and compensatory education has become increasingly necessary. New relationships with students have emerged. The ancient concept of *in loco parentis* has been largely abandoned and students have experimented with political activism. The finances of institutions have become more closely tied to enrollments. New modes of external governance for public colleges and universities have been established and institutional autonomy has been eroded significantly. Meanwhile, the participation of both students and faculty in institutional governance has changed—in some respects expanded and in others contracted. New educational technologies and practices have been adopted, vocational training has been greatly expanded—often at the expense of liberal learning—and curricula have proliferated. In this atmosphere the faculties have become more technical and specialized. Of the many changes, some have been constructive—particularly those which have been part of the process of opening opportunity to all classes of the society. Some changes may have been undesirable but probably unavoidable in a period of enormous growth.

During the period from about 1955 to 1970, higher education gained in public esteem and prospered financially. The academic profession then became an attractive magnet for capable and ambitious young men and women. But beginning in the early 1970s, the higher education community entered a period of prolonged financial

2. Cf. H. R. Bowen, 1982, p. 10.

stringency, the burden of which has fallen heavily on faculty compensation and working conditions. Since then, the attractiveness of the academic profession, both to those contemplating entry and to those already in, has diminished noticeably. It is not difficult to understand why this is so, for it is no exaggeration to suggest that nearly a century of progress for faculty, however uneven, seems now to have been interrupted, even reversed.

When Christopher Jencks's and David Riesman's *The Academic Revolution* was published in 1968, it was justly regarded as an insightful commentary on many developments in American higher education. The "revolution" in the book's title did not refer, as one might suppose from the vantage point of the mid-1980s, to student militancy in an era marked by the Free Speech Movement and campus demonstrations against escalating American involvement in Southeast Asia. Rather, the revolution of which they spoke was more properly an evolution. It was the gradual process by which the faculties of most institutions, over many decades of struggle and frustration, had gained increasing control over the academic core. While faculty frustrations, from causes real and imagined, had hardly vanished, the faculty nevertheless had "arrived." By the 1960s, the faculty clearly had become a dominant influence in academic matters. The authors were careful to distinguish between formal power, which the faculty had not obtained, and less formal influence. Their point was that the faculty had come to wield sufficient influence to enable them efficaciously to shape academic content and the selection of new colleagues.

Our view does not conflict with theirs for the period up to 1970, but about then, the influence of the faculty began to wane. The causes were multiple. What happened to faculty was the result of powerful demographic, economic, and political forces at work in society. We do not see the American faculties as victims of orchestrated attempts by others to claim control over faculty "turf." Rather it was the larger social forces, coupled with more than a little benign neglect, that contributed to the deteriorating condition of the American faculty.

Consider some of the changes since around 1970 that have had special pertinence to faculty and, accordingly, the attractiveness of academic careers. At that point, the momentum of rapid expansion throughout higher education was still much in evidence; academe was widely regarded as a "growth industry." Faculty compensation had risen sharply over the preceding decade. Faculty members were highly visible in U.S. presidential administrations throughout the 1960s; indeed, the sense was not uncommon that colleges and universities could be instrumental in helping to overcome deeply rooted social problems. Despite the battering many campuses took from public opinion and state executive and legislative leaders in the aftermath of persistent student demonstrations, most would regard 1970 as having been, on the whole, a good time for the American faculty: their real income was approaching historically high levels and they enjoyed unprecedented authority on their own campuses. A mere decade and a half later, the situation for faculty had in many ways deteriorated. Several developments help to illustrate the predicament of the professoriate.

Throughout the period from 1970 to 1985 persistent prophecies of sharp enrollment declines were assaulting the faculty with no letup in sight. Although by 1985 enrollments had barely begun to decline, the faculties at most institutions had been bracing themselves for substantial retrenchment. But it was the persistently gloomy outlook rather than the reality that cast a pall over faculty morale. As of 1985, however, a harsh reality seems more imminent and faculty members are deeply concerned about job security.

During the 1970s, real faculty earnings had been growing and in fact peaked in 1972–73. But, after that, real earnings (adjusted for inflation) not only slipped, they tumbled. Faculty fared worse than all but a handful of other occupational groups. While real earnings were falling during the period 1972 to 1985, the work environment was becoming less satisfactory in many respects. One reaction of faculty to these changes was to form unions and to engage in collective bargaining. In 1970 the faculty unionization movement had just begun to pick up momentum; by 1985 it had spread to over 700 campuses and embraced more than a quarter of all full-time faculty members.

Another development of concern to the corps of career faculty members was a startling increase in part-time faculty. Administrators, confronted with substantial enrollment shifts and economic constraints, welcomed part-timers for the financial relief and flexibility they afforded. In 1970–71 the ratio of full- to part-time faculty stood at 3.5 to 1; by 1982–83 the ratio had fallen to 2.1 to 1, and part-timers constituted about a third of the American faculty.

Finally, among the dramatic developments that have worried many faculty members have been massive shifts in enrollments among academic disciplines. Students, voting with their feet and presumably with images of their future bank accounts in mind, have abandoned the traditional arts and sciences in droves. The end of that flight was not yet in sight in 1985. This development put the traditional core of liberal arts faculty at many institutions in a defensive mode. At most campuses they have fought a rear-guard action in an often unsuccessful effort to prevent serious erosion of liberal education.

No one of these developments would suggest that the American professoriate was *in extremis*; considered together, however, they provide ample reason for serious concern. The picture has a brighter side, however. Over the period 1970 to 1985, American colleges and universities opened their doors to millions of minorities, women, older students, low-income persons, the handicapped, and other non-traditional students. During this period there was a great leap forward in the democratization of higher education. Also, part of this brighter side is that despite their manifest worries and dissatisfactions, the faculty have not defected in large numbers. Though the normal turnover has continued—some new faculty entering academe each year and some leaving for any of a variety of reasons—there have been no stampedes out of the academic profession. Faculty members have entered academe embracing certain values associated with academic life. And for the vast majority, even for those still retaining mobility in a tight labor market, they have opted to cling to the setting and lifestyle colleges and universities offer.

But the pressures on faculty have not relented. Nor are they likely to lessen for some years to come. In all, the changes pervading higher education have had profound effects upon the academic profession. They have brought about marked changes in the working conditions, attitudes, expectations, and possibly in the performance of the faculties. In the long run, they will affect the kinds of people attracted to and retained by the academic profession.

As of 1985, higher education is in an anxious period because of the downturn in the number of high school graduates and the uncertainty about prospective financial support. The financial outlook is less favorable than at any time since 1955, and the conditions and expectations of faculties are correspondingly bleak.

Our findings indicate that fewer and fewer persons, especially highly talented young students, are opting for academic careers. Indeed, there is serious risk that academic careers will become less attractive for highly able young people over the next ten years and more. That brings us to the nexus of the problem. This apparently sharp drop-off in interest among highly able persons, however understandable, comes at a time when we can foresee, perhaps a decade from now, numerous openings for new faculty. When enrollments begin to recover and retirements mount, colleges and universities, as always, will strive to hire the best available faculty. But unless the conditions of faculty employment improve and incentives for entering the profession become stronger, they may well be obliged to settle for the not very talented. The openings will be filled, one way or another; there has never been an insufficiency of persons—leaving aside the adequacy of their preparation—willing to enter upon academic careers. The question concerns the quality of the persons available at that time.

Given the history and the outlook, the purpose of this book is to address two questions relating to the future of American higher education over the next several decades. First: Will our colleges and universities be able to maintain an appropriate professoriate in the sense of attracting and holding people well qualified for the responsibilities they will be called upon to bear? Second: If not, what should be done to assure that the professoriate of the future is capable of meeting its responsibilities?

Many critically important issues are embedded in these broad questions. For instance, will higher education be able to attract its share of the inevitably tiny pool of exceptionally gifted and creative scientists, intellectuals, and artists who are important far beyond their numbers? And will higher education be able to make progress toward diversifying the faculties with respect to gender, ethnicity, and age?

The search for answers to these questions leads us to a wide-ranging study of faculty—what they do, who they are, how they are compensated, what their working conditions are, how well they do their work, and the circumstances of their entry to and exit from the profession. From these explorations, we hope to gain insights into the outlook for the professoriate and to offer recommendations for officials of federal and state governments, the trustees and administrative officials of colleges and universities, the professoriate itself, and the general public. Our intent is to approach these matters from the point of view of the public interest, not in terms of the narrow interests of the profession or of the colleges and universities they serve.

The academic profession stands at a crossroads. Our colleges and universities, and society at large, are faced with critical choices that must be made in the next few years. These decisions can lead toward a revitalized faculty fully capable of meeting its considerable responsibilities, or to neglect and perhaps irreparable damage to the nation.

Part One

THE CAST OF CHARACTERS

This section presents a profile of the American professoriate. It introduces the people who are members of the academic profession and thus sets the stage for the analysis that follows. Among the topics considered are the tasks and capabilities of faculty people, their personal backgrounds, their attitudes and values, emergent sub-groups within the faculty, and their effectiveness in terms of the quantity and quality of their work.

CHAPTER TWO

Faculty Tasks and Talents

Our purpose in this chapter is to introduce the people who are the subjects of this book. We speak of them individually as "faculty member" or "faculty person" and collectively as "the faculty" or "the professoriate." What we have to say about them is well known to a few academics but is perceived only vaguely in the consciousness of the rank and file of the professoriate and known scarcely at all to the general public. Even when it is known, it is too often left unstated.

The word "faculty" as we use it refers to a corps of professional persons of substantial learning who are employed within American institutions of higher education and are engaged directly in teaching, research, related public service, institutional service, or combinations of these. We include both those who serve full time and those who serve part time. In this connection, the term "institution of higher education" refers to colleges and universities that may be public or private, but are not-for-profit and serve mainly post-secondary students. This definition leaves out a great many learned people who teach or conduct research. It excludes teachers or researchers who serve in institutions other than colleges and universities, for example, in elementary and secondary schools, proprietary schools, the armed forces, government agencies, business corporations, churches, research institutes, and the media. It also leaves out freelance scholars, authors, artists, and others, and excludes unemployed persons who have been or aspire to be members of the professoriate. Associations between faculty members and intellectuals who work outside the academic community are often close, and there is constant movement of persons back and forth between colleges and universities and other institutions or occupations. But only those who are at any given time connected with colleges and universities are considered to be "faculty".

Within higher education, our definition includes only persons who are engaged directly in teaching, research, or public service and hold the rank of instructor or above. It excludes administrators, specialists such as librarians and technicians, graduate assistants, and non-professional people employed in clerical or secretarial work, food service, housekeeping, plant maintenance, and the like.[1] The faculty as we define it included in 1982 about 700,000 persons (Table 2–1), or about a third[2] of the total employees of our colleges and universities.

Frequently it is assumed, given the depressed conditions in higher education since the early 1970s, that the number of faculty members has been declining sharply either because vacancies, as they occurred, have not been filled or because numerous faculty members have been laid off. These assumptions are incorrect. In fact, from 1970–71 on, as shown in Table 2–1, faculty numbers have increased year after year almost without exception until 1983–84. Since then, however, faculty numbers have leveled off and probably declined slightly.

It is probable that the number of faculty members will decline gradually over the next few years. We expect the downward trend to continue until about 1995–96 (see Chapter 10 and Appendix D).

Within the ranks of faculty there is great variety. It is due in part to the individualistic personalities of faculty members. It is due also to the wide range of activities that different faculty members are called upon to perform. These vary with the type, size, governance, and affluence of the institutions they serve, and especially with the programs offered and the academic disciplines represented. The variety is also due to differences in gender, age, ethnicity, tenure, rank, and part-time versus full-time status. The professoriate is a mixture of many varied personalities and activities. And as we point out in Chapter 4 the faculty is becoming more diverse. Yet the vast majority of faculty members have much in common. Most have had a similar education culminating in advanced and extended study in perhaps 100 to 150 universities. They share an interest in teaching and learning and, with allowance for differences in disciplines, they conduct their teaching and research in comparable ways. The routines of campus life are remarkably similar over most of the American higher educational system. Faculty members are part of a nationwide communications network consisting of voluntary accreditation, professional associations, public licensing bodies, statewide coordination, multi-campus systems, consortia of institutions, and foundations and governmental granting agencies, and these impose or encourage considerable uniformity. Also there is much communication among institutions because of the mobility of faculty from one college or university to another. Finally, common attitudes and values arise in part because faculties in colleges of lesser prestige tend to emulate those in institutions of greater prestige.

1. Some of these people may engage part time in instruction and research and if so are counted as part-time faculty.

2. Computed from data in National Center for Education Statistics, *Digest of Education Statistics, 1983*, p. 102.

Table 2–1. Number of Faculty* in Institutions of Higher
Education, 1970–71 to 1982–83 (in thousands)

	Total	Full-Time Actual	Part-Time	Full-Time Equivalent
1970–71	474	369	104	402
1971–72	492	379	113	414
1972–73	500	380	120	417
1973–74	527	389	138	433
1974–75	567	406	161	457
1975–76	628	440	188	501
1976–77	633	434	199	501
1977–78	650	447	203	514
1978–79	647	445	202	513
1979–80	657	451	206	520
1980–81	678	466	212	537
1981–82	704	480	224	555
1982–83	713	485	228	561
1983–84	702	477	225	552
1984–85	680	464	216	536
1985–86	663	453	210	523

* Instructor and above.

SOURCE: National Center for Education Statistics, *Projections*, 1982, p. 88–89, and *Digest of Education Statistics*, 1983, pp. 103. Figures for 1979–80 to 1985–86 estimated by the National Center for Education Statistics.

On individual campuses, faculty members from different disciplines are drawn together through committees, senates, and social activities. Further, in the higher educational community there is a somewhat vague but influential conception of what the educated man or woman should be like and of which curricula, degree requirements, and campus extracurricular life are conducive to that ideal. Also the faculties in most colleges and universities engage in the consideration and analysis of values. They are involved in social and artistic criticism. They conduct philosophical systems and ideologies and they appraise existing social policies and recommend new ones. Through all these processes, the academic community creates an ethos. This ethos is not promulgated officially; it is certainly not shared by all the professoriate; it often differs from views prevailing among the general public, and it changes over time. Yet one can say that the weight of academic influence in any given period is directed toward a particular world outlook. Thus, though most faculty members enjoy considerable freedom in their work, and though there are substantial differences among them, it is not wholly outrageous to speak of an academic community as a nationwide (or even worldwide) subculture. And, despite the variety that exists in academe, it is appropriate for many purposes to treat the professoriate as a closely knit social group and not merely as a collection of disparate individuals or unrelated small groups.[3]

3. For a concise description of the academic career, see: Careers, Inc., 1984.

The remainder of this chapter and the three next chapters will be devoted to a review of the activities, characteristics, and attitudes of the American professoriate. We warn the reader to be slightly skeptical because the available data about faculty are spotty and sometimes out of date. Furthermore, they are difficult to interpret because they are unstandardized. Definitional categories are seldom clearly defined and vary from one source to another.[4]

The Tasks of the Faculty

The faculties in American colleges and universities mostly conduct their work inconspicuously and without much public notice or acclaim. Their work, and its significance, is not widely observed, understood, or appreciated. Faculty members seldom become celebrities. Even Nobel Prize winners, members of the several national academies, or acknowledged great teachers are usually unknown except among members of their own academic fields and among colleagues and students on their own campuses.

There are several explanations of this anonymity. Much of higher learning is specialized and technical and is incomprehensible to the general public. Teaching is usually conducted privately and not as a public event witnessed by many people. The scholarly and research work of faculty goes on in privacy, even solitude. Moreover, knowledge advances through frequent but small accretions. Authentic breakthroughs that command public attention are infrequent. Furthermore, the style of life of most faculty members is quiet and simple. Indeed, they could not afford a notorious style of life even if so inclined. The exceptions to the rule of anonymity are mainly those faculty members who distinguish themselves in literature and the arts, or become journalists and television personalities, or enter politics, or serve as athletic coaches, or become college presidents. And for this small minority, public fame is achieved by departing from conventional academic endeavors. For most of the professoriate, "fame" consists of the gratitude of students for good teaching and counseling, the respect of immediate colleagues and, above all, national or international recognition of scholarly attainments by members of one's own discipline. The anonymity of most faculty members by no means implies that what they do is unimportant or that it does not demand a high level of talent. Quite the contrary. What they do is extraordinarily important and demanding and should be more widely understood. Therefore, it may

4. Frequent problems are that different statistical studies cover different categories of institutions or different categories of faculty, without designating the specific coverage. For example, some include two-year institutions and some do not; some include part-time faculty or assistants or administrators and some do not. Some data are based on full-time equivalent faculty, others on head counts.

Table 2–2. Faculty Involvement in Four Tasks, by Type of Institution

	Percentage of Faculty Time				
	Instruction	*Research*	*Public Service*	*Institutional Governance and Operation*	*Total*
Private universities	55	25	4	16	100
Public universities	55	22	5	18	100
Elite liberal arts colleges	65	14	3	18	100
Public comprehensive institutions	68	10	5	17	100
Public colleges	68	10	4	18	100
Private liberal arts colleges	68	9	5	18	100
Community colleges	70	5	4	21	100
All institutions	64	14	4	18	100

SOURCE: Baldridge *et al.*, 1978, p.103.

be worthwhile to explore what they actually do and what qualifications are required.

The work of college and university faculties may be divided into four overlapping tasks: instruction, research, public service, and institutional governance and operation. Almost all of the work done by academicians may be assigned to one or more of these four categories. Table 2–2 shows the estimated percentage of time devoted to each of these four tasks.

Instruction. The main function of faculties is instruction, that is, direct teaching of students. All but a few engage at least to some extent in instruction, and the great majority consider it to be their primary responsibility (Warren, 1982, pp. 18–19). Even at universities where research figures very prominently in academic reward systems, faculty on the whole spend most of their time teaching. Instruction involves formal teaching of groups of students in classrooms, laboratories, studios, gymnasia, and field settings. It also involves conferences, tutorials, and laboratory apprenticeships for students individually. In addition, it requires time-consuming backup work. Faculty members must keep up with the literature of their fields, they must prepare for their classroom presentations and discussion, and they must appraise the work of their students.

Instruction also entails advising students on matters pertaining to their current educational programs, plans for advanced study, choice of career, and sometimes more personal matters. And this advisory work leads naturally to numerous student requests for letters of recommendation and assistance in placement either in advanced study or jobs. When a faculty member is teaching as many as five to eight courses a year, as is common, and is in personal contact in any one term with as many as 50 to 300 students (in some cases more), these tasks become time-consuming and arduous.

Finally, in recounting the instructional duties of faculty, one must include their responsibility to serve as exemplars whom students might emulate (Carnegie Foundation for the Advancement of Teaching, 1977, pp. 82–86). Serving in this capacity is not something that faculty members must consciously work at beyond merely being themselves; but it is a role that many of them occupy—whether they will it or not and whether their influence is positive or negative. In presenting themselves to students they become living representatives of what their particular colleges or universities stand for. Whether intended or not, the character they display represents the values the institution is setting before its students. As Cardinal Newman observed a century ago:

> The general principles of any study you may learn by books at home; but the details, the color, the tone, the air, the life which make it live in us, you must catch all these from those in whom it lives already.

Research.[5] Faculties contribute to the quality and productivity of society not only through their influence on students but also directly through the ramified endeavors we call "research." This term is used as shorthand for all the activities of faculties that advance knowledge and the arts. These include humanistic scholarship, scientific research in the natural and social sciences, philosophical and religious inquiry, social criticism, public-policy analysis, and cultivation of literature and the fine arts. These activities may be classed as "research" if they involve the discovery of new knowledge or the creation of original art and if they result in dissemination usually by means of some form of durable publication.[6] Only through dissemination do they become a significant advancement of knowledge or the arts. A brief comment on each of these activities may be in order.

The work of humanistic scholarship includes the discovery and rediscovery of past human experience, the preservation of texts and artifacts, the constant interpretation and reinterpretation of the knowledge acquired, and transmission of this knowledge from generation to generation. These activities are of inestimable value because they enable each generation to understand itself and its problems in the perspective of human experience.

Scientific research is divided into two overlapping categories: basic research which is intended to discover the laws of nature regardless of practical applicability; and applied research which is intended to discover ways of putting this knowledge to practical use. The distinction

5. This section draws substantially upon H. R. Bowen, 1977, chap. 10.
6. Or presentation in the case of sculpture, painting, and the like. It should be noted also that the results of research are sometimes disseminated by word of mouth.

Table 2–3. Faculty Involvement in Research, by Type and Quality of Institution

	Percentage of Faculty Members		
	with Heavy Involve-ment in Research	*with One or More Publications in Past Two Years*	*with Teaching Loads of 10 Hours or Less in Class per Week*
Universities			
High quality	50	79	91
Medium quality	40	72	85
Low quality	28	57	72
Four-Year Colleges			
High quality	26	54	72
Medium quality	12	37	52
Low quality	10	29	42
Two-Year Colleges	5	14	22
All Institutions	24	48	60

SOURCE: Fulton and Trow (in Trow, 1975, pp. 44, 46, 48).

between basic and applied is not sharp, but it is important in understanding the role of college and university faculty. They have historically emphasized the pursuit of basic knowledge in quest of "truth" rather than applied knowledge in search of results. Faculty members, on the whole, have been free to choose their research programs in terms of intrinsic scientific interest rather than practical outcome. Experience shows that this freedom of choice often produces outcomes of great though unplanned practical value. For example, penicillin, hybrid seed corn, the computer, polio vaccine, and the "pill" were direct results of basic research conducted in universities.

Higher education has, of course, no monopoly on research. It is conducted also by business, government, private research institutes, labor unions, libraries, museums, performing arts organizations, freelance scholars and writers, and others. Doubtless if colleges and universities should divest themselves of the research function, other organizations could fill the gap. Nevertheless, many members of the higher educational community have for nearly a century engaged in research as an activity complementary to and supportive of their main function, which is instruction. College and university faculties occupy a significant, probably dominant, position in the intellectual and artistic life of the nation.

It is sometimes assumed that academic research is confined largely to major universities. Though they are more heavily committed than other higher educational institutions, all sectors of higher education— even two-year colleges—are involved at least to some degree (see Table 2–3). The nature of this involvement is shown in Table 2–4. Also, the

Table 2–4. Median Percentage of Faculty
Involved in Research in Various Ways, 1972–73

Primarily committed to research (versus teaching)	24%
Published or edited books (over lifetime to date)	
none	69
1–2	22
3–4	6
5 or more	4
Published writings (in last two years)	
none	53
1–2	24
3–4	12
5 or more	11

SOURCE: Ladd & Lipset, 1975, pp. 349–53.

research function is shared among faculty members of all disciplines, though the amount in the humanities and fine arts and in professional fields is somewhat less than in the natural and social sciences. Nevertheless substantial research activity occurs in all fields (Ladd & Lipset, 1975, pp. 349–53).

Research as conducted by faculties in higher education is for the most part produced jointly with instruction. The two activities are carried on by the same persons and within the same facilities, and they complement each other so closely that it is difficult, both conceptually and statistically, to identify the relative amounts of faculty time and other resources devoted to each. A few faculty members devote all their time to research and a sizable minority spend most of their time on it; on the other hand, many faculty members are involved in research not at all or only slightly. On the average, it is likely that research as here defined claims not more than a fifth of all faculty time (H. R. Bowen, 1977, pp. 291–94). Nations differ in the degree to which research is carried on within higher education. In the United States the degree of responsibility for research is probably greater than in most other countries.

Many observers ask: Is it necessary for higher education to be so deeply involved in research? Would it be desirable to follow the example of the Soviet Union and some other countries by organizing research in institutes that are separate from the universities? Without trying to answer these questions definitively, we would support an opinion for which there is no proof but which is widely shared in academic circles, namely, that research, and public service as well, contribute to the intellectual aliveness of our universities and colleges by attracting many creative and stimulating faculty members and by encouraging and sustaining their intellectual vitality. It may be added that the success of American science and scholarship, which clearly leads the world (though others are catching up), suggests that it has not been a mistake to entrust the colleges and universities with considerable responsibility for research and public service. Moreover, on the much debated question of whether faculty pursuit of research interferes with teaching, our opinion is that it enlivens teaching.[7] There are undoubtedly instances of faculty members who neglect their teaching because of their keen interest in research. But the academic community is surely on the right track when it gives weight to intellectual and artistic achievement when judging faculty members for appointment and promotion. However, they should and generally do also give significant weight to performance in teaching. Fulton and Trow (in Trow, 1975, pp. 291) concluded from their detailed study of research in higher education that "only a very small minority of faculty are uninterested in teaching" and that "the normative climate in the United States, as reflected in academics' personal preferences, is far more favorable to teaching than most observers would have predicted." This view was strongly endorsed by Ladd (1979) who found that three-fourths of faculty respondents agree that their interests lean toward teaching (as contrasted with research) and agree that teaching effectiveness, not publications, should be the primary criterion for promotion of faculty.[8]

Public Service. The public service activities of faculties are perhaps less recognized and less understood than their other functions. Mostly, they are byproducts of instruction and research, but they are far from trivial (Crosson, 1983; Bonner, 1981). If public service is defined narrowly as specifically budgeted activities designed expressly for the benefit of the general public, the percentage of faculty time devoted to public service is quite small. But if public service is defined more broadly, as we think it should be, it looms up as a byproduct of much more than incidental or minor importance.

Some public services are performed by faculty in connection with their teaching and research. The most notable is health care delivered by faculty in university hospitals and clinics. Other examples are the operation by faculties of farms, dairies, hotels, restaurants, and other enterprises related to instruction and research. Similarly, as a byproduct of instruction, some faculty in most colleges and universities are instru-

---

7. Participation by students in faculty-guided research activities constitutes an important mode of student learning, particularly at the graduate level.

8. In a survey directed toward older faculty members, a question was asked as to whether the "growing emphasis on research and publication improved or diminished the quality of instruction." About one-third answered "improved"; one-third, "diminished"; nearly one-fifth answered "no change." AAUP *Academe*, Nov.–Dec. 1983, p. 9. Finkelstein (1984, p. 126) observed: "That good research is both a necessary and sufficient condition for good teaching . . . is not resoundingly supported by the evidence. Resoundingly disconfirmed, however, is the notion that research involvement detracts from good teaching by channeling professorial time and effort away from the classroom."

mental in providing recreational and cultural activities for persons in surrounding communities. These include dramatic and musical performances, library services, museum exhibits, broadcasting, recreational facilities, and spectator sports.

Faculties are also engaged in activities designed specifically to serve the public, usually in an educational and consulting capacity. The most important example is agricultural extension which combines adult education with consulting services for farmers, agribusiness, and consumers. Some universities offer similar organized consulting services for local government, school districts, public-health organizations, business firms, labor unions, lawyers, and other organizations (Mayville, 1980).

Perhaps the most important public service function of faculties is that they serve as a large pool of diversified and specialized talent available on call for consultation and technical services to meet an infinite variety of needs and problems. The faculty pool of talent would be of value on a standby basis, even if never used, just as hospitals or fire stations or auto repair shops have standby value. But in fact academic faculties are frequently called upon. Sometimes they serve on behalf of their institutions and sometimes on their own as freelance consultants and technical experts. The services may be in the form of technical consulting, public addresses, testimony before legislative committees or courts of law, arbitration, temporary employment on leave of absence, or preparation of studies. For example, if the State Department needs to know some details of Sino-Russian relations, if a physician needs advice about a rare disease, if a museum needs guidance in the restoration of a damaged painting, if a farmer needs assistance in overcoming an infestation of insects, if a community wishes information on land-use planning, faculty members somewhere will be able to help.

Society looks to the academic community not only for information and advice, but also in many cases for actual execution and administration of programs. Faculty members often go on foreign missions, administer governmental laboratories, organize research centers, provide statistical services, etc. Indeed, a new breed of professors has emerged who move easily between the academic world and business or government. Such professors are frequently called upon to serve in the state or federal civil service, they occasionally serve as high-ranking political appointees, they are sometimes elected to public office, they become members of special commissions, they help administer foundations, and they consult with business firms.

In the past decade or two the need for "knowledge transfer" from the discoverers to the users has been increasingly recognized. As Lynton (1982, pp. 21) has observed: "The traditional notion of the scholar focusing solely on the discovery of new knowledge and letting its application and dissemination trickle down to the user is simply no longer appropriate."

Institutional Governance and Operation. Faculties, individually and collectively, usually occupy a prominent role in the policies, decisions, and ongoing activities falling within the wide-ranging realm of institutional governance and operation. It is almost inevitable that faculty members would be thus called upon because the work of most colleges and universities is divided among many highly specialized and abstruse disciplines, each of which is well understood only by the faculty members engaged in it. In the largest universities, instruction, research, and public service may be offered in several hundred separate fields or identifiable sub-fields, and even in small institutions the number of disciplines may range from twenty to fifty or more. Under these conditions, no central authority would be competent to make decisions for all of these fields and sub-fields as to what should be taught, what research and public service projects should be undertaken, how the teaching and research should be conducted, or what requirements for admission, academic standing, and graduation should be established. Such decisions must be largely delegated to the various specialized faculties, usually organized into departments of professional schools, and within these units the content of particular courses must be decided by individual faculty members. Moreover, no central authority would be qualified to make decisions on the appointment, promotion, tenure, and firing of faculty members, or to decide what kind of building space and equipment each field should have, what books and journals should be in the library, and what kind of secretarial and research support for faculty members should be available. All of these decisions ultimately find their way to the central administration of any institution usually through the budgetary process. Yet in practice, at least among the most reputable institutions, the influence of the departments is substantial and often decisive.

The various specialized departments are interrelated in many ways. They teach each other's students, they compete for enrollment and for budget, they collaborate on research and public service, and they are keenly interested in the success of the whole institution to which they have committed their talents. As a result, there is need for the entire institutional faculty to serve as a deliberative, advisory, and even legislative body. The administration also needs access to faculty opinion and advice and therefore must have lines of communication to faculty. In most institutions there are various senates, committees, councils, task forces, and even kitchen cabinets which create two-way communications between faculty and administration. In a substantial minority of cases, faculty unions figure prominently in faculty-administration communications, and in some of these instances, unions displace the more traditional bodies. Institutions differ in the extent and form of faculty participation in governance, but all must have apparatuses for the delegation of decision-making, for communication and cooperation among the departments, and for two-way communication between central administrators and faculty. Substantial amounts of faculty time and effort are required for these activities which are not mere busy-work but an essential part of the task of making institutions manageable.

An important feature of faculty participation in the making of institutional policies and decisions is that it has a strong influence on faculty morale. Faculty members are intelligent and highly educated people who feel qualified to have opinions not only on matters affecting them personally and their departments, but also on matters pertaining to the institution as a whole. They also feel entitled to know about events and forces and decisions that are affecting the institutions. Therefore, reasonable involvement of faculty and communication with them are critical in the decision-making process of any college or university. This involvement is of special importance in connection with the appointment of administrative officers. Institutions vary, however, in the extent of faculty participation, and morale tends to vary accordingly.

Faculty members also contribute enormously to institutional success through their efforts to create and sustain a rich cultural, intellectual, and recreational environment on the campus. The list of their contributions to extracurricular life is long but it is worth recounting because it is so impressive and so important. On virtually every campus faculty members contribute their time and effort, and sometimes their money, to enrich the campus environment. They serve as impresarios or committee members in organizing outside lectures, concerts, art exhibits, and plays; they frequently give lectures or performances themselves; they help organize and sustain religious activities for students; they serve as advisers to student housing groups, student clubs, and student newpapers and radio stations; they befriend individual students; they open their homes to students or join students informally or at meals; they organize programs for foreign students; they take students on field trips; they serve on committees related to athletics, student social life, and student discipline; they help carry on an active social life for the faculty itself; they participate in commencement and other events; they keep in touch with alumni; and they perform many other extracurricular services which are often little noticed.

Not all faculty members do all these things but most do some of them some of the time. In doing so, they often express their individual personalities and their special interests, and they help to weld the campus into a meaningful community and to make of it an agreeable and civilized place for both students and faculty.

The role of faculty in the governance and operation of colleges and universities was recognized by the Supreme Court of the United States in the Yeshiva case (*National Labor Relations Board v. Yeshiva University*, 444 U.S. 672, 1980). In this case the Court ruled that faculty members—at least at some private institutions—are not employees in the ordinary sense but rather are members of "management." On the basis of this finding the Court decided that institutions where these conditions exist are not covered by federal legislation pertaining to collective bargaining. In several recent cases, the National Labor Relations Board has ruled that faculty members are managerial employees and may not bargain collectively under federal labor law (*Chronicle of Higher Education,* June 6, 1984) unless the employer chooses to permit them to bargain. The Yeshiva case, which is opposed by those faculty who are committed to collective bargaining, clearly recognizes the reality that in many institutions faculty are in effect part of "management." One may add that in institutions where this is not so, changes which will bring about meaningful participation of faculty in governance may be overdue (Strohm, 1983, pp. 10–15). We see no necessary inconsistency, however, between faculty participation and strong presidential leadership (Commission on Strengthening Presidential Leadership, 1984).

Learning. We have identified and described four basic functions of faculties—instruction, research, public service, and institutional service. These are based mostly on a single unifying process, namely, *learning.* Learning in this sense means bringing about desired changes in the traits of human beings (instruction), discovering and interpreting knowledge (research), applying knowledge to serve the needs of the general public (public service), and creating an environment that contributes to and facilitates learning (institutional service). Learning is the chief stock-in-trade of the professoriate. It occurs in all fields, it takes place in diverse settings, and it serves varied clienteles.

Institutions tend to be categorized largely on the basis of their relative emphasis on the four functions. Community colleges and regional four-year institutions focus on instruction and on public service for local or regional constituencies; major universities emphasize research, advanced study, and public service directed toward statewide, national, and international clienteles; liberal arts colleges and smaller universities lie between these extremes by emphasizing instruction and the creation of institutional environments favorable to human development. But most institutions engage in all four functions to some degree. Indeed, these functions are mutually supportive. Instruction may be enriched if it occurs in an environment of discovery, intellectual excitement, and contact with the outside world and its problems. Similarly, research and public service may be enhanced when they are combined with instruction. This does not imply that every community college or liberal arts college should become a great research center. Nor does it deny that the universities can overdo research and service to the neglect of instruction. It implies only that the spirit of inquiry and public service is appropriate in the academic enterprise.

The multifunctional role of a college or university may be illustrated by the work of a typical faculty member. In a given week, in addition to formal teaching, he or she may engage in a variety of activities related to learning. For example, the faculty person may advise several graduate students on their research, counsel with several undergraduates on their academic or personal problems, invite a group of students to a social affair, discuss an intellectual issue with a colleague, write testimony for a legislative committee, give a talk to a local professional society, read one or more professional journals, record data from a laboratory experiment in progress, attend a meeting of the academic senate or of a department, have lunch with a prospective colleague, block out a chapter in a new book, and attend a public lecture.[9]

Qualifications of Faculty

We have defined in a general way the tasks and responsibilities of faculty in American colleges and universities. We turn now to the question: What are the qualifications of persons who undertake these tasks and assume these responsibilities? Faculty members serve in varied institutions ranging from community colleges and small liberal arts colleges to huge universities, they cover a host of disciplines and subdisciplines, and they deal with students of diverse ages, backgrounds, interests, and temperaments. Moreover, the faculty members themselves present a variety of backgrounds, personalities, educational philosophies, interests, and temperaments. Given all this variety, is it possible to say anything useful about the qualifications of people who are to perform successfully the functions of faculty members? We think that to consider this question is possible and worthwhile because it sheds some light on the kinds of people the academic profession needs to recruit and retain, and undergirds the policy recommendations we present in later chapters. We shall begin by considering the minimal qualifications.

There are no precise formal standards for admission to the professoriate such as exist for physicians, lawyers, or elementary and secondary school teachers. Each institution is free to choose its faculty according to its own best judgment without external regulation. Faculty members, therefore, vary considerably in their formal credentials. The Ph.D. is the most frequent credential for faculty in four-year institutions, and the Master's degree for those in two-year colleges. These two degrees are generally regarded as the basic standard for the two types of institutions. But in both about a quarter of the faculty have other credentials. In such cases, it is usually assumed that the combination of formal education and experience is at least comparable to that of the Ph.D. in four-year institutions, and to the Master's degree in two-year colleges.

9. For an inventory of 71 activities related to professorial work, see Braxton & Toombs, 1982.

How intelligent are faculty members? Leaving aside controversies about the measurement of intelligence, one kind of evidence would be intelligence scores of some sort. Though little such evidence is available, some conception of the caliber of people who earn the Ph.D., and also those who become academics, is provided by the following average IQ scores (Berelson, 1960, p. 139. Cf., Herrnstein, 1973, pp. 111–39; Schwebel, 1968, pp. 207–14; Cole, 1979, pp. 68, 153, 155):

High School students	105	Ph.D.s:	
High School graduates	110	Natural Science	133
College entrants	115	Psychology	137
		Social Science	124
College graduates	121	Humanities	128
Graduate students:		Faculty members:	
Natural Science	128	Physics-Mathematics	143
Psychology	132	Sociology	130
Social Science	124	Psychology	135
Humanities	128	Chemistry	136
		Biology	129

These scores suggest that intelligence levels of academic people fall between two and three standard deviations above the mean for the general population.[10]

But the holding of a degree, or its equivalent, is not sufficient qualification for membership in the professoriate. This is particularly true of the Master's degree which often represents little more than eight or ten courses taken beyond the Bachelor's degree. But it is also true of the Ph.D. even though that degree usually involves several years of serious advanced study and the writing of an extended dissertation. But even this endeavor——often extending over a five- to ten-year period—does not guarantee all the competencies required of a faculty member. The characteristics one expects in considerable measure of the faculty member, whether in two-year or four-year institutions, can best be stated in terms of specific qualifications, not merely in terms of earned degrees. In our judgment, these qualifications are as follows (Cf. Levine, 1978, pp. 176–78):

1. Superior general intelligence: in the upper 5 or 10 percent of the population.

2. Sound general education:
   (a) Ability to communicate correctly and effectively in speaking and writing
   (b) Intellectual curiosity

10. In interpreting these figures, it should be noted that not all Ph.D.s become faculty members. Many enter other professions. In our opinion, however, higher education—in the past at least—has been able to compete strongly for intellectual talent.

(c) Open-mindedness and tolerance

(d) Breadth: intellectual and cultural interests beyond their special fields and ability to view the special field in a broad context

(e) Contemplative disposition.

3. Keen interest in and mastery of the special field:

   (a) Knowledge of the literature, ability to interpret it, and the energy and motivation to keep up to date

   (b) Ability to give cogent lectures, to discuss issues, to hold reasoned opinions, to write book reviews, and to prepare essays and memoranda

   (c) If the special field is practical, the ability and knowledge to be a successful practitioner.

4. Self-motivation:

   (a) Dedication to teaching as a vocation

   (b) Capacity for hard work in an environment that provides little direct supervision.

5. Rapport with students: patience, ability to elicit cooperation and respect, and plausibly to serve as exemplar to some.

6. A personage; not a nonentity.

These standards are minimal. They do not include the desirable trait of intellectual or artistic creativity, as evidenced, for example, by the ability to conduct important research, to produce significant works of art, to contribute to the leading journals, or to write learned books. Also these qualifications do not include such useful traits as exceptional entrepreneurial or administrative abilities, or extraordinary skills in the arts of persuasion, including public relations or fund-raising. But even though the above standards are minimal they are not trivial. Only a small minority of the general population can meet them. And surely not all faculty members conform to these criteria.

But the profession should, and in fact does, attract many persons whose talents exceed the minimum requirements. The profession ranges from a few submarginal persons to some of the most brilliant, the most learned, the most dedicated, the most creative, the most versatile people in our society. Many of these persons also have exceptional abilities in entrepreneurship, administration, or the arts of persuasion—talents which are useful within academic communities.

In judging the qualifications of faculty, it should be noted that the great majority of those in four-year colleges have had a rigorous and lengthy education. Though the minimal time for getting the Ph.D. is three years beyond the Bachelor's degree, most doctoral candidates spend much more time than that in formal study, ordinarily six to eight years (see p. 173). Moreover, there is ample evidence that the rank and file of the academic profession are intellectually a superior group (cf., L. Wilson, 1979, p. 22). They may be more contemplative than business executives, less combative than lawyers, and less affluent than doctors, but on the whole they are no less capable.

Whether in the future the professoriate will attract and hold its share of talent, ranging from the marginal to the most gifted, depends on the conditions and remuneration in the profession. If over a period of years these are not competitive, colleges and universities will not attract their share of the best talent and will be forced to operate with faculties of relatively marginal abilities.

Career Options

It is often assumed that members of the professoriate were predestined in youth for academic careers, and that after entering the profession were committed to it for a lifetime. On the contrary, most of the persons who meet the qualifications for the professoriate are versatile and have more than a single career option open to them. The specific characteristics that fit them for faculty positions also qualify them for various non-academic pursuits. Career options are open to them in youth, in mid-career, and even after retirement.

In their youth, before making a commitment to any vocation, most such people may choose among numerous professional careers. In doing so, they can weigh the pluses and minuses of academic life in comparison with other feasible careers.

Even after having prepared for an academic career by earning a Ph.D. or in other ways, various options would still be open to most of them. For example, those prepared for faculty positions in any of the natural sciences, the professional fields, and some of the social sciences might readily become independent practitioners or employees of government, corporations, or private non-profit organizations. Chemists might work for industry and economists for banks, lawyers and physicians might go into independent practice, social workers might be employed by government, and so on. A review of the disciplines included in the curricula of colleges and universities suggests that a large majority of all potential faculty members might, if given time for placement, find employment outside academe if they wished. Those qualified for academic work in all of the following fields would likely have such options:

Health professions: medicine, dentistry, pharmacy, nursing, hospital administration, various allied health professions

Natural sciences and related fields: physical, biological, and earth sciences, engineering, agriculture, home economics, mathematics, statistics, computer science

Social sciences, humanities, and related fields: psychology, economics, law, business administration, public administration, public policy analysis, library science, criminal justice, education, theology.

The only academic fields for which outside opportunities are seriously limited are the humanities such as literature and history and a few social studies such as anthropology. These are important fields educationally, but they do not provide many outside employment opportunities. But even these fields offer increasing options; moreover, some of the people trained in them can if necessary be retooled for employment within or outside academe. Others are more versatile and can adapt rather easily without formal retraining.

Mobility is also possible for faculty persons in mid-career. This is demonstrated by the considerable movement of people into and out of academe year after year, even in times of a depressed economy.[11] The net flow varies from year to year depending on compensation and working conditions in academe relative to that in outside employment. Fulton and Trow (in Trow, 1975, pp. 6–7) reported that 68 percent of all faculty had worked outside the academic profession for at least a year since obtaining their Bachelor's degrees. Freeman (1971, p. 177) found that one in four living Ph.D.s had made a shift from academic to non-academic or the reverse at least once in their careers. Toombs (1979, p. 10) found that 59 percent of a sample of academic people listed in *Who's Who* had had non-academic work experience, and that 19 percent of a comparable set of people in business or the professions had had an association with the academic world. And several studies (summarized in Chapter 12) reported that half to three-quarters of all faculty earn income for services rendered outside their institutions. Not all of this work is at a high professional level, but much of it is in the form of research, consulting, writing for newspapers and magazines, lecturing, professional practice of various kinds, temporary assignments for government, and other professional work. Even after retirement many professors find remunerative work outside academe.

The facts on mobility and on outside employment tend to verify that many academics are versatile, that their services are of value outside the ivy-covered walls, that they could be mobile if they wished to move, and that their career decisions are responsive to changes in the relative conditions of work and remuneration in higher education. These facts also suggest that colleges and universities must in the long run be able to pay salaries and provide working conditions that render higher education competitive with comparable occupations in government, business, the independent professions, and other occupations.

Concluding Comments

The work of faculty members is extraordinarily important to the economic and cultural development of the nation. To do this work well

---

11. See Chapter 9 and Appendix C.

requires people of considerable talent ranging perhaps from persons in the upper 5 or 10 percent of the population to the most brilliant and gifted persons at the very peak of human abilities. Most of these people have desirable career options. To attract and retain an adequate supply of such persons for our colleges and universities requires compensation and working conditions competitive with those in other occupations and industries for comparable talent. Since about 1970 the competitive position of American higher education has been slipping, and there is danger of continuing deterioration. Under these conditions, the caliber of faculties in our colleges and universities could be on the verge of serious decline with disastrous consequences for the nation. This is a matter we shall consider at length.

The faculties of America are a great national resource which has been subjected to deferred maintenance. Just as deferred maintenance of roads, water systems, dams, and campus buildings may go on for years without causing collapse—or indeed without the damage being noticeable to the casual observer—so the deferred maintenance of faculty may persist over long periods without widely noticeable effects. But eventually, such deferred maintenance of both physical plant and faculty will lead to a day of reckoning and exorbitant costs will be involved in restoration (cf. Lynton, 1982, p. 19). Serious impairment of the nation's academic talent may be closer than is generally realized.

CHAPTER THREE

Faculty Attributes

In this chapter we examine briefly the personal characteristics of the faculty including personal background, age, rank, tenure, attitudes, and values.[1]

Personal Background

Family. Historians of higher education have given us a clear picture of faculties in American colleges and universities. Up to the Civil War, they were overwhelmingly Protestant and male. They were often descendants of relatively prosperous New England families connected with business or the church. Their education was usually concluded with a B.A. in classics and in some cases topped off with a modest amount of postgraduate study. For many, the academic profession was a temporary way station between college education and careers in law, medicine, or the ministry. The duties included teaching of a wide range of subjects through the recitation method and the supervision of students *in loco parentis.* College teaching was not widely regarded as a profession devoted to advancing knowledge or requiring serious advanced study of specialized disciplines.

During the latter part of the nineteenth century, a major transformation in the profession resulted in drastic changes in the content and method of higher education and brought about the democratization

1. In the preparation of Chapters 3 and 4, we were greatly assisted by Martin J. Finkelstein's comprehensive and perceptive volume on the characteristics of the American professoriate. For anyone who wishes to delve into this subject more deeply than we have here, we can strongly recommend his *The American Academic Profession* (Columbus: Ohio State University Press, 1984). Other useful but less recent volumes on the American professoriate are: Caplow & McGee, 1958; Rudolph, 1962; Jencks & Riesman, 1968; Lazarsfeld & Thielens, 1971; Ladd & Lipset, 1975; Trow, 1975; Logan Wilson, 1979.

and professionalization of the faculties. This transformation, which continued through the early part of the twentieth century, was virtually complete by World War II. By that time, according to Finkelstein (1984, p.29), "Catholics and Jews constituted nearly one-quarter of a heretofore exclusively Protestant profession; the offspring of mid-Atlantic and upper midwestern states were supplanting New Englanders; the sons of farmers and manual laborers were increasingly joining the sons of businessmen and professionals; and *daughters* were now joining the sons." It could be added that in recent decades, higher education has been attracting increasing numbers of women, minorities, and white persons of modest socioeconomic origins. Yet, because of the close relationship between family background and level of education, a large proportion of faculty persons are still of relatively high socioeconomic origins.

The latest available figures on the social origins of academic people are not current; they refer to the years 1969 to 1975, when several splendid surveys were made. One of these revealed, as of 1969, that 39 percent of all faculty members were the sons or daughters of fathers who had attended or graduated from college (Ladd & Lipset, 1975, p. 343). In comparison, not more than 10 to 15 percent of all fathers in the same generation attended college. The same survey showed that 24 percent of the fathers were "manual workers" (Trow, 1975, pp. 6–7), which compares with the much higher percentage— about 50 percent—of the whole male labor force in manual occupations in the years prior to 1969.

A census report (cited in Ladd & Lipset, 1975, p. 172) indicated that about two-thirds of the faculty came from families in which the fathers were in professional, managerial, or entrepreneurial occupations; whereas in the general labor force only about one-fourth were in such occupations. Moreover, 20 percent of the faculty fathers were in elite or high-status occupations, a number which greatly exceeded the percentage for the entire population.

A 1975 survey (Ladd & Lipset, 1975–76, Sept. 22, 1975) confirmed on the basis of religious and ethnic data that the faculties contained a disproportionate number of persons of high socioeconomic status. This survey indicated overrepresentation of high-status Protestant denominations, including Presbyterians, Congregationalists, Episcopalians, and also of persons of British and North European ancestry. This survey also revealed the relatively large concentration of Jews in the academic profession. Whereas Jews make up 3 percent of the total population, they constitute 9 percent of the professoriate.

The conclusion is clear. Despite the steady flow of lower socioeconomic groups into the professoriate, the upper social strata continue to be heavily represented. However, it would be a mistake to conclude that the professoriate is an aristocratic body. The majority of the faculty derive from middle-class antecedents (Ladd & Lipset, 1975, pp. 172–73). Historically at least most have probably been upwardly mobile.[2]

The several surveys also provide interesting comparative data on the family backgrounds of faculties serving in different types of institutions and in different disciplines. The percentage of faculty whose fathers were manual workers varies inversely with the stature of the institutions in which the sons were employed. Similarly, the percentage of fathers and mothers who had attended college varies directly with institutional stature (Trow, 1975, pp. 6–7). The differences in family background among faculty of different disciplines is also striking. Whereas, on the average, 39 percent of all faculty had college-educated fathers, the percentage was 48 or more in the case of anthropology, law, and medicine; and it was 33 or less in the case of sociology, education, business, and agriculture (Ladd & Lipset, 1975, p. 343).

Religion. Another important aspect of faculty background is religious belief and observance. This was investigated in considerable detail in the surveys conducted in the early and mid-1970s. The religious origins of the faculty were found to be as follows (Ladd & Lipset, 1975, p. 170):

Protestant	66%
Catholic	18
Jewish	9
Other	4
None	3
Total	100

When these figures were classified by age of faculty, it became evident that the traditional position of higher education as a preserve of Protestants had been giving way to the entry of many Catholics and especially of Jews. The enormous success of Jews and their notable contributions to higher education are documented in Ladd and Lipset (1975, chap. 6).

Depth of faculty attitudes and observance toward religion is shown in Table 3–1. It indicates that about two-thirds report deep or moderate commitment, and about one-third as either indifferent or opposed. Clearly the American college or university as represented by its faculty is no longer the religious bastion it once was. Many decades ago it made the transition from a religiously oriented to a predominantly secular institution. However, in view of a resurgence in religious interest throughout the nation over the past fifteen years, the religious commitment of the faculties as shown in Table 3–1 may be understated. In any event, we believe that it is not vastly different from the commitment of the total population.

2. We suspect, however, that the percentage from upper-class origins may have risen during the 1960s and early 1970s and that it may have declined later in the 1970s, when many upper-class young people probably sought medicine, law, business, and science.

Table 3–1. Religious Commitment of Faculty, 1969–70.

	Protestants	*Catholics*	*Jews*	*Total*
I consider myself:				
deeply religious	16%	23%	5%	16%
moderately religious	52	52	28	48
Indifferent to religion	26	19	50	28
Opposed to religion	6	6	17	8
	100	100	100	100
I am presently not affiliated	20%	19%	26%	22%
I attend church or synagogue				
once a week	33%	62%	5%	25%
once a month	49%	67%	10%	48%

SOURCE: Steinberg, in Trow, 1975, p. 90

Education. Over the past century, the education of the faculties has changed substantially. In the mid-19th century, their formal education was mainly in the classics and theology and seldom advanced much beyond the Bachelor's degree. With the gradual secularization and professionalization of higher education, the education of the faculties has been channeled to many specialized disciplines and the doctorate has become the modal degree.[3]

The percentage of all faculty with doctorates has risen substantially over the years—from 38 percent in 1947–48 to 61 percent in 1977–78 (National Education Association, 1979, pp. 7, 19). It appears to be even higher today. Correspondingly, the percentage with only Bachelor's degrees has diminished almost to the vanishing point. A 1980 survey confined mostly to full-time faculty showed the following distribution of faculty members by highest degree awarded (H. Astin, 1980):

Bachelors or less	2%
Masters	18
Professional degrees (J.D., M.D., D.Ed., etc.)	10
Ph.D. or equivalent	70
Total	100

Table 3–2 shows the percentage of faculty members with doctorates by type of college or university. In all categories of institutions except two-year colleges and "other" liberal arts colleges, 65 percent or more of the faculties held earned doctorates. In interpreting all these percentages, it is well to remember that the doctorate is not appropriate for all disciplines or all faculty activities. For example, many institutions employ

---

3. The doctorate includes not only the Ph.D., but also Doctor of Education, Doctor of Arts, Doctor of Medicine, Doctor of Jurisprudence, and others.

Table 3–2. Percentages of Faculty Members with Doctorates, by Types of Colleges or Universities, 1975–76[1]

	Ph.D.s[2]	Other Doctorates[3]	Total
Two-year colleges	12	6	18
Selective liberal arts colleges	60	8	68
Other liberal arts colleges	41	9	50
Comprehensive colleges and universities	50	15	65
Doctorate-granting institutions	63	14	77
Research universities	66	19	85

SOURCE: Carnegie Foundation for the Advancement of Teaching, 1977, p. 85.
[1] The types of institutions are based on the classifications of Carnegie Council on Policy Studies in Higher education. For brevity some classifications have been consolidated.
[2] Includes Doctor of Arts.
[3] Includes doctorates in law, medicine, dentistry, education, and miscellaneous fields.

distinguished musicians and artists, persons of foreign backgrounds, self-educated people, athletic coaches, and others for whom the standard American doctorate is not necessarily *de rigeur*. Moreover, every institution employs some young faculty who are near the doctorate but have not yet completed all the requirements. American colleges and universities would lose a great deal if they required all prospective faculty members to have a doctoral degree in hand before employing them.

The Ph.D. is nevertheless the leading degree among faculty. It signifies on the average about seven years of advanced study, at a rigorous level, usually in a major university. It also suggests the ability to conduct an ambitious research project in a specialized discipline and to write an extended and original dissertation. When a great majority of the faculties in four-year American colleges and universities hold the Ph.D., the conclusion is that they constitute a highly educated and specialized professional corps. The process of acquiring a doctorate places an unmistakable imprint on most of those who experience it. It means that they have been strongly acculturated to the ways and values of the major universities, where great emphasis is placed on research and scholarship and on the duty of the academic person to pursue the life of the mind. This experience does not always prepare these doctorate recipients for life in the vast majority of colleges and universities, where the dominant task is undergraduate instruction. It sometimes leaves doctorate recipients with difficult problems of adjustment.

As shown in Table 3–2, the educational level of faculties in two-year colleges is quite different. There, only 18 percent hold a doctorate. This situation reflects partly the evolution of two-year colleges. They were in the beginning offshoots of secondary schools, and have over the years increasingly emphasized vocational skills rather than general education. A teacher of auto mechanics or cosmetology needs no Ph.D. On the other hand, there is considerable ambiguity concerning the

Ph.D., even for those community college faculty members who teach in the arts and sciences. One thoughtful study concluded that the degree seems "to hold simultaneously the possibility of status and scorn for community college faculty. Stories of pressure to seek terminal degrees for a sense of security in the college coincide with stories about the awkwardness with which those who hold the doctorate are treated by those who do not." (Seidman *et al.*, 1983; Burnett, 1982, pp. 18–23).

In an important study the American Council on Education (Atelsek & Gomberg, 1978, pp. 12–19) found that about one-third of all four-year colleges *expected* to increase the proportion of faculty with doctorates, about two-thirds expected to hold the percentage steady, and scarcely one percent indicated that they expected to reduce the percentage. Similarly, about two-thirds said they regard the doctorate as essential or strongly preferred for all new faculty. Not so, however, for the public two-year colleges. Only about one-third of them expected the proportion of doctorates to increase, more than one-third of them reported that they "do not prefer doctorates for any new faculty," and another one-third said they "slightly prefer doctorates for all new faculty." Only 7 percent said they strongly prefer persons with doctorates.

The data we have presented on the education of faculty reflect the period up to 1980. With an abundance of new doctorates being awarded and with slackness in the demand for faculty, employing colleges and universities have been able easily to recruit people with doctorates. On the other hand, under conditions of financial stringency, institutions have employed increasing numbers of part-time or temporary persons—many without doctorates. But without doubt the proportion of full-time faculty with doctorates in hand continues to rise. Coupled with the fact that numerous doctorate-holders under prevailing market conditions are obliged to settle for part-time positions, it is safe to assume that the proportion of all faculty holding doctorates still continues to edge upward.

Work Experience. Faculty members on the whole have had substantial and varied work experience both inside and outside academe. On the average, they have had at least 13 years of experience within higher education—those in four-year institutions have averaged 14 years of experience and those in two-year institutions 10 years (National Education Association, 1979, p. 7). Over two-thirds of the faculty have worked outside the academic profession for at least a year since obtaining the Bachelor's degree (Trow, 1975, pp. 6–7). In one study, 46 percent of those responding reported that they had been engaged in non-academic work in the job they held immediately prior to the present position in academe (Bayer, 1973, p. 29). Those with outside experience have worked in varied occupations, many in the armed forces or as teachers in elementary or secondary schools (National Education Association, 1979, pp. 3, 16). The percentage who have been school-teachers is particularly high in the case of faculty in two-year colleges.

Many faculty members engage in remunerative outside work while serving in higher education.[4] Some of this work may be carried on during vacations and other periods that may be regarded as "overtime," but much of it is fitted into the regular weekly work schedule which, as we shall show, is relatively long. There is no secret about this outside work. Indeed, many institutions openly encourage faculty members to take part to a reasonable degree in remunerative outside activities on the grounds that these activities enhance the skills and knowledge the faculty members bring to their teaching and that these activities serve society and enhance the prestige and the social usefulness of the institutions. *Curricula vitae* of applicants for academic positions often display outside activities prominently and proudly along with publications and other indications of professional attainments.

This outside work generates earnings from publications, inventions, sales of works of art, and fees for professional services such as lecturing, consulting, summer or part-time teaching in other institutions, serving as expert witnesses, research, and miscellaneous "moonlighting." These earnings supplement the non-competitive salaries common in academe. A few faculty members become wealthy from their outside earnings, especially from the sale of textbooks, from inventions, and from consulting.

Several studies made at various times indicate the amount of such outside work. One study reported that as long ago as 1961–62 well over one-third of faculty on 9- to 10-month contracts and one-half of those on 11- to 12-month contracts were engaged in work *outside* their institutions (Dunham et al., 1966). These figures excluded summer or overtime teaching within the home institution. Another study reported that 38 percent of all faculty engaged in paid consulting outside their institution, and also that substantial numbers of faculty were away from their campuses more than ten days a year for professional activities: 12 percent in the case of two-year colleges, 22 percent for four-year colleges, and 37 percent for universities (Bayer, 1973, pp. 18, 28). Another detailed study, referring to the year 1979–80, reported the following figures on outside employment (Minter, ca. 1981, pp. 91–92):

	Percent of faculty reporting income earned outside their home institutions	Average outside earnings as percent of base salary
Four-year private institutions	54.8	18.5
Four-year public institutions	50.9	13.0
Two-year public institutions	35.1	13.5

Even allowing that these figures reflect some proportion of faculty who are teaching part time at institutions other than their "home"

---

4. For a more detailed discussion, see pages 254–60.

institutions, these figures suggest that faculty people are in demand in the world outside academe and that the experience of many extends beyond the ivory tower. In Chapter 12 we shall return to the matter of outside earnings.

Age

The composition of the American professoriate by age is a matter of considerable contemporary interest. It is widely believed that the faculties are becoming older and that their effectiveness is therefore declining.

The distribution of the faculty by age, at any given time, is a result of "turnover," that is, of the entry of new faculty members and the departure of old ones. The process of faculty turnover is neither simple nor standardized. Faculty turnover occurs at all ages from the twenties to the seventies. There are many academic persons of all ages who are or might be qualified for jobs in government, business, professional practice, and other pursuits. There is in fact a constant outflow and, under some conditions, more might leave higher education. Similarly, there are numerous persons employed outside who are, or could readily become, qualified for faculty positions and who might under some conditions enter higher education. Some people move back and forth *within* academe between faculty appointments and administrative positions. Some drop out for reasons of illness or death, because of the transfer of a spouse, or because of other changes in family circumstances. Some are temporary (often part-time) appointees who exist around the edges of the academic world and move in and out frequently. Even the concept of retirement at a fixed age is becoming far less certain than it once was. Because the entry and exit of persons can occur at all ages, the makeup of the faculty by age groups is not a simple function of the numbers entering in youth and the numbers leaving at a specified retirement age.

Data on the age distribution of faculties over the past several decades are available from various sources. These data are not wholly comparable. They employ different definitions of faculty (or fail to specify the definitions), they use different age brackets, and they cover different periods of time. Nor do these data yet reflect the effects of federal law raising the standard age of retirement from 65 to 70. Yet they probably reveal the approximate trend despite lack of precision. Some of these data are presented in Tables 3–3 and 3–4.[5]

It is worth noting that in Table 3–3, the data for 1973–74 and 1980–81 are derived from comparable surveys conducted by Helen

5. For an interesting account of trends in the aging of faculty in Great Britain, see Flather, 1982.

Table 3–3. The American Professoriate by Age Groups as Reported in Various Studies, 1962–63 to 1980–81

Age Group	1962–63[1]	1968–69[2]	1972–73[3]	Age Group	1973–74[6]	1977–78[4]	1979–80[5]	1980–81[6]
Under 31	10.3%	15.4%	7.2%	Under 30	6.1%	4.0%	6.0%	1.1%
31–35	15.7	17.0	17.8	30–34	14.2	14.3	8.0	5.2
36–40	17.9	16.4	17.1	35–59	17.9	21.0	25.0	17.1
41–45	15.7	14.4	16.3	40–44	16.2	16.1	16.0	18.1
46–50	13.6	12.9	14.0	45–49	14.2	15.5	15.0	16.6
51–55	10.5	9.6	11.7	50–54	12.7	12.3	12.0	15.5
56–60	7.9	6.8	8.3	55–59	9.3	9.3	10.0	13.2
61–65	5.6	5.3	5.6	60–64	6.0	5.5	6.0	9.4
Over 65	2.9	2.2	2.0	Over 64	3.3	1.9	2.0	3.9
Total	100.0	100.0	100.0	Total	100.0	100.0	100.0	100.0
Estimated Mean Age	44	42	44		44	44	44	48

[1] Dunham *et al.*, 1966; four-year institutions only.
[2] Cartter, 1976, p. 165.
[3] Bayer, 1973, p. 27.
[4] National Education Association, 1979, p. 7.
[5] National Science Foundation, 1981a, p. 60. Includes only full-time faculty in science, engineering, and social science.
[6] H. Astin, 1973, 1980. Includes mainly full-time faculty in four-year institution.

Astin. They show a moderate but perceptible aging of the faculties. On the other hand, the figures for 1979–80 collected by the National Science Foundation show no substantial change from surveys of earlier years. The explanation may be that the National Science Foundation data cover only faculties in science, engineering, and social science—disciplines for which student demand has remained relatively strong— whereas Astin's data cover all disciplines, including those with weak demand.

An additional consideration is that the faculties in two-year colleges are somewhat younger than those in four-year institutions. As of 1977–78, the mean age of the two-year group was 42 and of the four-year group as 44 (National Education Association, 1979, p. 7). Thus, the rapid expansion of the two-year institutions may have slowed the rise in the average age for all institutions combined.

From these figures, we draw the tentative conclusion that, over nearly four decades, the age distribution of the faculties, at least until quite recently, has been remarkably stable. The faculties continue to be widely spread over the various age brackets. However, the mean age may have risen a bit recently, and a slow but steady increase seems likely over the next decade and a half.

In our judgment the "aging" or "graying" of the faculties that is so often referred to with dread has not yet occurred to an alarming extent. However, this does not rule out such aging in the future. The shift to a steady or declining state in the number of faculty would, unless preventive measures are adopted, continue to reduce the intake at the early ages while those now in the prime ages of 35 to 60 would be steadily getting older. In that event, the nation's faculties would surely become more heavily populated with older persons. For example, Table 3–5 presents two projections of the age distribution of the professoriate over the years 1980–81 to 2000–2001. The first of these, which refers only to tenured faculty, was computed by the Carnegie Council on Policy Studies in Higher Education. The second is a modification of those figures to include non-tenured persons. Both estimates show an increase in average age, but certainly nothing like the drastic increase often postulated.

The faculties are now widely distributed over the middle-aged groups from 35 to 60, and over the next several decades the number of retirements will probably increase, thus creating more positions for younger persons. Moreover, institutions can plan and manage their faculty personnel policies toward a balanced age distribution of their faculties. This they can do, for example, by facilitating timely retirement, by recruiting people selectively by age groups, and especially by filling the few vacancies that arise with great care so that every appointment will add strength regardless of the age of the incumbent. These and other policy implications associated with the graying of the professoriate, including early retirement policies and the relationship between age and productivity, we shall consider in later chapters.

Effects of Aging. The opinion is widespread that recent and impending changes in the age structure of the professoriate are or will be seriously detrimental to higher education. The argument is in two parts. First, it is held that educational quality and research productivity

Table 3–4. The American Professoriate by Age Groups as Reported in Various Studies, 1947–48 to 1977–78[1]

	1947–48[2]	*1968–69*[3]	*1969–70*[4]	*1972–73*[5]	*1975–76*[6]	*1977–78*[7]
Under 35	32.0%	33.4%	27.6%	23.0%	25.4%	18.3%
35–49	45.6	43.2	46.6	47.7	45.7	52.6
50 and over	22.5	23.4	25.8	29.2	28.9	29.0
Total	100.0	100.0	100.0	100.0	100.0	100.0
Estimated Mean Age	42	42	43	44	44	44

[1] These data are from different surveys and are not necessarily comparable. Source: National Education Association, 1979, p. 19.
[2] U.S. Office of Education.
[3] American Council on Education.
[4] National Education Association.
[5] American Council on Education.
[6] National Education Association.
[7] National Education Association.

Table 3–5. Age Distribution of the Professoriate, 1980–81 to 2000–2001

Age Group	Tenured Faculty[1]			Total Faculty (including non-tenured persons)[2]		
	1980–81	1990–91	2000–2001	1980–81	1990–91	2000–2001
26–35	3.5%	2.1%	2.6%	15.0%	13%	20%
36–45	40.1	15.7	18.7	35.3	15	15
46–55	34.4	46.1	24.7	30.3	41	21
56–65	20.2	30.9	45.5	17.8	27	37
over 65	1.8	5.2	8.5	1.6	4	7
Total	100.0%	100.0%	100.0%	100.0%	100%	100%
Mean age	48	52	55	45	49	49

SOURCE: Carnegie Council on Policy Studies in Higher Education, *Three Thousand Futures*, 1980, p. 26.
[2] Estimated by the authors by adjusting the Carnegie Council figures to include non-tenured faculty. The assumption underlying the adjustment is that the recruitment of young faculty will decline during the late 1980s and early 1990s, but will accelerate toward the end of the century, when there will be many retirements.

will deteriorate if the number of fresh, vigorous, imaginative, younger faculty declines relative to the number of tired, disillusioned, older faculty. This may be true, other things being equal, though evidence on this matter is scarce. Common sense tells us that in the interests of variety and continuity it would be good to have faculties representing a wide range of age groups. No one would seriously advocate the extremes of a faculty predominantly under 30 or over 65. But between these extremes no one knows precisely where the optimum lies. Within a fairly wide range, the distribution by age is probably not among the more important factors in distinguishing a great faculty from a mediocre one. In the near future, it may well be that an older faculty, most of whom chose the academic profession in a period when higher education was buoyant and attractive, may prove to be more capable than young people recruited in the future at a time when higher education may be in the doldrums.

Some evidence on the matter of age and academic proficiency has been gathered by Warren. His findings are summarized in the following passage (1982, p. 19):

> As faculty members grow in experience, their contacts with other faculty members as well as their own experiences might be expected to bring about a gradual change in perspective. Among slightly more than 300 faculty members in California colleges and universities, those over 50 years of age were more likely than younger faculty to be concerned with the general intellectual growth of their students and to value strict academic standards. Those under 30 were least concerned with students' general intellectual growth, valuing the fixed structure and knowledge of the discipline more highly, and were most likely to be flexible in their standards for student performance.

Gaff and Wilson (1971) presented evidence along these same lines and R. C. Wilson et al. (1975) found no relation between the age of faculty persons and their nomination by students or colleagues as effective teachers. Warren (1982, p. 19) continued his commentary by asserting:

> The effective teachers were those who were most involved with the students in educationally relevant ways, which might be expected to parallel the older faculty members' concern for general intellectual growth. . . . High social involvement with students was not associated with student or faculty judgments of effectiveness as a teacher. . . .

McKeachie (1983, pp. 8–10), a distinguished psychologist and scholar of higher education, has assembled the most complete and most reassuring information on the process of aging among academic people. His opening paragraph reads:

> We hear a great deal these days about mid-life crises, academic menopause, the crisis of the aging faculty, the terrible consequences of the tenure system, and the need to get rid of dead wood; all of these represent stereotypic beliefs about a downward course of vitality and productivity among mature faculty. [p. 8]

He then described the aging process indicating that intelligence holds up reasonably well; that student ratings of teachers are relatively unrelated to age; that there are not large numbers of ineffective teachers of any age; individuals have the ability to change at all ages; that when continuity and stability are found they are the result of a constant environment rather than built-in rigidity; and that productivity and renewal are best enhanced by freedom, diversity of opportunities, risk-taking, complexity, and time pressure (pp. 9–10). Thus, the most important enemy of continued vitality and productivity is sameness of environment, monotony of tasks, and lack of challenges. McKeachie concludes (p. 10):

> Faculty members, like other human beings, can do a great deal to construct their own environments. They are curious; they like to feel competent; they will work hard when they feel they are making a contribution. A sense of being appreciated for those contributions, a feeling that the future has at least some bright spots, a sense of being valued by one's colleagues and administrators, these are the things that induce performance.

An important paper by Shulman (1983) echoes McKeachie's thesis that freedom, variety, new experiences, and the like are needed to sustain vitality.

Any sensible person knows that the productive lifetime of any individual has ultimate limits, and that these limits are not the same for all individuals. But a great deal of productivity resides in many persons who have passed the now outmoded retirement age of 65.

Some evidence on the matter of age and scientific productivity as distinct from teaching effectiveness was collected in a carefully documented essay by Barbara F. Reskin (1979, pp. 189–207). It began with the following words, "Because almost all the empirical studies of this question have been flawed methodologically, unequivocal evidence on the existence, strength and form of the relationship between age and scientific productivity is still lacking" (p. 189). Her concluding comment (p. 203) was that "in no case did productivity show simple negative relationship with age . . . . Multivariate analyses for specific scientific disciplines that permit non-linear effects of age are necessary to learn whether age *per se* has any independent effect on scientists' performance."

Klitgaard, in an important paper (1979, p. 67) reviewing the literature on age and scientific and scholarly productivity, concluded:

> What would happen if new [faculty] hirings decline? The discussion leads me, at least, to doubt many of the negative effects that are widely assumed. Only one finding falls on the other side: current methods for identifying academic stars before graduate school are surprisingly ineffective. If the decline leads to greatly reduced graduate enrollments, graduate schools may not be able to avoid screening out some of the very best potential researchers—even if the very able are proportionally more likely to apply.

Ladd and Lipset, in their 1975 survey, made interesting comparisons between the attitudes and values of older and younger faculty members (1975–76, May 24, 1976). Faculty persons over 55 expressed views, as one would expect, that were more conservative than those of their younger colleagues. The following is a sampling of these differences:

	Under 35	Over 55
Republican	18%	37%
Nationalize big corporations	35	25
Ban homosexual teachers	14	40
Opposed to laws against pornography	66	40
Legalize marijuana	71	44
Expel disruptive students	56	72
Student demonstrations have no place on the campus	13	42
Favor collective bargaining	79	57
Attend religious services at least once a month	34	50

But on matters of academic policy, as shown by the following figures, the opinions of young and old were quite similar:

	Under 35	*Over 55*
Meritocracy is a smokescreen for discriminatory practices	46	45
Salary increases should be based on merit	57	49
Tenure should be awarded on basis of national standards	64	77
Benign quotas for hiring minority faculty are justified	35	32
Reduce admission standards to raise minority enrollments	63	59

Ladd and Lipset found, in addition, that older faculty members are somewhat more likely than their younger colleagues to prefer teaching over research. They are also a little less likely to publish, are inclined to prefer "soft" philosophical approaches to their disciplines rather than "hard" scientific approaches, and are likely to be less enamored of radically new and "wild" ideas.

Another aspect of faculty age distribution pertains to money. Because of traditional seniority, older faculty draw larger salaries than younger faculty. It is often asserted, therefore, that a relative increase in the number of oldsters will raise the cost of operating our colleges and universities. This argument would be valid if it could be assumed that as the proportion of older faculty increased, total funds available for higher education would increase, or that the salaries of young faculty would be reduced, or that funds would be shifted from non-salary items to salaries for oldsters. We think that a result more likely than any of these would be a change in the salary scales such that the salaries of older faculty members would be constrained in comparison with those of younger faculty members.

In considering various implications of the age structure of the professoriate, we do not mean to argue that the aging of the faculty would be a good thing. A regular turnover of faculty with a steady inflow of new talent is clearly desirable. Rather, we mean to point out that the problem of age may not be as drastic or as critical as is often assumed. But if it turns out that a drastically aging faculty becomes a reality, it will be up to the leadership of higher education to learn how to organize and motivate people of rising average age to conduct excellent education, and at the same time to recruit in each generation a significant inflow of the most talented young people. It would not be surprising if, over the next ten years, the mean age would rise by several years from recent levels around 44 or 45. But that need not be regarded as a calamity.[6]

6. For an extended discussion of the aging of faculty, see Calvin, 1984, pp. 8–12.

Table 3–6. Percentage Distribution of Full-Time Faculty, by Academic Rank, 1974–75 to 1981–82

	1974–75	1975–76	1976–77	1977–78	1978–79	1979–80	1980–81	1981–82
Professors	23.0%	23.3%	23.2%	23.9%	24.7%	25.8%	26.5%	27.0%
Associate Professors	22.9	23.4	23.6	24.2	24.4	24.8	24.5	24.5
Assistant Professors	28.9	28.2	27.1	26.5	25.8	25.2	24.4	23.9
Instructors and Others[1]	25.3	25.1	26.1	25.3	25.1	24.1	24.6	24.6
	100.0%	100.0%	100.0%	100.0%	100.0%	100.0%	100.0%	100.0%

SOURCE: National Center for Education Statistics, *Digest of Education Statistics*, 1976, p. 107; 1979, p. 104; 1980, p. 107; 1982, p. 107; 1983, p. 103.
[1] Includes lecturers and persons of undesignated rank.

Rank

In past decades, the professoriate has been divided about equally among the four ranks,[7] and the distribution by rank has been fairly stable. However, since the mid-1970s, as the number of new entrants into the faculties has leveled off, the percentage in the two upper ranks has crept up and the percentage in the two lower ranks has declined (Table 3–6). Thus, by 1981–82, the faculties had become slightly more top-heavy than they had been a decade earlier. That the proportion in the upper ranks did not increase more rapidly in recent years was due to the cautious regulation of appointments and promotions so as to prevent faculties from being overloaded at the top. As indicated in Chapter 4, the distribution of the professoriate by academic ranks is quite different for women as compared with men.

Tenure

Available statistics on trends in the proportion of faculty members on tenure are difficult to interpret. Many of the estimates are based on differing methodologies or include differing categories of faculty. An estimate for 1981–82 is presented in Table 3–7.

Our judgment, based on a review of this and many other sources, is that the percentage of the full-time faculty on tenure in the 1960s was of the order of 50–55 percent, that this percentage rose to 63–68 percent by

7. Professor, associate professor, assistant professor, instructor (or lecturer or adjunct).

1980, and that it has since held steady at about that level—perhaps inching up a little. Probably, as of 1985, over 70 percent—perhaps even 75 percent or more—of the full-time faculties are on tenure, the percentage being greater for public than for private institutions.

While reviewing tenure, it is well to consider also those faculty members who are on "tenure track" appointments. These are persons who are on probationary status and under consideration for tenure. Some are young persons seeking their first appointment to tenure; others are more senior persons who have relinquished a tenured appointment at one institution and must undergo another, usually shorter, probationary period at another institution. In either case, under the "up or out" rule, failure to receive tenure ordinarily means outright discharge. There is little information on the numbers of persons on tenure track appointments, but two surveys conducted in 1979–80 provide some information. One of these covered full-time science, engineering, and social science faculty in doctorate-granting institutions (National Science Foundation, 1981a, pp. 10–19), and the other covered full-time humanities faculty in all types of institutions (Atelsek & Gomberg, 1981a, p. 24). These two surveys suggest that in 1979–80 around 15 to 20 percent of full-time faculty were in the ambiguous and uncertain status of "on the tenure track." Most of these people were probably hoping for tenure and many were undoubtedly living in a state of nagging anxiety about their future status. In earlier times, these fears would have been less acute because of the many alternative opportunities in a growing profession and a strong economy. But in recent years alternative oppor-

Table 3–7. Percentage of Faculty on Tenure, by Type of Institution and Academic Rank, 1981–82

	Public				Private			
	Univer-sities	Other Four-Year Insti-tutions	Com-munity Colleges	All Insti-tutions	Univer-sities	Other Four-Year Insti-tutions	Com-munity Colleges	All Insti-tutions
Professors	96.9%	95.5%	98.0%	96.3%	96.6%	92.3%	82.7%	94.0%
Associate Professors	84.9	82.4	92.9	84.5	75.6	73.3	81.6	74.1
Assistant Professors	17.1	35.5	61.6	32.0	10.0	18.3	32.4	16.1
Instructors	5.9	10.2	22.6	12.3	3.1	2.6	7.1	2.8
Lecturers	3.7	0.1	26.8	2.6	1.7	1.1	0.0	1.5
No Academic Rank	1.0	29.7	80.6	78.8	0.0	48.4	55.0	47.9
All Ranks	66.1	65.2	76.3	68.1	60.8	53.4	49.2	55.7

SOURCE: National Center for Education Statistics, as reported in *Chronicle of Higher Education*, June 16, 1982.

tunities have been uncertain, and the issues surrounding the status of these people have been painful. These same uncertainties and fears apply also to those persons who are on short-term contracts outside the tenure system.

Appointment to tenure is not an iron-bound contract. It may be annulled in cases of serious malfeasance on the part of individual professors and in cases of financial exigency on the part of institutions. But most institutions and most professors regard it as a contract not to be terminated except in grave circumstances. The piling up of tenure in the 1970s, therefore, was seen by administrators and trustees as an ominous addition to financial liability. Hence many institutions undertook to slow up the rate of both faculty promotion and appointment to tenure. As a result the rapid rise in the percentage of faculty on tenure leveled off.

As shown in Table 3–7, the differences in tenure percentages among types of institutions are considerable. Comparing the universities and other four-year institutions, the latter often grant tenure to persons in the lower ranks, whereas the universities usually reserve tenure for the upper two ranks. The striking difference, however, is between the two-year and four-year public institutions. The two-year institutions have a different faculty structure with relatively more persons in the lower ranks or with "no rank," and they often grant tenure to these persons—sometimes with only a brief waiting period. The overall tenure ratio of the public community colleges greatly exceeds that of any other institutional group.

The tenure percentage also varies considerably by gender. Because relatively few women have attained the senior ranks, the percentage of women on tenure is considerably less than the percentage of men. On a comparison by academic ranks, however, the differences between the proportion of men and of women who hold tenure are small and not all in favor of men.

Since the 1970s the slowing of the rate of promotion in rank and of the rate of advancement to tenure, combined with the actual dismissal of tenured faculty in a few institutions, has placed a severe strain on the higher educational system. This is especially so because of the "up or out" rule that obtains in many institutions. Under this provision a denial of promotion from assistant professor to associate professor, after a long probationary period of usually six or seven years, often means not merely no advancement in rank but loss of a job, and in some cases loss of a promising career. Legitimate expectations built up over decades are disappointed, the level of apprehension of the whole faculty—even those who may never be affected—is raised, a general pall of uncertainty and injustice pervades the campus, and not least, the administrators and faculty committees involved are faced with disagreeable tasks and placed in vulnerable positions. We shall give further consideration to tenure from a policy standpoint in Chapter 12.

Faculty Attitudes Toward Their Careers

Most faculty members express favorable attitudes toward their careers. In a 1973 survey (Bayer, 1973, p. 30) an overwhelming majority indicated that if they could retrace their steps, they would again choose the academic life. Only 14 percent responded negatively; 18 percent, however, said that if given another chance they would choose a different discipline. These responses were roughly the same for faculty members in universities, four-year colleges, and two-year colleges. Other studies from the mid-1970s confirm these results. A huge majority of faculty members said they were moderately or strongly satisfied with their career choices. Only 10 or 12 percent indicated that if they were to begin their careers again, they would definitely or probably not want to be professors (Ladd & Lipset, 1975–76, May 3, 1976; Carnegie Foundation for the Advancement of Teaching, 1977, pp. 82–84). Moreover, most faculty expressed satisfaction with the particular institutions to which they were attached. When asked, "In general, how do you feel about this institution?" half responded, "it is a very good or fairly good place for me" and only around 9 percent that "it is not a good place for me" (Trow, 1975, p. 28; Ladd & Lipset, 1975–76, May 3, 1976). Responses for different types of institutions were also quite similar. It should be noted, however, that the job satisfaction of part-time faculty, who now constitute about one-third of all faculty members, likely is much lower. (See, for example, Tuckman et al, 1978).

 Boberg and Blackburn (ca. 1983, p. 9) report on the basis of their surveys that a major factor in faculty satisfaction with their jobs is "their concern for quality—in their students, in their colleagues, in their work environment." During the past decade or more, faculty members in many but not all institutions have perceived a diminution of quality and this has been a major source of discontent and poor morale. What Boberg and Blackburn learned (p. 11) is "that faculty liked their career choice. They want to be professors. What they are unhappy about is the conditions of work. Tending to these is one way to improve the quality of life for an institution's most vital personnel."

 A 1980 survey by Helen Astin (Table 3–8) also suggests that faculty attitudes toward their jobs were generally up-beat.

 Our own 1984 interviews with faculty members and administrators (discussed in Chapter 8) tend to corroborate the data in Table 3–8, although widespread concern about faculty morale suggests that an additional half-decade of financial stringency likely has reduced the proportion of faculty who would claim to be "very satisfied" with their careers.

 Turning to more specific matters, a 1975 survey revealed that 62 percent of faculty members thought the intellectual environment in their departments was excellent or good, and only 9 percent thought it was poor. Three-quarters of the faculty thought most undergraduates to

Table 3–8. Career Satisfaction Expressed by Faculty Members, 1980, by Academic Ranks

	Professor	Associate Professor	Assistant Professor	Instructors & Others	All Ranks Combined
Very Satisfied	50%	37%	35%	42%	43%
Satisfied	36	43	38	39	39
Marginally Satisfied	11	16	21	16	15
Not Satisfied	3	4	6	3	3
	100%	100%	100%	100%	100%

SOURCE: H. Astin, 1980.

be basically satisfied with the education they are getting, and two-thirds expressed the opinion that most faculty are strongly interested in the academic problems of undergraduates (Trow, 1975, pp. 28–29). On the other hand, in another study 52 percent agreed "that most colleges reward conformity and crush student creativity," and 57 percent, "that undergraduate education would be improved if education were less specialized" (Ladd & Lipset, 1975, pp. 355–57). A large majority (74 percent) spoke well of the degree of academic freedom prevailing in their institutions, though 23 percent felt there were some important constraints and 3 percent felt that there was little academic freedom (National Education Association, 1979, p. 12).

Similarly, Baldridge and others (1978) found high morale as measured by various indicators among only 50 to 60 percent of the faculty as shown by the responses recorded in Table 3–10.

Finally, in an informal survey of elderly persons who have been members of the American Association of University Professors for more than fifty years, about two-thirds expressed the opinion that both tenure and academic freedom have become stronger over the years and only about one-tenth thought they have become weaker (*Academe*, Nov.–Dec. 1983, p. 9).

Though the general mood of the faculty seemed on the whole positive (at least at the time of the various available surveys), large numbers also offered criticisms or suggestions about the state of higher education. Prominent among these were complaints about due process in decisions affecting faculty and about inadequate opportunities for the faculty to receive information about the affairs of their institution and to participate in policy decisions (National Education Association, 1979, p. 13). Other areas for which a majority thought improvement was needed included faculty compensation, financial support for professional growth, promotion policies, evaluation of faculty, and teaching loads (pp. 14–15).

Preponderantly favorable attitudes of faculty toward their profession do not guarantee high morale. Evidence for this is found in faculty responses gathered by the National Education Association in 1977–78 when depressed conditions in higher education were actually being felt or were widely anticipated. They show one-fourth or more of the faculty experiencing low morale (see Table 3–9). These responses were about the same for faculty in four-year institutions and in community colleges. Our 1984 interviews (discussed in Chapter 8) also revealed perceptibly weakened morale. The cause appeared to be not only adverse trends in compensation and working conditions, but also a pervading sense of insecurity for the future and a sense of the declining status of the profession. We find no inconsistency between the relatively high levels of job satisfaction reported previously and the shaky condition of faculty morale encountered in our recent campus visits. The general mood in 1984 was somber; faculty spirits had sagged. However, the problem had not yet reached the proportion of a crisis. The great majority of faculty were carrying on as usual.

Attitudes and Basic Values of the Professoriate

In this section we consider the attitudes and values held by American faculty members, classified by various disciplinary groups and by different types of institutions.

Disciplinary Groupings. The American professoriate is in many ways a homogeneous professional group with shared interests and values. At the same time, it is composed of persons identified with hundreds of different disciplines and sub-specialties. And people in the various disciplines tend to differ as to personal and educational backgrounds and world outlooks. Each discipline attracts individuals of particular talents and interests, and the experience of working in each field places its mark on their personalities. Many studies have shown that faculty members in different fields exhibit significantly different personal characteristics and attitudes. Ladd and Lipset (1975–76, Sept. 15,

Table 3–9. Faculty Responses to Questions About Their Morale, 1977–78

	My morale is:	The morale of other faculty seems to be:	Compared with that of five years ago, the morale of other faculty seems to be:
Very high	22%	5%	9%
Fairly high	50	53	40
Fairly low	23	37	35
Very low	5	5	16
Total	100%	100%	100%

SOURCE: National Education Association, 1979, p. 12.

Table 3–10. Indicators of Faculty Morale, by Type of Institution, 1978
(Percentage of Respondents Expressing the Attitude)

	High Trust	High Satisfaction	High Institutional Identification
Private Multiversity	66%	80%	46%
Public Multiversity	62	76	48
Public Comprehensive Institutions	54	56	47
Elite Liberal Arts Colleges	73	80	47
Other Liberal Arts Colleges	63	60	56
Community Colleges	56	54	56
Private Junior Colleges	53	48	62
All Institutions	60	59	53

SOURCE: Baldridge *et al.*, 1978, p. 41.

1975) in their extensive studies of the professoriate conclude, " . . . we commonly find greater differences of opinion among the various scholarly disciplines than we can locate among the most grossly differential groups in the general public, such as rich and poor, young and old, and white and black." They show that both in academic matters and in politics, the several major divisions of the faculties vary systematically. The most liberal are those in social sciences; next in order are faculty members in the humanities and the natural sciences; and the most conservative tend to be those in the applied professional fields. Table 3–11, which presents a 1975 sampling of faculty opinions on political and academic matters, shows the considerable variations by academic disciplines.

Institutional Groupings. Given the considerable differences among types of institutions, it might be supposed that faculty attitudes and values would differ as much among the types of institutions as among the various disciplines. But this is not the case. There are differences to be sure, but they are not as pronounced as might be imagined. For example, Ladd and Lipset (1975–76, Oct. 27, 1975) found that faculty members in major universities were inclined to somewhat more "liberal" positions on political issues such as busing to achieve school desegregation, capital punishment, or collective bargaining on the campus than their counterparts in "lower tier" institutions, but the differences were not striking. Even on the question of the relative importance to be given to teaching and research, the differences were small—not because faculty in the lower-tier institutions favored heavy emphasis on research but because faculty in the high prestige institutions placed pronounced importance on teaching (ibid., April 12, 1976).

In a survey of faculty in 1972–73, Bayer presented a large sample of faculty members with a list of goals and asked them to indicate which were "essential" or "very important" in their teaching of undergra-

Table 3-11. Percentage of Faculty Members Expressing Opinions on Various Political and Academic Issues, by Academic Discipline, 1975

	Cut welfare spending[1]	Oppose capital punishment[1]	Favor busing[1]	Favor policies to reduce income differences[1]	Favor collective bargaining in academe[2]	Oppose any restrictions on freedom of inquiry[3]
Social Science	22%	64%	65%	78%	79%	57%
Humanities	26	56	62	73	77	47
Fine Arts	—	46	49	55	—	—
Physical Science	41	41	46	59	66	37
Biological Science	41	36	36	52	61	45
Business Administration	56	27	30	33	65	36
Engineering	—	26	28	43	53	30
Agriculture	58	19	26	32	34	30
Education	—	40	47	60	74	41
Medicine	—	35	46	56	50	—
Miscellaneous applied fields	—	23	31	39	—	25

[1] Ladd & Lipset, 1975–76, Nov. 3, 1975
[2] Ibid., Feb. 7, 1976
[3] Ibid., March 15, 1976. This question was related specifically to research on the heritability of intelligence.

Table 3–12. Responses of Faculty to Questions Regarding the Goals of Their Undergraduate Teaching

	Percent Indicating Goal Is Essential or Very Important			
	Two-Year Colleges	*Four-Year Colleges*	*Universities*	*All Institutions*
Goals for which there was close agreement:				
To master knowledge in a discipline	92	92	91	92
To increase the desire and ability to undertake self-directed learning	86	90	89	89
To develop ability to think clearly	97	98	96	97
To prepare students for employment after college	69	61	60	62
To provide tools for the critical evaluation of contemporary society	55	62	53	56
To develop religious beliefs or convictions	11	13	5	10
Goals for which there was less agreement:				
To convey a basic appreciation of the liberal arts	55	65	46	55
To develop moral character	55	52	38	47
To provide for students' emotional development	51	44	33	41
To develop responsible citizens	72	62	53	60
To prepare students for family living	35	24	16	23

SOURCE: Bayer, 1973, p. 25.

duates. The responses were quite similar among three types of institutions for goals relating to the intellectual development of undergraduates, but differed considerably for goals pertaining to the personal and moral aspects of human development. In view of common perceptions about how much community colleges differ from four-year institutions, perhaps Bayer's most startling discovery was the extent to which the values of faculty members at the two types of institutions were commonly held.

From available data, one can only conclude that basic attitudes and values are not sharply divergent among colleges and universities of various types, but that significant differences among faculty members are present on each campus and are more closely related to discipline than to type of institution.

Basic Values. Our review of the attitudes of faculty members would be incomplete without reference to their basic values. These values are derived from long academic tradition and tend to be conveyed from one generation to the next via the graduate schools and also through the socialization of young faculty members as they are inducted into their first academic positions. These values may be subsumed under three main categories: the pursuit of learning, academic freedom, and collegiality.

The pursuit of learning and its dissemination are regarded in the value system of faculties as the main functions of colleges or universities. Institutions of higher education may have other legitimate functions such as helping students achieve broad personal development and providing agreeable experiences for them, but the central task is the advancement and diffusion of learning. The primary responsibility of each faculty member, therefore, is to be a learned person and to convey this learning through discussion, teaching, and publication. The work of a faculty member is evaluated primarily on the basis of performance in discharging these duties. In this context, learning refers to the *truth*— that which, so far as possible, is judged to be true. Faculty members are expected to be loyal to the truth wherever it leads, even when the truth is inconvenient, unpopular, or contrary to widely accepted dogma. The pursuit of learning is sometimes intended for practical ends and sometimes for knowledge valued for its own intrinsic interest. In either case, it may lead to unexpected findings or applications or consequences. Colleges and universities should not limit themselves to knowledge intended only for practical application.

Academic freedom includes the right of faculty members to substantial autonomy in the conduct of their work, and to freedom of thought and expression as they discover and disseminate learning. This freedom is essential to the advancement of learning. To ensure this freedom, faculty members ordinarily should be evaluated by their peers who may include local colleagues and sometimes outside specialists. To reinforce academic freedom, faculty members should be given lifetime tenure subject to safeguards relating to a probationary period, dismissal for cause, and financial exigency on the part of the employing institution. To further reinforce this freedom, employing institutions should be able to operate with a minimum of detailed supervision from outside agencies such as the central offices of multi-campus universities, statewide coordinating bodies, state budget offices, legislative committees, federal agencies, and private donors and sponsors.

Collegiality is a many-faceted concept. It includes faculty participation, through committees and senates, in the affairs of the institution, especially in educational matters such as admission of students, curricula, degree requirements, and faculty appointments and promotions. Generally, faculty members believe they should be informed at least, if not consulted, on other matters of departmental or institutional significance, for example, campus building plans, finances, appointments of presidents and deans, and the like. Collegiality refers also to membership of faculty persons in a congenial and sympathetic company of scholars in which friendships, good conversation, and mutual aid can flourish. Collegiality contains in addition the ideal that knowledge within any one field is worth as much as knowledge in any other field, and therefore that no faculty member should receive preferment over any other simply on the basis of academic field.

The ideal academic community from the point of view of faculty is a college or university in which the three values—pursuit of learning, academic freedom, and collegiality—are strongly held and defended. However, these values are seldom achieved totally. Colleges and universities are beholden to governments, private donors, foundations, religious organizations, and to their own students for the resources needed to carry on their functions, and the values of these groups cannot be ignored. Yet the three basic academic values as we have defined them are ideals that are widely held by faculty members, and deviations from them are often bitterly resisted. Only when the influence of these values is understood does the behavior of faculty become explicable.

The three values are generally well developed in the graduate schools, where prospective faculty members prepare for their careers. These are usually mature and well-established institutions where academic tradition has taken root. The concepts of learning, collegiality, and academic freedom are sometimes, though not necessarily, less firmly entrenched in the less mature institutions where most young academic persons become faculty members. There is sometimes a mismatch between the socialization of these persons, many of whom were trained at research-oriented universities, and the conditions in the colleges and universities where they serve on the faculty, with resulting disillusionment and dissatisfaction (see, for example, Drew, 1985). The mismatch has probably been worsening in recent decades as institutions have grown larger and more impersonal, as administrative decisions have moved up from local campuses to system offices, as legislatures have taken a keener interest in the detailed budgets and administrative decisions of colleges and universities, and as an increasing proportion of higher education takes place in urban settings which are often less hospitable to collegiality than rural or small-town environments.

CHAPTER FOUR

Emergent Groups Within the Faculty

We turn now to a consideration of three particular sub-groups of the professoriate: women, minorities, and part-time faculty. The emergence of these groups has contributed new characteristics and diversity to the makeup of American faculties. First, we consider the entry of women and minorities into the professoriate. Then we examine the growing phenomenon of part-time faculty members.

Women[1]

Men have long dominated the professoriate. They have held a majority of the positions and their dominance has been even greater in the senior ranks. However, at least since the Civil War, women have occupied a significant place on higher educational faculties. It has been estimated that as long ago as 1869–70 about 12 percent of the faculty were female. Just prior to World War I, in 1909–10, about 20 percent of faculty positions were occupied by women. From 1920 to 1960, the percentage fluctuated around 25 percent, and from 1960 to 1981 it rose rapidly to 34 percent by 1980–81 (National Center for Education Statistics, *Digest of Education Statistics*, 1983, p. 101).

Table 4–1 provides another set of estimates of the percentages of faculty positions held by women during the years 1960–61 to 1982–83. This was a period when a dramatic increase in the number and proportion of women in faculty positions occurred. Their share of the full-time jobs increased from about 17 to 27 percent and of the part-time positions from 23 to 36 percent. In terms of total number of positions

1. In preparing the sections on women and part-time faculty we had the benefit of valuable literature reviews by, respectively, Barbara Koolmees Light of Whittier College and by Eileen Heveron of The Claremont Graduate School. These reviews are available on request. For details see Preface.

held by women, the increase was 101,000 full time, 61,000 part time, and 121,000 full-time-equivalent[2] positions. Much of the increase was directly attributable to affirmative action and civil rights legislation (H. Astin and Snyder, 1982). However, as of 1985 the struggle for full equality of women in academe has not yet been fully won. Not only do they hold far fewer than half the faculty positions, they have not yet achieved equality in other respects. Although found in all types of institutions, they are most heavily represented in community colleges, and least represented in universities (National Center for Education Statistics, *The Condition of Education*, 1980, p. 122). When comparisons are made for tenured faculty, differences by type of institution are even more pronounced. Similarly, the percentage of women has been lower in institutions judged to be of high educational quality than in those of lesser quality (Trow, 1975, pp. 6–7). Other studies have shown that women tend to be concentrated in the lower academic ranks (see, for example, Tuckman & Chang, 1983, p. 20). Only 10 percent of professors and 20 percent of associate professors are women, whereas 35 and 52 percent, respectively, of assistant professors and instructors are women.[3] Along the same line, women are relatively more numerous among part-time faculty than among full-time faculty. In 1981–82, women occupied 36 percent of the part-time positions as compared with 27 percent of the full-time positions (Table 4–1). Finally, women's salaries have been lower than men's at comparable academic ranks (ibid., p. 107. See also Astin & Snyder, 1982, p. 29; Gappa & Uehling, 1979, pp. 41–44; Menges & Exum, 1983, pp. 123–25).

Nothing so clearly presages the increasing participation of women in the nation's faculties as the spectacular rise in their attendance at graduate schools and in the number of advanced degrees received. They have caught up with men in attendance at graduate schools and in the number of master's degrees awarded, and as of 1984–85 they were receiving more than one-third of all the Ph.D.s awarded and one-fourth of the first professional degrees. By no means do all of those with advanced degrees, as with their male counterparts, wish or intend to seek academic careers. Highly educated women are fanning out into many fields. There is virtually no professional calling from which they are excluded. But many of them do seek academic careers.

Whereas the number of Ph.D.s awarded to men has been declining steadily since 1972–73, the number awarded to women has been increasing sharply: from 1969–70 to 1984–85 the number of Ph.D.s awarded to men declined by about 4700 (an 18 percent drop) while those earned by women shot up by about 8400 (a 212 percent increase).

2. "Full-time-equivalent" is defined here as total full-time positions plus one-third of part-time positions. It represents the number of full-time persons who would be needed if only full-time persons were employed.

3. National Center for Education Statistics, *Digest of Education Statistics*, 1983, p. 103. See also Carnegie Council on Policy Studies in Higher Education, 1980, p. 310.

Table 4–1. Percentage of Faculty Positions Held by Women,
Selected Years 1960–61 to 1982–83

	Full-Time Faculty	Part-Time Faculty	Full-Time and Part-Time Faculty	Full-Time-Equivalent Faculty
1960–61	16.9%	23.2%	19.1%	17.7%
1965–66	18.2	25.0	20.0	19.0
1970–71	20.9	30.8	23.0	21.8
1975–76	24.1	34.0	27.1	25.5
1980–81	26.0	35.9	29.1	27.4
1981–82	27.0	36.4	30.0	28.2

SOURCE: Estimates of the authors. The basic data on number of instructional staff of the rank of instructor or above in institutions of higher education are shown in National Center for Educational Statistics, *Digest of Education Statistics*, 1983, p. 103. The figures on the distribution of faculty between men and women were derived from many sources including: National Center for Education Statistics, *Digest of Education Statistics*, 1983, table 94, p. 103; table 90, p. 102; table 92, p. 103; table 6, p. 11; and comparable tables in previous annual issues of the *Digest*; American Council on Education, *Fact Book on Higher Education*, 1984–85, pp. 119–21.

Consequently, the overall total has been just about stable since 1970–71. The figures in Table 4–2 on Ph.D.s awarded to men and women show unmistakably what has been happening: in a decade and a half the women's share of the Ph.D. market mushroomed from little more than one of every eight Ph.D.s awarded to almost three of every eight.

Though women have been represented in higher education for more than a century, they have only recently aspired in large numbers to sustained academic careers and only recently have received the support of affirmative action and of new liberal attitudes toward their entry into the academic profession. Many of the career faculty women who have entered the profession within the last decade are still in relatively junior positions. Given the substantial progress in the numbers of women in academic careers, and in their acceptance and promotion, one may reasonably expect that persisting discrimination in status and rewards for women will be largely eliminated in the years ahead. In concluding a careful study of the progress of women in higher education, Astin and Snyder provided a balanced overview (1982, p. 26):

> ... the women's movement and affirmative action legislation have prompted higher education institutions to take significant steps to remedy a history of neglect of women. Academic men and women today are treated more nearly on a par. Women are better represented on campus. Even though there are still discrepancies in rank and salary, the gap has diminished considerably. There is a better balance between teaching and research for both men and women. Women engage in research and publication to a greater extent now than they did in the early 1970s.
>
> This response by the higher education community is encouraging, but perhaps most important is the fact that academic men and women are coming to know and respect each other.

Table 4–2. Number of Ph.D.s Awarded to Men
and Women, 1969–70 to 1984–85

	Men	Women	Total	Percentage of Women
1969–70	25,890	3,976	29,866	13.3%
1970–71	27,530	4,577	32,107	14.3
1971–72	28,090	5,273	33,363	15.8
1972–73	28,571	6,206	34,777	17.9
1973–74	27,365	6,451	33,816	19.1
1974–75	26,817	7,266	34,083	21.3
1975–76	26,267	7,797	34,064	22.9
1976–77	25,142	8,090	33,232	24.3
1977–78	23,658	8,473	32,131	26.4
1978–79	23,541	9,189	32,730	28.1
1979–80	22,943	9,672	32,615	29.7
1980–81*	22,700	10,200	32,900	31.0
1981–82*	22,600	10,700	33,300	32.1
1982–83*	22,100	11,200	33,300	33.6
1983–84*	21,700	11,800	33,500	35.2
1984–85*	21,200	12,400	33,600	36.9

SOURCE: National Center for Education Statistics, *Projections of Education Statistics,* 1982, p. 70.

* Estimated

Minorities

Just as women are assuming an increasing role in higher educational faculties, so are minorities, though available information on minority participation is less abundant than it is for women. Separate (not necessarily comparable) surveys made between 1973 and 1980 suggest an upward trend in the percentage of all faculty members who are of minority background as follows:

1973 6.2%[4]
1975 6.9 [5]
1976 7.2 [6]
1980 9.1 [7]

Another survey covering only doctoral scientists, engineers, and social scientists employed in higher education revealed that the minority percentage had risen from 5.5 in 1973 to 7.6 in 1979 (National Science Foundation, 1981a).

4. H. Astin, 1973.
5. Carnegie Council on Policy Studies, 1980, p. 310.
6. National Center for Educational Statistics, *The Condition of Education,* 1978, p. 194.
7. H. Astin, 1980.

A study by the National Research Council showed in some detail the trend in employment of minorities by racial and ethnic groups. These data are displayed in Table 4–3. They refer only to persons who hold the Ph.D., are specialized in the sciences, social sciences, or humanities, and are employed in higher education. With these limitations, the data compared three groups: (1) those who had received their Ph.D.s in the decade of the 1960s before new attitudes and affirmative action had strongly affected the employment of minorities; (2) those who had received their degrees in the 1970s; and (3) all Ph.D. recipients in the included fields regardless of the date of the degrees. As shown in Table 4–3, minorities as a percentage of each relevant group rose from 8.2 to 9.8 percent for the science and social science group and from 4.8 to 7.2 percent for the humanities group. These figures suggest an upward trend of minority employment in academe. Lest one be carried away by these numbers, one should note that they represent an increase of only about 35,000 in the number of minority persons holding academic positions. The greatest gains appear to have been achieved by blacks and

Table 4–3. Percentage Distribution of Persons Who Hold Ph.D.s in Science, Engineering, Social Science, and Humanities and Who Are Employed in Higher Education, by Ethnic and Geographic Origin[1]

	Recipients of the Ph.D. between 1960 and 1969		Recipients of the Ph.D. between 1970 and 1978		All Recipients of the Ph.D. Regardless of When Degree Was Received	
	Sciences[2]	Humanities	Sciences[2]	Humanities	Sciences[2]	Humanities
Non-minorities						
U.S. born	80.4%	84.8%	83.1%	84.9%	81.8%	85.0%
Foreign-born	11.4	10.4	7.1	7.9	10.1	9.0
Sub-total, non-minorities	91.8	95.2	90.2	92.8	91.9	94.0
Minorities						
U.S. born						
Hispanics	0.5	1.4	0.8	1.8	0.6	1.6
Blacks	0.5	1.1	1.5	1.6	1.0	1.4
Asian	0.8	0.2	0.5	0.2	0.6	0.2
American Indian	0.5	0.2	0.4	0.4	0.3	0.4
Foreign-born						
Hispanics	0.2	0.9	0.4	1.5	0.2	1.3
Blacks	0.1	—	0.6	0.4	0.3	0.2
Asian	5.7	0.9	5.7	1.2	5.1	1.0
Sub-total, minorities	8.2	4.8	9.8	7.2	8.1	6.0
GRAND TOTAL	100.0%	100.0%	100.0%	100.0%	100.0%	100.0%

[1] Includes Ph.D.s in all subjects except professional fields such as theology, business administration, home economics, journalism, speech and hearing sciences, jurisprudence, social work, library science, and education.
[2] Includes sciences, engineering, and social science.
SOURCE: National Research Council, *Employment of Minority Ph.D.s: Changes Over Time*, 1981, p. 22.

foreign-born Hispanics. Foreign-born Asians are notable for their pre-eminence in the sciences even though they showed only a modest gain in their overall percentage.

The overwhelming conclusion from all the data on minorities in higher education is that they continue to be severely underrepresented. (Menges & Exum, 1983, pp. 125–27). The minority groups combined constitute 20 to 25 percent of the American people, but they probably occupy not more than 8 to 10 percent of academic positions. And this number is strongly influenced by the relatively heavy participation of Asian minorities without whom the minority showing would be even more depressing. The rate of gain will probably continue to be slow at best as it is controlled not only by the rate at which minority persons become qualified for academic positions but also by the decrease in the overall rate of growth of faculties. But even slow growth in the propor-tion of minorities is far from assured. The prospect of actual decline—possibly sharp decline—is altogether real, as discussed in Chapter 8.

Recent data indicate that growth in black enrollments in higher education has been sluggish. From 1970 to 1980, college enrollments of blacks increased only from 26 to 28 percent of black high school gradu-ates 18 to 24 years of age. Moreover, black undergraduate enrollment as a percent of all college students, though it increased from 1970 to 1976 from 6.8 to 10.3 percent, leveled off thereafter and has actually begun to decline slightly. Also, the number and percentage of full-time graduate students who are black has not grown since 1972 and in fact has declined slightly. And awards to blacks of advanced degrees have held fairly steady from 1976 to 1981 with the number of doctoral degrees increasing only slightly (from 1,200 to 1,300). It is clear that much of the momentum has left the affirmative action movement as it affects blacks.[8] We were unable to find comparable data for other minority groups.

Part-Time Faculty

A subgroup of increasing size and visibility is the part-time faculty.[9] Indeed, few features of the faculty have changed as markedly in recent years, or bear such significant implications, as the use of part-time

8. The figures are derived from National Center for Education Statistics, *Participation of Black Students in Higher Education*, 1983, pp. 4, 6, 7, 8, and 14. For a valuable discussion of minority faculty see Commission on Higher Education of Minorities, *Final Report*, especially pp. 37–38.

9. Another category of workers usually employed fewer hours than regular faculty is that of graduate assistants. They are not included here among the part-timers because under our definitions they are apprentices and not members of the faculty proper. It gives us pause to realize, however, that if the teaching assistants were included as part of the faculty, about one-fourth of all the teaching in our colleges and universities could be said to be performed by part-time persons.

faculty. It is a highly variegated subgroup, and generalizations are risky. Part-time people are employed, usually, for a quarter or less up to a half of a full load. There is nothing new about the presence of part-timers in American higher education. Colleges and universities have long employed local professional people—such as practising lawyers, doctors, or accountants—as adjunct teachers. Frequently, they also have employed women who have left the full-time labor market, semi-retired persons, and others who happened to be qualified and available on a part-time basis as needed. In recent years, however, there has been a surge in number of part-time persons. As shown in Table 4–4, the number has increased nearly threefold, from 82,000 in 1960 to 220,000 in 1982. Given such substantial growth in absolute numbers, it is not surprising to find a major swing in number of part-timers as a percentage of the total academic labor force. In 1960–61 they accounted for 35 percent of all faculty persons and 15 percent of full-time-equivalent faculty. By 1970–71, their numbers had declined to 22 percent of all faculty and 9 percent of full-time-equivalents. Thereafter, the percentages rose to 32 percent of all faculty and 13 percent of all full-time-equivalents. What accounted for this extraordinary swing in the relative number of part-timers?

We believe that the heavy use of part-timers in the early 1960s was a carryover from the financially depressed 1950s and was due also to a shortage of qualified full-time faculty during the 1950s. The decline in relative numbers of part-timers during the 1960s occurred partly because of the prosperity of higher education during this period and partly because of an increasing flow of new Ph.D.s to the academic profession. The subsequent increase in the relative number of part-timers in the 1970s and early 1980s was a result of at least five factors: (1) the declining prosperity of higher education, (2) the effort of institutions to achieve flexibility of staffing in a time of uncertain and rapidly shifting enrollments, (3) the availability of numerous persons with advanced degrees who have been unable to obtain, or have not sought, full-time teaching positions, (4) the growth of the community colleges which have become heavy users of part-time people, and (5) the expansion of life-long learning programs.[10]

There have been some predictions that the rise in the percentage of part-timers will level off or even decline during the late 1980s and 1990s. It is hypothesized that when institutions are forced to cut faculty, the largely non-tenured part-timers will be the first to go. On the other hand, if the financial position of higher education becomes more precarious, part-timers may be hired in larger numbers as an economy

---

10. An unpublished 1979 report of the Chancellor's Office of the California Community Colleges indicated that in 1979–80 part-time faculty accounted for 64.5 percent of total faculty in community colleges of the state and 36.4 percent in California State University. A similar percentage was reported for the state of Maryland (Albert, 1982, p. 2). These figures refer to head count, not to full-time-equivalent.

Table 4-4. Full-Time and Part-Time Faculty, Instructor and Above, 1960–61 to 1982–83

Year	Number of Persons (000) Full-Time	Part-Time	Total	Number of Full-Time-Equivalents[1]	Part-Timers as percent of Number of Persons	Full-Time-Equivalents
1960–61	154	82	236	181	35	15
1965–66	248	92	340	279	27	11
1970–71	369	104	474	404	22	9
1975–76	440	188	628	503	30	12
1980–81	466	212	678	536	31	13
1981–82	475	220	695	548	32	13
1982–83	475	220	695	548	32	13

[1] Three part-timers are assumed to equal one full-time-equivalent.

SOURCE: National Center for Education Statistics, *Digest of Education Statistics*, 1983, p. 103.

measure. In any event, the possibility of hiring part-time faculty weakens the job market for potential faculty who are available and qualified for full-time work. A disciplined retreat from the practice of employing numerous part-timers would create openings for thousands of well-qualified full-time faculty.

Tuckman and his associates (1978a, p. 23) have classified part-time faculty into six categories. These categories, together with the percentages of part-timers assigned to each, are as follows:

Semi-retired persons	3%
Graduate students teaching in institutions other than the one where they are pursuing their studies	24
"Hopeful full-timers" who are seeking a full-time job but unable to find one	19
"Full-moonlighters" who are serving as teachers simultaneously in two or more institutions	32
"Homeworkers"	7
"Part-time moonlighters" who have a part-time or full-time job outside academe	15
Total:	100%

Tuckman provides detailed data on the characteristics of part-timers (see Tables 4–5 and 4–6). They indicate that most part-timers achieve little job security and far less than full participation in faculty affairs. Regarding their compensation, Tuckman and his associates found conflicting evidence. If earnings of part-timers and full-timers were compared without reference to rank, the part-timers were clearly underpaid, often substantially. But if the comparison were made with rank controlled, then the pay of the two groups was about equal. However, most part-timers have been relegated to the lowest ranks. Tuckman's conclusion was that if the comparative compensation were adjusted for differences in personal characteristics, the part-timers would probably prove to be underpaid.

There has been much discussion of the influence of part-timers on quality of instruction. Everyone concedes that many of them are highly capable and add to the quality and diversity of available talent. Some of the most brilliant and capable physicians, lawyers, scientists, and public figures in the nation may be counted among them. On the other hand, many are of mediocre talent and training. The range of ability among part-timers is undoubtedly wider than among full-timers. One suspects that the average ability level among them is lower than that for full-timers, though there is no hard evidence on this matter. Part-timers have the disadvantage also that they usually do not become part of the

Table 4–5. Personal Characteristics of Part-Timers

Percent with Ph.D.	19.7	Percent previously full-time	28.9
Percent with M.A.	49.7	Average age	40.0
Percent female	38.7	Average years taught full-time	2.5
Percent married	76.5	Average years taught part-time	4.6
Percent with spouse in academe	13.7	Percent who have published an article	19.5
Percent with (resident) children	58.8	Average number of articles published	1.1
Percent Black	3.1	Earned income, 1976	$14,826
Percent Caucasian	91.7	Total household income, 1976	$23,410

SOURCE: Tuckman *et al.*, 1978a, p. 24.

academic community. They are not available to bear their share of student advising, of participation in educational policy-making, and of the intellectual discourse of a campus. By their not participating fully, the burden of maintaining the institutions falls increasingly on the full-time resident staff. So long as the part-timers are few in number, the problem is not serious. But when part-timers become numerous (in some community colleges they exceed the number of full-time faculty) the problem approaches the intolerable. In a 1977 survey, about 19 percent of the faculty respondents in four-year institutions gave the opinion that their institutions were employing too many part-timers, while in the two-year colleges this percentage was 43 percent (National Education Association, 1979, p. 13). The growing dependence of academe on part-time faculty is widely regarded as a serious problem (Cohen & Brawer, 1977 and 1982; Friedlander, 1983; Moodie, 1980; Mayhew, 1980). We agree. We believe that the time has come for institutions to review the part-time phenomenon and to accomplish an orderly but partial retreat from the practice of employing part-timers. In particular, the candidates for elimination would be those who are poorly paid, basically dependent on their academic jobs, and professionally on the periphery of their departments.

Our review of part-time faculty leads to a consideration of what might be called "the marginal faculty." This subject is frequently discussed and is a favorite topic of the public press. The "marginal faculty" consists of persons with various academic credentials who have, or hope to have, a connection with higher education but who have not become fully integrated into the academic community to their satisfaction. Some are underemployed; some work full time but in non-academic settings which they find unpalatable; some are serving full time in college teaching or research but on contracts of short duration with few prospects of continuous appointments; some are engaged in part-time college teaching or research with temporary appointments; a few are totally unemployed. "The Ph.D. who drives a taxi" or the "lost generation of scholars" or "the gypsy faculty" are phrases that are frequently used to describe these people. It should be underscored that not all part-time faculty are in the "marginal" category. Some of them are quite satisfied with their part-time appointments, but some are marginal in the sense that they hope for full integration into academe. The extent of the marginal faculty problem must be judged in relation to the general rates of unemployment among Ph.D.s. Careful studies have revealed that these unemployment rates are remarkably low. Though many academic people may be placed in various insecure or non-preferred situations, the vast majority are, in fact, employed, even those in the humanities. We explore this matter further and in detail in Chapter 9.

Concluding Comments

In this and the previous two chapters, we have tried to produce a portrait of the American professoriate. We have described the tasks and responsibilities of professors, we have identified their roles in American society, and we have sketched their leading characteristics. We have done all this with some sense of change or trend over time.

The functions of the faculties, which once were almost exclusively the teaching of undergraduates in a traditional classical curriculum, have expanded and include research, public service, and participation in the governance of the colleges and universities. They have also expanded to include a much broader range of subjects as the natural sciences, the social sciences, and numerous professional and vocational fields have been introduced and have taken root. In the course of these changes the higher educational system has been transformed from a predominantly religious to a largely secular orientation. New types of institutions have been devised to accommodate all these changes, notably, the research university, the land-grant university, the regional

Table 4–6. Job-Related Characteristics of Part-Timers

Percent hired to meet enrollment overload	10	Percent with rank below assistant professor	91
Percent in permanent position	17	Percent with regular vote in faculty affairs	24
Percent in evening or continuing education division	60	Percent who feel they are paid at least proportionately to full-time faculty	28
Percent at:		Average contact hours	5.0
two-year institution	53	Average courses taught	1.5
four-year college	34	Average total hours devoted to part-time job	13.5
university	13	Average level of satisfaction on scale of 100	29.9

SOURCE: Tuckman *et al.*, 1978a, p. 27

university, and the community college. The liberal arts college, the prevailing institution of the nineteenth century, still survives and prospers but as a much smaller part of the higher educational system. Yet even it has undergone substantial modification.

Meanwhile, these far-reaching changes in the higher educational system have been accompanied by pronounced trends in the origins and characteristics of the faculties. They now derive from broader strata of society—more from modest family backgrounds, more women, more members of minority groups, and more who serve part time. The faculties also have become more specialized, more rigorously educated, more worldly, and more experienced. They have also become more numerous as enrollments have climbed. The result of these developments has been to transform the American professoriate into a profession that is much more diverse than ever before. In 1915, when John Dewey and his colleagues formed the American Association of University Professors to try to represent more effectively the interests of the "professionalized" professoriate, the faculty, though far from monolithic, nevertheless shared much in common. All in all, the same might be said, though with less force, into the post-World War II era. But the increasing diversity of American postsecondary education means that less and less can we speak of "the faculty" as though it were a totally homogeneous group. The American faculty today is complex and pluralistic. Nevertheless, as we have repeatedly asserted, the socialization that occurs in graduate school and on the job in colleges and universities welds the faculties into a coherent profession with which they have a strong sense of identification and which to a considerable degree transcends differences of background and discipline.

From the information we have assembled and from our own inquiries about faculty people across the country, we conclude that the academic profession is reasonably sound and healthy. On the whole, its members are intelligent, well educated, cultivated, hard working, productive, and honorable. They are mostly of a vigorous age, though the average age may be slowly rising. They place a high value on autonomy, academic freedom, and collegiality, and do sacrifice financial reward for these values. Presidents and deans often report that their present faculties are the best ever. Even so, the outlook for this extraordinary cohort of faculty members, as subsequent chapters suggest, is problematic.

Part Two

PROFESSORS AT WORK

In this section we consider trends in the rewards and working conditions of the profession. We consider first work load, time use, and performance of faculty. Next, we turn to faculty compensation and conditions in the work environment. Finally, we report our findings based on extensive campus visits.

Though the trends we find are mostly disturbing, not yet have they caused major defections from the profession or have they impaired performance drastically. But there is already evidence of diminished morale and a downturn in the entry of promising young people into the profession. A continuation of the downward slide would almost surely reduce the competitive power of the profession to attract and retain talent.

CHAPTER FIVE

The Quantity and Quality of Faculty Work

In this chapter we focus on the work of faculty. We raise and attempt to answer two centrally important questions: How do faculty use their time? What is the quantity and quality of their work?

The Workload

Over recent decades, the basic duties of academic people have remained largely unchanged. As shown in Chapter 2, these tasks are: instruction, research, public service, and institutional governance and operation. In performing these tasks, faculty members must also continue to maintain and build their intellectual capital. That is to say, they must "keep up" in their disciplines and in broad intellectual discourse as well. Keeping up is done partly in connection with preparing for teaching, partly as a byproduct of specific research projects, but mainly through the constant study of the literature and the learning of new skills. These tasks are almost limitless. There is no end to the amount of time and effort that can usefully be devoted to them. All competent faculty members live with the sense that they are dealing with infinity—that they can never fully catch up. The actual amount of their working time and effort depends on their motivation and energy, and these tend to be quite high on the average. The many studies of the faculty work week indicate that it is generally well above forty hours.

In allocating their time and energy to the several tasks, faculty members have characteristically felt pulled in different directions. For example, a curricular change or a growth of enrollments may call for increased attention to teaching; the desire to earn promotion or to achieve professional recognition may call for greater effort in research; attempts of one's institution to correct a decline in student enrollment may demand more attention to student advising; or the opportunity to consult with a government agency may pull the faculty member in still another direction. And then to all these demands must be added the time that is due to families and to religious and civic organizations. These pressures are partly inner-directed. Faculty members on the whole want to do a good job of teaching—to be active in research; to put in a respectable performance in student advising; to carry their weight in public service activities; to be active in campus and civic life; and above all to keep abreast of their disciplines. Most faculty members learn to cope with the pressures by establishing an allocation of time and effort that yields a tolerable total workload. When changes occur that produce increasing demands, allocations are shifted so that the overall time and effort does not increase drastically or permanently. Therefore, changes in the demands on faculty tend to affect allocations more than overall totals. For example, if an influx of underprepared students causes an increase in the teaching load, faculty members may compensate by reducing the time and effort devoted to "keeping up" with their fields. But if the pressures come simultaneously from all directions, as they sometimes do, faculty members may adjust simply by reducing the quality or intensity of effort all round. Sometimes, however, the pressure from all sides cannot be contained within the preferred work week, and the time and effort will have to be increased. Faculty attitudes toward changes in workload will tend to be negative not only if the pressures require increased time and effort but also if they prevent optimal allocation of time and effort, even if the total is constant. For example, if an increase in enrollment can be accommodated only by encroaching on the professors' research time, there may be as much upset as though the adjustment required an increase in total time and effort. The dilemmas relating to the use of time are particularly severe for younger faculty who are candidates for tenure. They are often expected to establish themselves as teachers, carry heavy classroom loads, and produce publishable research—all in a few years—while meeting obligations to a young family. We shall turn now to the question of whether, and to what extent, the overall workload has increased over the years from 1970 to 1985.

One type of relevant information would be changes in the working hours of faculty. Have they actually increased? Though there have been many one-time reports on working hours we know of no studies with consistently comparable data over time and covering a substantial number of institutions. Our review of the spotty data suggests at least that working hours have probably not decreased since 1970. That conclusion, however, is speculative. Another source of information is the responses of faculty and administrators to more specific questions about the faculty workload. Surveys including questions of this type were conducted during the period 1975 to 1982 (Minter & Bowen, 1977, pp. 29–32; 1978, pp. 45–48; 1980a, pp. 44–47; 1980b, pp. 41–43; May 26, 1982, p. 8). The respondents indicated, for example, that the number of academic programs and courses offered was growing rapidly as institutions were responding to changes in student demands (Bowen & Minter, 1975, pp. 28–31; 1976, pp. 50–52).[1] In 1980 and 1982 the respondents reported that faculty performance in teaching and research was improving, partly because of stricter standards for appointment, promotion, and tenure. They also indicated that teaching loads expressed in number of classroom hours, size of classes, and load of student-advising were rising. These responses, if taken at face value, suggest either that the overall workload was becoming heavier or that some elements of the workload such as "keeping up" or "institutional service" were being squeezed out by increasing demands for teaching and research.

Another partial indicator of changes in the faculty workload is the ratio of students to faculty. For this purpose, it is appropriate to count both students and faculty as full-time-equivalents in order to avoid giving excessive weight to part-time persons. Our calculations (see Table 5–3) show that the ratio fell from 14.7 in 1969–70 to 14.2 in 1982–83, suggesting that the load relating to students might have fallen slightly.[2] These figures take into account only the student-related workload and omit the part of the load related to research, public service, and institutional service. We found that expenditures for research and public service grew more slowly than expenditures for instruction, thus suggesting at least that if research and public service had been included in the calculations, the decline in faculty load would have been greater than the figures above show. We conclude that the calculated workload probably did not increase since 1970, and that the impact of changes occurring since then has been primarily a rearrangement of the content of academic work. Faculties have been forced to spend more time and effort on obligatory tasks which have expanded and less time and effort on discretionary activities. The result may have been the cutting of corners in teaching, the slighting of committee work, and above all the reduction of time for "keeping up" intellectually. As a result, faculty persons, like the buildings and equipment around them, may have themselves become afflicted with deferred maintenance. Their intellectual capital may have gradually diminished and their effectiveness over the long run eroded (H. R. Bowen, 1980c). These conclusions do not apply to every institution or to every professor, but they are probably pervasive enough to be of wide concern.

1. Surveys for the years 1975 to 1978 were confined to private colleges and universities; those for 1980 and 1982 included both private and public institutions.

2. The same calculation when the student load was adjusted to give greater weight to graduate and other advanced students also yielded a slight decline—20.5 in 1969–70 and 19.8 in 1982–83. Cf. H. R. Bowen, 1980b, pp. 37–43.

Time Use

Faculty members are largely autonomous in the use of their time. They are specifically required to be present only for meetings of their classes. On the average, these involve three class sections or twelve hours a week in four-year colleges and universities, and four class sections or sixteen hours a week in community colleges (National Education Association, 1979, p. 11). The required hours vary markedly among institutions of different types and also among faculty members with different commitments to research, departmental duties, and public service (Trow, 1975, pp. 7–8; Ladd & Lipset, 1975, p. 348). Faculty members are also under some obligation to spend at least several hours a week in office hours and faculty committee meetings. Beyond these minimal obligations they are free at their own discretion to work when and as much as they wish. They are judged by overall performance rather than by number of working hours. Indeed, one of the priceless benefits of the academic profession is that it gives the individual worker considerable flexibility in the use of time. In this respect, the professor is more like a freelance author or artist than a slave to a timeclock.

This latitude is based on the reality that people cannot be forced to think or to be creative by controlling their hours of work. People may sit for hours in an office without ever having an idea or being creative, and they may get their best ideas while having lunch with a colleague, walking in the woods, or vacationing in Europe. And the best teaching may take place over coffee or on a picnic. In the academic world, the distinction between work and leisure is inevitably fuzzy. The discretionary nature of academic work does sometimes result in unconscionable shirking and occasionally in outrageous abuses related either to plain laziness and indifference or to excessive engagement in outside remunerative work. Yet, on the basis of innumerable studies of the use of faculty time, one can only conclude that for the great majority the time expended is at least comparable with that of persons who have rigidly enforced work schedules.

The many studies of faculty work loads show almost uniformly that faculty members on the average put in considerably more than the 36-hour week that has become the average for all workers in non-agricultural employment. Many report average hours per week in the range of 55 to 62, and some of course exceed these averages. Many of the reports, however, refer only to the academic year and do not include the lengthy vacation periods when working hours may be fewer or even non-existent. According to tradition, academic vacations were intended to be used in part at least for "keeping up" with one's discipline, engaging in scholarship and research, and achieving personal development and refreshment, travel, professional work, and other rewarding experiences. They were regarded as opportunities for self-development and

Table 5–1. Average Weekly Hours of Work, Full-Time Faculty Members in
Science, Engineering, and Social Science, by Type of Activity, 1978–79

	All Activities	Instruction	Research	Public Service	Remunera- tive Outside Work	Professional Enrichment
All faculty in all institutions	45.8	17.8	11.0	8.6	3.7	4.7
By type of institution:						
Universities	48.2	14.9	15.6	9.5	4.2	4.1
Other 4-year inst.	42.7	21.6	5.0	7.4	3.1	5.5
Public inst.	47.1	17.6	11.9	8.9	3.8	4.8
Private inst.	42.6	18.4	8.7	7.8	3.3	4.4
By highest degree of faculty:						
Doctoral faculty	46.5					
Non-doctoral faculty	42.4					
By academic rank of faculty:						
Professor	47.2					
Associate professor	43.7					
Assistant professor	46.2					

SOURCE: National Science Foundation, 1981b, pp. 1, 22, 35, 56.

professional achievement, and not solely as times for rest or non-professional work. However, the pace of activity undoubtedly slows down
on the average during holidays.

A study conducted in 1978–79 by the National Science Foundation
dealt with the ambiguities related to vacations by recording time-use of
faculty throughout the calendar year (Table 5–1). The basic finding from
these data is that science, engineering, and social science faculty in
four-year institutions work, on the average, 45.8 hours per week year-
round. During the academic year, the average goes up to 50 hours a
week, falling off during the summer to 35 hours a week. The division of
this work among the several major activities of faculty is shown in the
top row of Table 5–1. According to these figures, 17.8 hours per week are
devoted directly to instruction, about 11.0 hours to research, and 8.6
hours to public service. Instruction, research, and public service are the
major outputs of faculty members' efforts and together they account for
37.4 hours, which is slightly more than the average weekly hours of all
workers in private, non-agricultural employment. Faculty on the average also put in 4.7 hours a week on professional enrichment. This
includes a wide variety of activities such as keeping up with the literature of their academic fields, attending formal courses in quest of
advanced degrees, attending professional conferences and seminars,
and visiting museums and libraries. These activities are essential to the

maintenance of faculty members' intellectual capital and are indispensable if the flow of teaching, research, and public service is to be sustained throughout their lives. When the 4.7 hours for faculty enrichment are added to the 37.4 hours in direct teaching, research, and public service, then the weekly total time devoted to the mission of the employing college or university rises to 42.1 hours. Finally, there is the item of remunerative outside work which averages 3.7 hours weekly of faculty time. Even this part of the work week cannot be construed as solely in the interests of faculty members themselves, and not for the benefit of the employing institutions. As we indicate elsewhere (see Chapter 2, pp. 27–28 and Chapter 12, pp. 254–60), most institutions encourage outside work within reasonable limits on the ground that it contributes to the knowledge and worldly sophistication of the faculty and provides a valuable public service. Indeed, one interpretation of the social role of academic faculties is that they are a pool of specialized talent available to society for dealing with wide-ranging problems and needs. The question may be asked whether the faculty members or the colleges or universities which employ them should receive the income from this outside work. The practice has become widespread, though not universal, that the income should go to the faculty member, the probable reasons being that faculty salaries have been low for the caliber of talent involved, and, judging from the figures on working hours, because outside work is usually performed on overtime.

Table 5–2 provides information on weekly hours of faculty work by academic discipline. The figures apply only to faculties in sciences, engineering, and social sciences and only to those in universities. It shows a basically consistent pattern but many variations in detail. The overall work week is shown to be similar for all the disciplines except for mathematics and psychology, where it appears to be inexplicably low.

Another study covering all institutions of higher education and all academic disciplines reveals that there are substantial differences in the work week between community colleges and four-year institutions (National Education Association, 1979, p. 10). The average work week was found to be 49 hours in four-year colleges and 41 hours in community colleges. Even in the community colleges, however, the weekly hours were considerably above the average for all workers in non-agricultural private employment.

Our reading of the facts, based on data from the sources we have cited and many others as well, is that the American professoriate as a whole works at least as long as comparable workers in other industries. Because faculty members are not subject to rigidly specified working hours and because they are not bound to carry on their work at specified work places but may work at home, in the library, in the laboratory, and at other places, they are peculiarly vulnerable to the charge of loafing. As in all organizations, some individuals do not carry their weight.

Table 5–2. Average Weekly Hours of Work, Full-Time Faculty Members in
Science, Engineering, and Social Science, Universities Only, by Type of Activity, 1978–79

	All Activities	Instruction	Research	Public Service	Remunera- tive Outside Work	Professional Enrichment
All faculty	48.2	14.9	15.6	9.5	4.2	4.1
By Academic Discipline						
Physical Science	49.6	13.8	20.5	8.0	3.9	3.4
Mathematics, Statis- tics, Computer Science	40.6	16.7	9.6	7.2	2.9	4.2
Engineering	49.1	15.5	14.7	10.0	5.8	3.0
Life Science	50.6	13.4	18.8	11.0	3.6	3.9
Environmental Science	49.7	13.7	18.8	10.8	1.9	4.4
Psychology	39.5	13.0	9.1	8.8	5.6	3.0
Social Science	48.1	18.2	10.5	8.4	4.8	6.1

SOURCE: National Science Foundation, 1981, p. 3.

But faculty members by and large work long hours and, it may be added, they work hard.

Some critics of higher education do not dispute the claim that faculty members work long hours, but hold that some or much of this work is not productive. Special targets for criticism are research, faculty and committee meetings, and outside activities with and without remuneration. These, it is often alleged, are carried on excessively and at the sacrifice of student learning. Charges of misallocation of time raise a whole set of complex issues pertaining to the role of colleges and universities in research and public service and pertaining also to the relationship between research and public service and the quality of education. These critical matters were considered in Chapter 2.

The Quantity of Faculty Work[3]

Here we shall try to marshall some of the limited evidence on the quantity of work accomplished. We then examine the quality of this effort. The opinion is widespread that both quantity and quality have deteriorated in the years since 1970, when higher education entered a depressed period, and that if the depressed conditions persist, productivity will decline still further. We found this opinion hard to verify or to refute because of the scarcity of evidence on both quantity and quality.

3. In the preparation of this and the following section, Roberta Stathis-Ochoa, California State University, San Bernandino, was a major contributor.

Table 5–3. Trends in the Ratio of Students to Faculty, Selected Years, 1959–60 to 1982–83

	1 Full-Time-Equivalent Students Unadjusted[1]	2 Full-Time-Equivalent Faculty[2]	3 Ratio of Students to Faculty Unadjusted[3]	4 Full-Time-Equivalent Student Units Adjusted[4]	5 Ratio of Student Units Adjusted to Faculty
1959–60	2,777,000	209,000	13.3 to 1	3,756,000	18.0 to 1
1969–70	6,319,000	430,000	14.7 to 1	8,822,000	20.5 to 1
1979–80	8,487,000	605,000	14.0 to 1	11,885,000	19.6 to 1
1982–83	9,221,000*	651,000*	14.2 to 1	12,873,000*	19.8 to 1

* Projected.
[1] National Center for Education Statistics, *Projections of Education Statistics*, 1982, p. 58, and corresponding data from previous issues. See also American Council on Education, *Fact Book*, 1981–82, p. 115.
[2] National Center for Education Statistics, *Projections* . . ., p. 89; and American Council on Education, p. 115.
[3] Column 1 divided by column 3.
[4] Adjusted by giving greater weight to more advanced students relative to freshman and sophomore students. The weights, which are intended to allow for the greater teaching load connected with advanced students, are as follows: freshmen and sophomores 1, juniors and seniors 1.5, and graduate and advanced professional students, 3.0. For a full discussion of this weighting system, see H. R. Bowen, 1980b, pp. 39–42, 263–66.
[5] Column 4 divided by column 2.

We have already considered the quantity of work performed in terms of the use of time by faculty members. From our analysis, we were unable to find any evidence that faculty members are working fewer hours today than they did in former times.

Another approach to the quantity of faculty work is to review trends in the ratio of students to faculty.[4] Table 5–3 presents some figures on student-faculty ratios covering the years 1959–60 to 1982–83. Column 1 of the table shows the numbers of full-time-equivalent students, and Column 2 the numbers of full-time-equivalent faculty. In calculating these numbers, each part-time student was counted as a third of a full-time student, and each part-time faculty member as one-third of a full-time faculty member. Column 3 presents the ratios of students to faculty. It shows that there was a substantial increase during the 1960s when enrollments were rising more rapidly than the growth in faculties. In the 1970s, faculty numbers partially caught up with enrollments and the ratio fell—but not to the 1959–60 level—and has been about stable since.

Columns 4 and 5 of Table 5–3, using a different and perhaps more refined method of computation, also show student-faculty ratios. The full-time-equivalent students shown in Column 4 are adjusted to give greater weight to advanced students relative to freshmen and sophomore students. This is done on the grounds (a) that the time and effort

4. Measurement of both students and faculty presents serious statistical difficulties relating to the treatment of part-time students and faculty, whether to include teaching assistants as faculty as well as students, and whether to give different weights to students enrolled at different academic levels.

required of faculty members is greater for advanced students than for beginners, and (b) that more research is associated with advanced students than with beginners. As the calculations turned out, the *trend* of the adjusted ratios shown in Column 5 was very similar to the *trend* of the unadjusted ratios shown in Column 3. Our conclusion from these data is that the teaching load, as measured by student-faculty ratios, has been heavier in recent years than it was in 1959–60 prior to the great enrollment growth, though not quite so heavy as it was in 1969–70 at the height of the enrollment expansion.

The Quality of Faculty Work

We turn now to the qualitative aspects of faculty productivity. To begin, we draw upon annual reports based on surveys of a representative sample of about 100 private and 100 public colleges and universities conducted during the period 1980 to 1983.[5] As part of these surveys, chief academic officers and chief student affairs officers were asked to express their opinions about changes over the past year in the qualifications, competence, and performance of the faculty in their particular institutions. The 1982 responses as tabulated in Table 5–4 suggest that the performance of the faculties was in most respects either holding steady or improving. Other surveys in the same series for earlier years and also for 1983 produced similar results.

The authors who conducted the surveys commented on Table 5–4 as follows:

> These favorable reports are plausible because of the unique conditions in the faculty labor market over the last two decades. In the late 1950s, the 1960s, and the early 1970s—the boom period of American higher education—great numbers of gifted young people were choosing the academic profession. As they emerged from the graduate schools, they got jobs in the leading colleges and universities throughout the country. By the early 1980s, these people were at the prime of life and had become the cream of the professoriate. In the late 1970s and up until the present, many of the same kinds of people were still coming out of the graduate schools, but the academic labor market had deteriorated. Many of them were forced to seek jobs at second-tier institutions. As a result, those institutions were able to employ people of a quality that had not been generally available to them before. Thus, on the whole, the quality of faculties now at the nation's colleges and universities is probably as high as it has ever been. How many of those people the institutions will be able to hold to retirement, and whether they will be able to replace them with equally gifted people, remains to be seen.[6]

5. W. J. Minter, 1983; and H. R. Bowen, 1980a, 1980b, 1982.
6. Minter & Bowen, 1982, p. 8.

The authors also noted that the responses on quality of faculty members were somewhat more favorable at private institutions than at public institutions; that the only negative response pertained to "commitment to the institution"; and that the response relating to "competence of new faculty" was much more positive than the corresponding response on "quality of overall faculty performance" (p. 8).

The data in Table 5–4 are based on a small (though representative) sample of institutions, and they express not facts, but rather the perceptions of institutional officials some of whom may be biased or poorly informed. However, the validity of the data is enhanced by repetition. Similar surveys have been made over the year since 1975 with responses from presidents, deans, student personnel officers, senior faculty members, and student leaders. The results have been consistent over several years and among the different categories of respondents. In our opinion, the results in Table 5–4 strongly suggest at the least that the performance of faculty has held up reasonably well.

President Donald Kennedy of Stanford University recently observed that the leading private universities in the United States are beginning to face new competition for research funding. The graduate programs at the leading institutions, he said, have been turning out more people with doctorates than are able to find employment in top-ranking universities. Capable scientists are therefore joining institutions which in the past offered little competition. He concluded:

> The average quality of faculty at a broad spectrum of colleges and universities, both public and private, had been growing significantly in recent years. Principal investigators at relatively unknown institutions are becoming increasingly competitive for sponsored research support.[7]

One may ask: How are these conclusions possible when, as we will show (Chapters 6 and 7), the compensation and the work environment of faculty have been generally deteriorating for more than a decade? It would be unreasonable to believe that these changes would not exert adverse effects on faculty productivity with respect to both quantity and quality. They would almost surely reduce to some extent the ability of the faculty to perform well and would almost surely impair motivation. How, then, can the opinions recorded in Table 5–4 be correct? We believe that the explanation lies in the high caliber of persons attracted to the academic profession in the golden years between roughly 1955 and 1975, and who now constitute a faculty corps of unprecedented ability. At the same time, however, there has been a steady erosion of conditions in the profession and this is an increasing drag on the effectiveness, motivation, and morale of the faculty corps. If the time

---

7. *The Times* [London] *Higher Education Supplement*, Feb. 18, 1983. See also Drew, 1985.

Table 5–4. Changes over the Past Year in Qualifications, Competence and Performance of Faculty as Reported by Chief Academic Officers and Chief Student Affairs Officers for Their Institutions, 1982

	Increase	No change	Decrease	Don't know	Total
Public Institutions					
Percentage of faculty with Ph.D. or equivalent	43%	49%	3%	5%	100%
Concern for teaching	25	61	6	8	100
Concern for advising students	36	45	14	5	100
Productivity in research and scholarship	15	63	8	14	100
Willingness to innovate	23	56	15	6	100
Commitment to institution	17	52	25	6	100
Competence of new faculty members	47	40	5	8	100
Rigor of assessing student performance	35	54	4	7	100
Rigor of academic standards	32	57	6	5	100
Quality of overall faculty performance	26	61	7	6	100
Overall quality of learning environment	46	43	6	5	100
Private Institutions					
Percentage of faculty with Ph.D. or equivalent	55	40	0	5	100
Concern for teaching	37	58	2	3	100
Concern for advising students	58	35	5	2	100
Productivity in research and scholarship	23	69	2	6	100
Willingness to innovate	30	56	10	4	100
Commitment to institution	30	51	15	4	100
Competence of new faculty members	61	32	3	4	100
Rigor of assessing student performance	36	55	3	6	100
Rigor of academic standards	30	60	4	6	100
Quality of overall faculty performance	42	51	2	5	100
Overall quality of learning environment	44	50	1	5	100

SOURCE: Minter & Bowen, 1982, p. 8.

comes when the profession cannot attract highly capable people and at the same time the drag on motivation and morale persists, higher education will become qualitatively a disaster area. But that time has not arrived, at least not yet.

Our conclusions on faculty performance are far from definitive. In particular our inquiry did not touch on the subtleties of educational performance. It did not consider changes in the lives of students as a result of their contact with faculty, or the advancement of learning as a result of faculty efforts. So far as it went, however, it did not uncover drastic deterioration of performance. In fact, faculty performance from all indications (including our own field studies) appears to have held up well.

CHAPTER SIX

Changes in Faculty Salaries

BY W. LEE HANSEN[1]

What explains the dramatic decline from 1970 to the early 1980s in the relative earnings position of college and university professors? What forces altered the structure of faculty salaries in this same period? To what extent were the changes linked? These three sometimes baffling questions provide the focus for this chapter. Its ultimate purpose is to yield insights about the determinants of academic salaries and to suggest what might be done to improve academic salaries and thereby strengthen the future health and quality of American higher education.

The questions are by no means new. Numerous attempts have been made to answer the first question ever since the salary position of academics began to deteriorate in the early 1970s. So many plausible explanations have been offered that it is difficult to sort them out. We do know that the evidence supporting the explanations has been sketchy, partial, and at best inconclusive. Considerably less attention has been given to answering the second question about the determinants of changes in the structure of academic salaries.

We are aware of efforts to reduce disparities between the salaries of men and women faculty members. We know that the expansion of demand in certain sectors of the private economy spilled over into public-sector labor markets, including those for academics. But how strong these various effects have been remains unclear. We still need to understand whether or not, and if so how, changes in salary levels and salary structure interact. Has one change influenced the other? What has been the direction of the linkage? And what causal forces account for this linkage?

1. Professor of Economics, University of Wisconsin. This chapter was commissioned by the authors of this book. The chapter will be supplemented by an extended monograph which is to be published by the American Association for Higher Education with the support of the Teachers Insurance and Annuity Association and College Retirement Equities Fund.

Whether having answers to these questions will help alleviate the stresses in academe resulting from the sharply different economic prospects for faculty members and for prospective faculty members now, as contrasted to a decade or more ago, remains unclear. But without a better grasp of what happened and why it happened, we will be handicapped in our ability to deal with the serious salary and compensation problems that have developed and that threaten the future vitality of higher education.

The conclusion of this chapter can be summarized briefly. The most important determinant of the deteriorating salary position of college and university professors is the action of elected officials—state legislators for public universities and four-year institutions and local education board officials for two-year colleges and technical schools. Perhaps, unaware of the consequences of their decisions, they have systematically appropriated smaller percentage salary increases for instructional staff in public institutions of higher education than for other public employees performing functions unrelated to education. An exception is non-instructional personnel in higher education, who have fared about as well as other state and local personnel.

This conclusion rests on the dominance of public institutions. And this dominance casts the public institutions in the role of wage-setter. Because private institutions as a group are in too weak a position to play the role, they seek to maintain their salary positions relative to those of comparable public institutions.

The most important changes in the structure of faculty salaries have come about through the growth of the private sector of the economy and the demand this has created for individuals with highly specialized skills. This has resulted in a bidding up of salaries in academic departments whose members are most vulnerable to outside offers. Institutions have been forced to raise salaries in high demand fields at the expense of salary increases for faculty members in other fields.

These two developments are closely linked. Had the salary position of faculty members not deteriorated so seriously, faculty salaries would have remained more nearly competitive with those in the most vigorously growing portions of the economy. At the same time colleges and universities have come to be recognized increasingly as repositories of sizable amounts of intellectual talent, with the consequence that many faculty members are seen as potentially valuable contributors to the private sector of the economy. While the results of salary deterioration can be ascertained with some precision, the strength of the inherent attractiveness of the private sector of the economy to faculty members is difficult to ascertain.

The plan of this chapter is to review changes in the salary position of college and university professors and then examine various explanations for the changes. Subsequently, changes in salary structure are examined as are the reasons put forth for the changes. This is followed

Table 6–1. Average Faculty Salary by Category, Type of Affiliation, and Academic Rank, 1984–85

Category and Ranks	All Combined	Public	Private Independent	Church Related
Category I — Doctoral-Level Institutions				
All Ranks	$34,830	$33,860	$39,020	$34,640
Category IIA — Comprehensive Institutions				
All Ranks	$29,770	$29,930	$29,200	$29,170
Category IIB — General Baccalaureate Institutions				
All Ranks	$25,600	$26,880	$27,790	$23,360
Category IIC — Specialized Institutions				
All Ranks	$28,860	$32,520	$28,450	$22,470
Category III — Two-year Institutions with Academic Ranks				
All Ranks	$26,120	$26,460	$20,370	$17,440
Category IV — Institutions with No Academic Ranks				
No Rank	$24,550	$24,830	$16,240	$18,010
All Categories Combined				
All Ranks	$30,960	$31,240	$32,950	$26,250

SOURCE: "Starting the Upward Climb?" Annual Report on the Economic Status of the Profession, 1984–85, *Academe*, March–April 1985.

by a brief attempt to show the linkages between the two types of changes. The concluding section comments on the implications of the findings.

Faculty Salary Levels

To set the stage for the discussion that follows, it is useful to have in mind the absolute levels of faculty salaries and how these levels compare with those for other comparable groups of workers.

Salaries of faculty members, based on academic year salaries for full-time employees of colleges and universities, averaged $30,960 in 1984–85, as shown in Table 6–1. With fringe benefits equal to an average of $6,902 or 22.2 percent of average salary, total compensation averages $37,810. Overall average salary levels by rank are as follows: professor, $39,870; associate professor, $29,910; assistant professor, $24,610; instructor, $19,150; and lecturer, $22,020.

Differences exist in average salary levels by category of institution and type of affiliation, ranging from a high of $39,020 for faculty members at private, doctoral-level institutions to a low of $16,240 at private independent institutions without standard academic ranks (two-year colleges). Salary levels at private institutions exceed on average those at public institutions, with church-related institutions lagging considerably behind. Differences within categories are more varied, being quite wide for categories IIC and III and relatively narrow in categories IIA and IIB.

The level of academic salaries relative to those of workers in comparable professional and managerial occupations for the calendar year 1982 is shown in Table 6–2. Based on data for male year-round, full-time workers from the Current Population Survey, we find that the median earnings of post-secondary teachers is $26,608 and the mean is $28,145.

Among the sixteen occupations listed, post-secondary teachers rank tenth in both median and mean money earnings. The earnings position of professors is even less favorable when we consider that most of them have completed Ph.D. programs requiring upwards of four years of advanced graduate training, often with marginal incomes. By contrast, workers in such occupations as engineering and administration typically hold only bachelor degrees but they earn considerably more, as indicated by their positions in the earning distribution: third for federal administrators and officials at $38,991, fourth for administrators and officials in manufacturing at $35,894, fifth for engineers at $32,616, sixth for those in finance, insurance, and real estate at $32,542, and so on. Administrators in state and local governments are only slightly below faculty members in the earnings distribution.

Table 6–2. Total Money Earnings of Male Civilian Year-Round, Full-Time Workers Age 15 and Over by Occupation of Longest Job in Managerial and Professional Specialty Occupations in 1982

Occupation	Median	Mean
Health Diagnosing	$49,915	$55,953
Lawyers and Judges	41,858	44,810
Federal Administrators and Officials	38,991	39,704
Salaried Administrators and Officials—		
Manufacturing	35,894	39,545
Engineers	32,616	34,478
Salaried Administrators and Officials—Finance,		
Insurance, and Real Estate	32,542	35,413
Salaried Administrators and Officials—Other		
Industries	30,607	34,277
Natural Scientists and Mathematicians	30,319	31,769
Accountants and Auditors	26,745	28,858
Teachers, Post-secondary	26,608	28,145
State and Local Administrators and Officials	25,948	27,647
Health Assessment and Treating	22,377	23,550
Teachers, except Post-secondary	21,284	21,597
Salaried Administrators and Officials—Sales	20,486	23,513
Other Professional Specialty	20,223	23,513
Other Administrators and Officials—		
Self-employed	15,824	19,112

SOURCE: *Money Income of Households, Families, and Persons in the United States: 1982*, Consumer Income, Current Population Reports, Bureau of the Census, Series P–60, No. 142, pp. 178–79.

Table 6–3. Percentage Changes in Average Faculty Salaries, the Consumer Price Index, and Average Real Faculty Salaries, 1970–71 to 1984–85

Comparison	1970–71 to 1976–77	1976–77 to 1984–85	1970–71 to 1984–85
Average Faculty Salary	33.1	68.4	124.1
Consumer Price Index	47.7	80.4[a]	166.4[a]
Average Real Faculty Salary	−9.9	−6.7[a]	−15.1[a]

[a] Based on estimated inflation rate of 4 percent for 1984–85.

SOURCE: Faculty salary data based on annual percentage changes in academic-year salary levels of full-time faculty at institutions reporting comparable data from one year to the next, for All Ranks Combined, for All Categories Combined, from Committee Z Report on the Economic Status of the Profession, *AAUP Bulletin* and *Academe: Bulletin of the AAUP*, various issues. Consumer price index from the Bureau of Labor Statistics; converted to academic year basis.

Recent Changes in Salary Levels

This section offers a brief review of changes in salary levels. We focus largely, though not exclusively, on the period 1970–71 to 1984–85. Because of subsequent comparisons of changes in salary structure that must begin with 1976–77, we shall use the two sub-periods 1970–71 to 1976–77 and 1976–77 to 1984–85.

The real salary position of college and university faculty members, after adjustment for price level changes, improved considerably in the 1960s from its depressed level in the mid-1950s but suffered a sharp erosion in the 1970s that has persisted to the present.

From 1970–71 to 1984–85 the real salaries of college and university professors declined by 15 percent as shown in Table 6–3. This decline is split about evenly before and after 1976–77, with a 10 percent decline in the first period and an almost 7 percent decline in the second period. The declines were concentrated in two shorter sub-periods (see Figure 6–1): 1972–73 through 1974–75, and 1977–78 through 1981–82. In both subperiods inflation averaged 10 percent per year and salary increases lagged substantially behind unprecedentedly large increases in price levels. Since 1981–82 the pace of inflation has slowed dramatically. As a result real salaries for academics have increased over the past four years but not enough to offset any appreciable part of the longer term decline.

The earnings position of college and university professors also declined substantially relative to that of workers in comparable, competing occupations and in the rest of the economy (see Table 6–4, which shows data up to 1983–84 only). Beginning in 1970–71, salaries in the private sector (line 2) started to increase more rapidly and continued doing so until the early 1980s, when the rate of increase slackened. As a result academic salaries fell steadily relative to those in the private sector over the 1970s. The actual decline in relative earnings position from 1970–71 to 1983–84 was about 17 percent. The fact that both real and relative salaries of academics fell by about the same percentages simply means that salaries in other occupations just about kept up with inflation over this period.

A comparison of the earnings of academics with all employees provides another benchmark. The average annual wage and salary per

Figure 6–1. Percentage Changes in Average Faculty Salaries, the Consumer Price Index, and Average Real Faculty Salaries 1969–70 to 1983–84

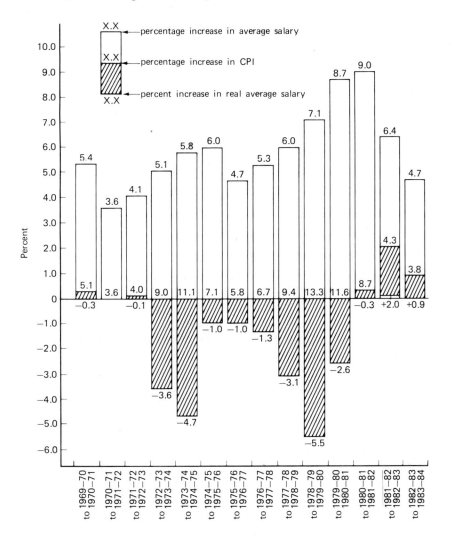

Table 6–4. Percentage Changes in Average Real Faculty Salaries and in Average Real Salaries, Earnings, or Income of Other Comparison Groups, 1970–71 to 1983–84

Comparison Groups	1970–71 to 1976–77	1976–77 to 1983–84	1970 71 to 1983–84
1. All Faculty Members	– 10.6	– 9.0	– 18.7
2. Private Sector Employees in Jobs Equivalent to GS Grades 11–15 (Group C)	+ 0.1	+ 1.7	+ 1.8
3. Wage and Salary Income per FTE Worker, All Domestic Industries	+ 1.5	– 3.5	– 2.0
4. Manufacturing	+ 4.1	– 0.8	+ 3.3
5. Government	+ 2.8	– 4.0	– 1.3
6. Disposable Personal Income per Employed Member of Civilian Labor Force	+ 3.0	+ 0.1	+ 3.0
7. Disposable Personal Income per Capita	+ 9.2	+ 5.8	+ 15.5

SOURCE:
  Line 1 — See Source, Table 6–1.
  Line 2 — Bureau of Labor Statistics, U.S. Department of Labor, *National Survey of Professional, Administrative, Technical and Clerical Personnel*, various issues. Based on March survey data (March 1971, 1976, and 1984).
  Lines 3–5 — Department of Commerce, *Survey of Current Business*, July issues.
  Line 6 — Disposable personal income divided by employed members of the civilian labor force, *Economic Report of the President, 1985*. Appendix Tables B–24 and B–29.
  Line 7 — *Economic Report of the President, 1985*, Appendix Table B–24.

NOTE: All salary and income figures are adjusted for price level changes. Data in lines 1 and 2 use the consumer price index converted to an academic year basis. Data in lines 3–7 are for calendar years 1970, 1976, and 1983.

full-time-equivalent employee in domestic industry fell by 2.0 percent in real terms over the 1970–71 to 1983–84 period. Employees in manufacturing achieved a 3.3 percent real increase. Employees in government registered a small 1.3 percent real decline in earnings. From this evidence it is quite clear that the salaries of academics fell substantially relative to a wide range of other wage and salary workers in the economy.

Still another indication of the changing economic position of academics is revealed by movements in disposable personal income. Disposable personal income per employed member of the civilian labor force rose in real terms from 1970 to 1983 by 3.0 percent, with all of the gain occurring from 1970 to 1976. A comparison with disposable personal income per capita is also revealing. This measure indicates the capacity of the economic system to support its various public and private sector activities. Real disposable personal income per capita rose in real terms by 15.5 percent from 1970 to 1983, with most of the gain occurring from 1970 to 1976. The increases in disposable personal income per capita are greater than in personal income per worker because they take into account the rising proportion of the total population employed as a result of the maturing of the post-World-War-II baby boom as well as the sharp drop in the proportion of children as a result of the birth rate declines of the 1970s.

All of this adds up to a discouraging picture for the future vitality of the colleges and universities. Able young people who might aspire to join the ranks of their professors will be discouraged by the dismal salary picture reflected in these data. (Evidence presented in Chapter 11 suggests that such a trend is already noticeable.) Without any apparent prospects for substantial improvement, many of the most talented young people who in the past chose to become academics will find it even easier to opt for other more financially attractive careers. Meanwhile, younger members of the academic profession will be more susceptible to attractive offers from outside the academic sector. Not all academics will have such opportunities because of the limited transferability of their skills to other sectors, but substantial numbers will, and some of them can be expected to move. Unfortunately for the academic profession, those likely to leave the profession are also likely to be among the most talented, and they will be difficult to replace.

Explanations

What explains the deteriorating salary position of college and university professors? A host of explanations has been offered to account for the precipitate decline in faculty salaries. These explanations can be divided into categories that we label as market forces, institutional constraints, and societal preferences. Each of these explanations is outlined more fully below, followed by an effort to evaluate the force of each explanation.

Market forces
1. Faculty salaries were too high in 1970 and thus the relative decline since then represents the return to a more normal situation that reflects looser labor market conditions.
2. The abundant supply of new Ph.D.s in the 1970s reduced the labor market tightness of the 1970s and has since flooded the market with new Ph.D.s so that it is not surprising that salaries have declined relatively.
3. The slowing growth rate of enrollments in American colleges and universities during the 1970s and early 1980s weakened the demand for new Ph.D.s and thereby depressed the rate of growth of salaries.

Institutional forces
1. The imposition of various wage control and wage guidelines policies of the 1970s artificially held down the rate of increase in faculty salaries relative to other groups of workers.

2. The slow and cumbersome salary-setting process in public higher education, which dominates the wage-setting process, renders it incapable of responding quickly to sharp inflationary shocks.
3. The academic sector found itself unable to invent an effective method for recommending prospective salary increases that would maintain its salary position.

Political and social forces
1. The less buoyant economy of the 1970s and early 1980s limited the revenue-generating powers of states and hence restricted the funds available to support higher education and with it faculty salary increases.
2. The higher priorities for new public programs in the health, welfare, and environmental fields cut into funds that might otherwise have supported higher education.
3. Public support for higher education eroded relative to that which prevailed in the 1960s.
4. Faculty members were not in fact awarded salary increases equivalent to those awarded other state and local employees even though state and local officials frequently claimed they gave the groups equal treatment.

This is a complex list of explanations that are by no means independent of each other. The difficulties of evaluating them individually and collectively are formidable. Nonetheless, we try to do this in reasonably brief fashion.

Market Forces

Academics Were Overpaid in 1970. One can imagine two kinds of evidence that would support the explanation that because faculty members were overpaid in 1970 they might have reasonably expected some salary erosion subsequently. One is a dramatic relative improvement in the salary position of academics in the years leading up to 1970. The other is that owing to high prevailing salaries over this same period, excessively large numbers of young people were attracted into the academic profession or were preparing to enter it.

We do find evidence for the first. In the 1960s salary increases for academics exceeded those of private sector employees doing roughly comparable work. Based on comparable data for 1960–61 to 1970–71, real salaries rose 23.0 percent for faculty members as compared with 18.0 percent for private sector employees doing similar types of work (see line 2, Table 6–5). This represents a modest 5-percentage-point improvement. Before jumping to the conclusion that academics were overpaid by 1970, it is important to remember that their larger salary increases were designed to accomplish two goals. One was to attract more people into the profession so as to cope with the baby boom population entering college in the mid-1960s; the size of the 18-year-old

Table 6–5. Percentage Changes in Average Real Faculty Salaries and in Average Real Salaries or Earnings of Other Comparison Groups, 1960–61 to 1983–84

Comparison Groups	1960–61 to 1970–71	1970–71 to 1983–84	1960–61 to 1983–84
1. All Faculty Members	+ 23.0	− 18.7	0.0
2. Private Sector Employees in Jobs Equivalent to GS Grades 11–15 (Group C)	+ 18.0	+ 1.8	+ 20.2
3. Wage and Salary Income per FTE Worker, All Domestic Industries	+ 22.5	− 2.0	+ 20.0
4. Manufacturing	+ 17.4	+ 3.3	+ 21.2
5. Government	+ 28.4	− 1.3	+ 26.8
6. Disposable Personal Income per Employed Member of Civilian Labor Force	+ 26.4	+ 3.0	+ 30.1
7. Disposable Personal Income per Capita	+ 33.2	+ 15.5	+ 53.8

SOURCES AND NOTES: Same as Table 6–4 except line 1 for 1960–61 to 1970–71 increases which are based on data for a selected group of 36 institutions drawn from AAUP data.

group would increase by more than 35 percent (from 2.7 to 3.7 million) from 1964 to 1965. The other was to offset the relatively low salaries of the 1950s. President Eisenhower's 1957 Commission of Education had called for a doubling of faculty salaries by 1970. Salaries did not in fact double in real terms. From 1960–61 to 1970–71 salaries rose by 64.1 percent in nominal terms, which translates into an increase of only 23.0 percent in real terms.

In view of the Commission's goal, the real salary gain for academics is hardly impressive. In fact, the gain for academics was about the same as that for workers in the rest of the private sector whose real earnings rose by 22.5 percent over this same period. Surprisingly, government workers did substantially better with a 28.4 percent gain. Workers in manufacturing lagged further behind.

When we move to broader comparisons, we find that disposable personal income per employed worker over the 1960–61 to 1970–71 period rose somewhat more than academic salaries. Meanwhile, the fiscal capacity to support higher education increased substantially as indicated by the 33.2 percent growth in disposable personal income per capita (see Table 6–5, line 7).

To sum up, increases in average real salaries for faculty members in the 1960s were modest in view of the need to attract large numbers of new faculty. This makes it difficult to claim that academics were over-paid in 1970. Nor did the high salaries produce the massive influx of people into the academic profession that one might have reasonably expected these allegedly high salaries to do. More on this point follows.

The Abundant Supply of New Ph.D.s. During the 1960s academic

Table 6–6. Doctorate Projections, Recipients, and Post-Graduation Plans, 1957–58 to 1983–84

	1965 Cartter Projections	1974 Cartter Projections	Total Doctorates	Percent Seeking Employment at time of Ph.D.	Percent with Definite Employment Plans in Academe at time of Ph.D.
	(1)	(2)	(3)	(4)	(5)
1957–58	—	—	8,773	14.3	41.5
1959–60	—	—	9,733	13.9	43.6
1961–62	—	—	11,500	13.9	41.0
1963–64	—	—	14,325	13.0	42.3
1965–66	16,100	—	17,949	11.7	43.4
1967–68	18,000	—	22,936	13.7	44.4
1969–70	21,400	—	29,498	19.4*	43.1
1971–72	26,000	—	33,843	21.6	40.2
1973–74	31,500	34,600	33,047	23.9	35.2
1975–76	35,700	36,600	32,947	25.7	33.5
1977–78	40,400	39,100	30,873	25.7	30.7
1979–80	46,000	42,000	31,013	23.7	29.7
1981–82	49,700	43,200	31,048	24.7	27.7
1983–84	53,200	42,100	n.a.	n.a.	n.a.

SOURCES:

Col. 1. A. M. Cartter and R. Farrell, "Higher Education in the Last Third of the Century," *Educational Record* (Spring 1965), 119–28.

Col. 2. A. M. Cartter, *PhDs and the Academic Labor Market* (New York: McGraw-Hill, 1976), p. 143.

Col. 3. National Research Council. *Survey Report 1982: Doctorate Recipients from United States Universities* (Washington: National Academy Press, 1983), p. 16.

Col. 4. *Ibid.*

Col. 5. *Ibid.*

*Note: Definitions changed.

employment rose sharply with the rapid expansion in the number of doctorates completed, particularly in the second half of the decade (Table 6–6). But contrary to expectations, doctorate production stabilized early in the 1970s. This occurred at about the same time that the real salary position of academics began to deteriorate, thereby casting doubt on the thesis that the large output of Ph.D.s slowed salary increases.

Projections made by Cartter and Farrell in the mid-1960s and the mid-1970s of the number of Ph.D.s to be awarded each year showed a steady upward trend to 1985 when doctorate production was expected to reach almost 55,000. Cartter's later projections based on data through 1972 and on the assumption that doctorate production would respond to labor market conditions indicated that the number of Ph.D.s would continue to rise through the 1970s and then peak at slightly over 40,000 in the early 1980s. Even these more cautious projections were far higher than the actual number of Ph.D.s awarded which peaked at 34,000 in 1972 and since then stabilized at about 31,000. While academic salary increases had started to slow by 1970, as already noted, young people contemplating graduate training must have been either very sensitive to this change or highly responsive to Cartter's predictions of an over-supply of Ph.D.s.

Evidence of an ever worsening market since the 1972 peak in Ph.D. production is not clear. The percentage of Ph.D. holders still seeking employment at the time they received their degrees has remained relatively constant even with the sharp decline in the percentage of new Ph.D.s taking academic jobs. While one cannot rule out the impact of the continued larger numbers of Ph.D.s in depressing academic salaries, the case is not as clear as some observers have suggested.

The Declining Growth of Demand. Widely publicized reports in the early and mid-1970s about prospective enrollment declines in higher education apparently led many to believe that enrollments were actually declining and that the need for faculty members was therefore diminishing. College enrollments did not decline in the 1970s; indeed, they rose by almost 40 percent from 1970 to 1980 (see Table 6–7). Even in the first half of the 1980s' enrollment growth was still positive. Clearly, the pace of enrollment growth slowed progressively since the 1960s, but slowing is not decline.

This slowing of enrollment growth and the prospect that overall enrollments are likely to fall in the late 1980s and/or early 1990s obviously mean that the overall market for new Ph.D.s has weakened. This development is already reflected in the declining proportions of new Ph.D.s taking academic appointments, as noted earlier. But these

Table 6–7. Higher Education Enrollments, Instructional Staff, and Full-Time Instructional Staff, 1960 to 1995

Year	Total Enrollment (in thousands)	5-Year Percentage Increase	Total Instructional Staff (Instr. & Above) (in thousands)	Full-time Instructional Staff (Instr. & Above) (in thousands)
	(1)	(2)	(3)	(4)
1960	3,789	56.3	236	154
1965	5,921	44.9	340	248
1970	8,581	30.3	474	369
1975	11,185	8.2	628	440
1980	12,097	0.6	678	466
1985 (proj.)	12,174	− 0.3	—	—
1990 (proj.)	12,139	− 0.3	—	—
1995 (proj.)	12,101	—	—	—

SOURCE: Col. 1: Table 255, p. 161 and Table 207, p. 134, Bureau of the Census, *U.S. Statistical Abstract 1984.*
Col. 2: Calculated from Col. 1.
Cols. 3 and 4: Table 257, p. 161, Bureau of the Census, *U.S. Statistical Abstract 1984.*

declines have been largely offset by the growing numbers of new
Ph.D.s finding employment in the non-academic sector. Thus far, there
is little or no visible unemployment of Ph.D.s though there could be rising underemployment. The principal result of these market shifts has
been to give employing institutions much greater choice about whom
they hire; and they can now insist on hiring people who have completed their Ph.D.s. Moreover, many institutions now find themselves
able to upgrade the quality of their faculty as a result of the availability
of well-trained Ph.D.s from the leading graduate departments.

The net effect of these market forces is not clear. Though it is possible to argue that market forces are dominant and do explain what happened, the evidence is mixed. Moreover, still other explanations must
be examined.

Institutional Forces

Wage Controls and Guidelines. President Nixon's wage controls in
the early 1970s and President Carter's anti-inflation program in the late
1970s appear to have had a relatively minor effect on faculty salary
increases, despite widespread concern at the time. In 1971–72 and
1972–73 average faculty salaries rose by less than the average for federal
employees whose salaries were controlled. But in 1973–74 and 1974–75,
as wage controls were relaxed then ended, faculty members made up in
part for losses sustained earlier. In 1978–79 and 1979–80 average salary
increases for academics exceeded the targets set by the Carter administration and imposed on federal employees. Overall, then, these programs did not seem to affect the rate of increase in faculty salaries
through the 1970s. This does not rule out the possibility that controls
had some limited effect on particular institutions.

More important is the likelihood that faculty members and their
institutions were at times constrained from securing larger increases,
either because slated increases were deferred or it was difficult under
the circumstances to press for increases that exceeded the standards
being enforced. In some cases state legislatures used the standards as an
excuse for not offering larger increases. It is difficult to establish any
overall substantial impact for these controls except to report that a
knowledge of the experiences of particular states suggests that controls
did put a damper on higher education's hopes of larger salary increases
that would have helped them keep pace with inflation over the 1970s.

Inflationary Lag of Salaries. Because faculty salaries are typically set
well before the academic year to which they apply, and because salary
increases are ordinarily based on the experience of the recent past, it is
almost inevitable that salary increases lag behind price level increases.
This conclusion emerges clearly from Figure 6–1, which shows the pattern of increases in the Consumer Price Index and average faculty sala-

ries over the 1970s and the early 1980s. The most obvious examples of lags are in the early 1970s and late 1970s when price increases soared but salary increases lagged. In the first case the salary increases received after the rate of inflation dropped were still not large enough to produce any real salary gain even though the nominal salary increases were greater than ever before. This situation did not recur after the late 1970s' burst of inflation. Partly because this inflationary episode was longer and more severe, nominal salary increases became progressively larger, rising steadily from 1976–77 to 1981–82. These eventually resulted in small real salary gains because of precipitous declines in the rate of increase in prices.

Inability To Target Needed Salary Increases. College and university faculty members have always had difficulty making an effective case for increased remuneration. The difficulty arises in part from the character-istics of academics themselves and in part from the peculiar institu-tional setting that surrounds higher education. Faculty groups have attempted to cope with these difficulties but there is as yet no evidence that any particular approach holds the key to success.

The push for higher salaries in higher education has usually been a diffident one. Academics typically find considerable satisfaction in their work and as a result are less inclined to focus their attention on securing better pay. Even if they wanted better pay and were willing to press aggressively for it, they would encounter difficulties. Private, indepen-dent colleges as well as those with religious affiliations are generally in a weak financial position that tempers any strong effort toward higher wages. Their faculty members are well aware of the great difficulties of raising funds for support of academic salaries. Public institutions typi-cally have a separate governing board and a network of employment practices that differentiate them from other governmental units. This makes it difficult to press for salary increases with the same vigor as other public employees who increasingly are unionized. The unioniza-tion of some college and university faculty members may have altered these relationships somewhat but still leaves a wide gap in the approach taken to secure salary increases.

In an effort to surmount these difficulties the American Association of University Professors in the late 1950s initiated a major effort to raise faculty salaries. Impetus for this effort came from President Eisen-hower's 1957 Commission on Education Beyond the High School which, as noted, proposed a doubling of the level of real faculty salaries within the decade. Several approaches were employed. The first set out a system for grading the salary levels of colleges and universities, employing the same grading scale of letter grades A through F used to evaluate student performance. The idea was to stimulate competition among institutions to raise their salaries and thereby improve their grade in the salary standings, in the hope that average salaries would be pushed upward for the profession as a whole. Another approach initiated in the mid-1960s tried to improve on the grading system. Each year the AAUP published projected salary levels over the next two years for each letter grade level. This approach proved to be helpful because it eliminated any uncertainty for institutions about what magnitudes of increases would produce higher grades.

There has been no evaluation of the effectiveness of these two approaches. It seems probable that they would have had some impact on salaries because no college or university would want to receive low grades in the salary competition. At the same time the novelty and effectiveness of these approaches would be likely to have diminished over time. Thus, average salaries may have been raised somewhat during the 1960s because of the AAUP's efforts.

To the extent that salaries did rise, it seems more plausible to believe that the tremendous growth in demand for new college faculty members in the 1960s accounted for the increase. College and universities had to offer higher salaries to lure people to their institutions. Supply was desperately tight, and there was extensive talk about faculty shortages. Some indication of the difficulties in hiring is provided by Table 6–7 which shows enrollment and faculty growth by five-year periods beginning in 1955.

How to separate out the impact of growth of demand from the impact of the efforts by the AAUP to raise average faculty salaries is unclear. It is reasonable to assume that growth was the dominant factor in accounting for salary increases over the decade of the 1960s. And yet the projected salary scales were intended to show how compensation should increase if the economy grew at the projected rate and the profession maintained its relative position among U.S. income earners.

What actually happened is that in the 1960s nominal salaries increased almost without exception by more than the increase used to set the compensation scales for the coming year. This continued until 1969–70 when the actual increase in salary fell just short of the percentage increase in projected compensation. Because actual increases started to fall behind the projected increases, the AAUP shifted its approach slightly to take into account expected inflation and also average long-term productivity increases. By 1977–78 the AAUP shifted its ground again, arguing that the increase in the projected scale should include a threefold adjustment to take into account the expected increase in the cost of living, an increase in real compensation to reflect seniority, and participation in the overall increase in national productivity. However, in the following year the AAUP reverted to the use of expected inflation and the average productivity rise.

With the approach of 1979–80, real salaries had declined so substantially that it seemed fruitless to continue publishing projected scales. It was not evident that the scales had had any long-term continuing impact. Consequently, they were quietly dropped. Attention was shifted from compensation to salary, and greater emphasis was given to the cumulative effect of salary deterioration over the preceding 10-year period and in some cases over even longer periods.

It would appear then that the AAUP's approaches may have had some impact in their initial years of operation by pushing particular institutions to advance their salary ratings, and hence to raise average salary levels. Later, because of the changing economic climate of the 1970s their effects were probably less important.

Political and Social Forces

Less Buoyant Economy. The state of the economy during the 1970s was considerably less favorable to most employee groups than that of the 1960s. As economic conditions worsen, state and local governments experience pressures to increase their expenditures just as their revenues are contracting because of the reduced level of economic activity. They are then faced with having to raise taxes and cut expenditures. Neither approach is painless. A favorite target is to cut salary increases because salaries constitute the largest component of state expenditures. And it has often been easier to hold down salary increases in higher education than in other sectors of state government. Clearly, then, this is an important consideration.

Erosion of Public Support. Public support for higher education, so strong in the 1960s, weakened in the 1970s. The image of higher education was tarnished by the student unrest and other events of the late 1960s and early 1970s thereby making it easier for legislators to hold down expenditures on higher education. Assessing the strength of public support for higher education is not easy because no ready measures are available. But it does appear that financial support dropped and has not yet recovered.

Higher Priority for Other Programs. The growth of spending for social programs, combined with the assumption of broader responsibilities by the states, heightened competition for state tax dollars throughout the 1970s. The strength of groups wanting to spend on social programs and the lack of effective organization by higher education and its constituents resulted in more resources flowing to these other programs. There is no simple way to quantify the change in priorities except to note that from 1970 to 1980 the proportion of all state spending devoted to public higher education declined.

Equity in Pay Increases. This leaves one last matter for consideration, namely, the extent to which state governments in a period of growing unionization of state and local employees (largely but not completely excluding those in higher education) did in fact provide, as they often claimed to provide, salary increases to faculty members in public institutions that were equivalent in percentage terms to those provided other state workers.

There is, however, reason to believe that overall salary increases

were not equivalent. The across-the-board salary increases were frequently equivalent, with faculty members and other state employees
being awarded similar increases, of say 5 percent. But they did not work
out that way. For faculty members the 5 percent had to be used to provide for a wide range of salary increases, including those associated
with promotions, rewarding merit, and frequently some across-the-
board component to offset increases in the price level. For individuals
in state and local civil-service systems with explicit salary schedules,
their 5 percent typically was distributed across the board. But in
addition, these civil servants benefited from salary increases associated
with movement up the steps in their salary schedules (reflecting longevity) as well as with reclassifications of their positions and promotions to new positions. These components of salary increases were
embedded somewhere else in the budget. This means that the different
salary structures and associated rules for granting and financing these
increases generally guaranteed that state and local employees outside of
higher education gained larger salary increases than was indicated by
their well-publicized across-the-board increases.

Empirical support for this view is available from several sources.
One is an inspection of compensation plans and their associated salary
schedules. Given the diversity of these plans as well as the difficulty of
assembling them for study, we must turn to another approach. That
approach utilizes data collected each year by the U.S. Bureau of the Census for its reports on public employment which record as of October
each year full-time-equivalent employment, total payrolls, and average
full-time-equivalent monthly earnings by function for state governments and for local governments.

These data indicate, as shown in Table 6–8, that for the periods
1970–71 to 1976–77, 1976–77 to 1983–84, and 1970–71 to 1983–84, the percentage increases in average full-time-equivalent earnings in state-
funded public higher education lagged well behind the gains achieved
by state and also by local employees engaged in functions other than
education. In real terms the earnings of instructional staff fell by 11.7
percent from 1970–71 to 1983–84. State employees in other than education meanwhile experienced a slightly greater than 2.0 percent decline
in real earnings. By contrast, the real earnings of local employees in
other than education neither rose nor fell to any appreciable extent; in
other words, they kept up with inflation.

(It should be noted that the percentage changes reported in Table
6–8 differ for faculty members from those presented in Tables 6–4 and
6–5. These differences arise because the percentage changes are calculated from overall salary averages reported in the census data as compared with salaries standardized for changes in the mix of faculty, as is
the case for the AAUP data. Changes in the overall averages thus reflect
the combined effect of changes in salary levels by rank and changes in
the mix of faculty by rank. As the faculty has aged, the proportion of

Table 6–8. Percentage Changes in Real Average Earnings of Full-Time Equivalent State and Local Employees in Education and in All Other Activities, 1960–61 to 1983–84

Type of Employee	1960–61 to 1970–71	1970–71 to 1976–77	1976–77 to 1983–84	1970–71 to 1983–84	1960–61 to 1983–84
State					
Public Higher Education					
Instructional Staff	+ 27.4	− 4.6	− 7.4	− 11.7	+ 12.6
Noninstructional Staff	+ 28.8	+ 3.9	− 3.3	+ 0.4	+ 29.3
All Other Functions					
Total Staff	+ 34.6	+ 0.9	− 3.0	− 2.1	+ 31.6
Local					
Public Elementary–					
Secondary Education					
Instructional Staff	+ 27.4	− 5.8	− 6.8	− 12.3	+ 11.8
Noninstructional Staff	+ 38.8	+ 1.5	− 6.1	− 4.7	+ 32.3
All Other Functions					
Total Staff	+ 28.6	+ 3.5	− 3.1	+ 0.3	+ 28.9

SOURCE: U.S. Bureau of the Census, *Public Employment*, annual issues.

lower-paid assistant professors has declined, thereby pushing up the overall salary level faster than that for salaries by rank. In addition, the census data report information only for public institutions.)

Both state and local employees did better in the 1970–71 to 1976–77 period, with real earnings increases of 0.9 and 3.5 percent, respectively; meanwhile, real salaries of instructional staff in public higher education fell by 4.6 percent. From 1976–77 to 1983–84 both state and local employees experienced real declines of about 3.0 percent as compared with a 7.4 percent drop for instructional staff in higher education. It is interesting to note that the earnings of people in public higher education other than instructional staff experienced much the same pattern of change as did state and local employees; overall they maintained their real earning power during the 1970–71 to 1983–84 period. This is not surprising because most employees in public higher education other than instructional staff are under civil service and thus subject to the same kind of compensation plan as other state employees.

The evidence is quite clear that salary decisions by state governments were biased in favor of state employees other than instructional staff in higher education. This leads us to inquire what kinds of evidence would support the view that salary increases of this magnitude were justified for other state employees.

Several factors come into play. Perhaps the most obvious is the impact of collective bargaining which has given other state employees a stronger voice. Another may have been the force of pay comparability legislation adopted in many states and patterned after the Federal Pay

Comparability Act. Whether the strength of market forces had much bearing is not clear. Generally, state employment is viewed as attractive, well-paid, and secure. Turnover is low and, though no data are readily at hand, one gets the impression that there is intense competition for state jobs, with a high ratio of applicants to openings, and queues of qualified applicants for new openings. Moreover, an abundance of both new high school and college graduates has been available to fill these jobs, not to mention the many women who entered the labor force over the decade. In short, the supply of people to state government has been ample, making it relatively easy for state employers to fill most of their open job positions. One also suspects that the task of hiring the full complement of state workers would have been no more difficult had nominal wages advanced more slowly than they did.

An Interim Appraisal

Many explanations for the declining real and relative salary position of faculty members have been offered here along with limited evidence in support of or against each. Additional information is needed to round out the evidence, of course. But in the meantime it appears that the decisions of state legislators have been most critical in accounting for the deteriorating salary position of faculty members.

We are left with a puzzle, however. On the one hand legislators have often prided themselves on having given equal percentage increases to both faculty and non-faculty personnel. But in fact the increases have not been equal. To argue that faculty members received smaller percentage increases because of a perception among legislators that the labor market for Ph.D.s had weakened is at odds with the claims of legislators.

The central issue in the puzzle revolves around the fact that state governments typically fund positions, which are filled by individuals who are advanced through a series of positions over time. Colleges and universities typically fund individuals rather than positions, and this means that individual faculty members cannot gain comparable salary increases because no additional funding is provided by position money. Put another way, when a highly paid state employee retires, someone is advanced to that salary position, somebody else replaces the person advanced, and so on down the line. But in the academic sector the retirement of a highly paid faculty member usually is followed by the hiring of a new young Ph.D., with no opportunities for other faculty members to advance up the ladder to replace the individual who retired. And as a consequence the salary base is eroded by the difference between the salary of the retiree and that of the assistant professor replacement. This is the only explanation I can offer to resolve the puzzle.

Whatever the judgment about the reasons why faculty salaries fell so far behind those of a wide range of other members of the labor force over the period 1970–71 to 1983–84, the financial impact has been significant on those individuals who were already faculty members in 1970–71.

We can illustrate this in another way besides that of showing how faculty salaries fell progressively further behind those of other groups. Instead, we estimate the cumulative effect of falling behind by summing the shortfalls of salary over past years to the present. To do this properly we must allow for the fact that had these shortfalls not been experienced, faculty members would have derived satisfaction from being in control of additional increments of salary each year. The easiest way to put a value on this satisfaction is to assume that the amount of satisfaction denied is equivalent to what these funds might have earned had they been invested to add to a person's retirement savings.

We can estimate the financial impact of salary shortfalls by thinking of ourselves performing the mental experiment of comparing what did happen with the counterfactuals' reflecting what might have happened under other circumstances. We have several possible counterfactuals. One is the assumption that faculty salaries would have advanced at the same rate as the Consumer Price Index, i.e., academics would have experienced no decline in real salaries over the 1970s and early 1980s. Another is the assumption that faculty salaries would have advanced at the same rate as those of state employees performing other functions than those in education. After using these two alternatives to estimate the size of each year's shortfall of salary, we apply an appropriate interest rate for compounding to 1983–84 each increment of salary shortfall. For this purpose we use a long-run real interest rate of 2.5 percent.

The nature of this calculation is shown in Table 6–9. The results are presented in the lower part of the table. If we assume that faculty salaries might have been expected to rise at the same pace as the Consumer Price Index, then the cumulative value of salary shortfalls since 1970–71 equals 1.71 times the 1983–84 average salary level. Given that the average 1983–84 salary for full professors (all categories combined) was $37,400, the value of salary shortfalls—the cost to the average full professor at that rank since 1970–71—is almost $64,000 ($34,700 × 1.71). If instead we assume that faculty salaries would have risen apace with state salaries, the 1983–84 value of the loss—the cost to the average full professor at that rank since 1970–71—is about $60,000. If one were to assume a different interest rate for compounding, the results would of course differ.

It is clear that the gradual whittling away of faculty salaries relative to the Consumer Price Index or to the salaries of other groups is far more dramatic when expressed this way as compared with showing declining salary relatives.

The implications of this analysis are important because the results

Table 6–9. Estimation of Financial Impact of Salary Shortfalls for Faculty Members, 1970–71 to 1983–84

Year	Index of Average Faculty Salary: 1970–71 = 100.0 (1)	Counterfactuals: Faculty Salary Would Have Risen at Same Rate as: Consumer Price Index 1970–71 = 100.0 (2)	Index of Average Salary of Other State Workers 1970–71 = 100.0 (3)	Counterfactual Salary Estimates: Col. 1 ÷ Col. 2 1970–71 = 100.0 (4)	Col. 1 ÷ Col. 3 1970–71 = 100.0 (5)	Salary Shortfalls Cumulated in Percent: 100.0 − Col. 4 (6)	100.0 − Col. 5 (7)	Amount of $1.00 Compounded from Each Year to 1983–84 (8)	Present Value of Salary Shortfalls: Col. 8 × Col. 6 (9)	Col. 8 × Col. 7 (10)
1970–71	100.0	100.0	100.0	100.0	100.0	0.0	0.0	1.41	0.00	0.00
1971–72	103.6	103.6	103.6	100.0	100.0	0.0	0.0	1.38	0.00	0.00
1972–73	107.7	107.7	107.9	100.0	99.8	0.0	0.2	1.35	0.00	0.00
1973–74	113.3	117.4	117.2	96.5	96.7	3.5	3.3	1.31	0.05	0.04
1974–75	119.9	130.4	129.8	91.9	92.4	8.1	7.6	1.28	0.10	0.09
1975–76	127.1	139.7	139.5	91.0	91.1	9.0	8.9	1.25	0.11	0.11
1976–77	133.1	147.7	148.4	90.1	89.7	10.3	10.3	1.22	0.12	0.13
1977–78	140.2	157.6	156.7	89.0	89.5	11.0	10.5	1.19	0.13	0.13
1978–79	148.6	172.4	170.1	86.2	87.4	13.8	12.6	1.16	0.16	0.15
1979–80	159.2	195.4	190.8	81.5	83.4	18.5	16.6	1.13	0.21	0.19
1980–81	173.1	218.0	212.7	79.4	81.4	20.6	18.6	1.10	0.23	0.21
1981–82	188.7	236.9	232.8	79.7	81.1	20.3	18.9	1.08	0.22	0.20
1982–83	200.8	247.1	244.2	81.3	82.2	18.7	17.8	1.05	0.20	0.19
1983–84	210.2	256.2	250.7	81.0	83.8	18.0	16.2	1.03	0.19	0.17
									1.71	1.60

Summary of Results

1983–84 Present Value of Cumulative Salary Shortfalls since 1970–71, Expressed as a Multiple of 1983–84 Salary Level.

- Assuming Faculty Salaries would have risen as fast as CPI: 1.71

- Assuming Faculty Salaries would have risen as fast as salaries of state employees in activities other than education: 1.60

indicate that even if salaries for 1983–84 were adjusted on a one-time basis, to "catch up" up to the CPI or other salaries, this would in no way compensate for past salary losses.

Recent Changes in Salary Structure

This section provides a brief review of changes in the structure of faculty salaries. It begins by focusing on the traditional comparisons made possible by the published salary data from the AAUP. These data show salary averages by rank, institutional category (universities, four-year institutions, and two-year institutions), and type of control (public, private, and church-related). Attention is then turned to differences in salaries between men and women faculty members, again based on AAUP data. This is followed by an examination of changes in and differences among average salaries by academic disciplines based on the Oklahoma State University salary survey of public universities.

The comparisons presented here are for the 1976–77 to 1983–84 period. The year 1976–77 was chosen as a benchmark because of the nature of the available data. Beginning in 1976–77 the AAUP salary reports cover a larger number of institutions because the AAUP began to use the faculty salary data collected by the National Center for Educational Statistics. The annual salary data by academic discipline collected by the Oklahoma State University did not become fully representative of the universe of institutions and departments within them until 1976–77 even though the survey began two years earlier. Several other sources of data are available but these two are sufficient to provide a comprehensive picture of recent changes in the structure of academic salaries.

The structure of salaries by rank has been surprisingly stable. From 1976–77 to 1983–84 average salaries at all institutions reporting comparable data from year to year rose by 57.8 percent. Differences in the increases by rank were exceptionally small, ranging from a high of 58.6 percent for assistant professors to a low of 56.3 percent for associate professors; full professors gained 57.4 percent. When these calculations are repeated for the various institutional categories we find essentially the same result. Even in Category I (doctoral-level institutions), where competition for faculty members is likely to be most keen, we find that the average salary of assistant professors rose only marginally more than that for the other ranks, by 59.7 percent for assistant professors versus 57.7 percent for full professors, and 56.1 percent for associate professors who had the smallest increase. These differences are so small that it is difficult to say that any change in the rank structure of salaries occurred.

We do find some change in the structure of salaries across institutions by type of control. For the early part of the period from our base year of 1976–77 through 1978–79, the relative increases in salaries at public institutions reporting comparable data from year to year exceeded those in private independent institutions, which in turn exceeded those in church-related colleges and universities. Later, from 1979–80 through 1983–84, average salaries for all ranks combined and overall by categories rose relatively more in the private independent sector than in the church-related sector, and the latter group's average rose relative to the public sector. Over the entire period 1976–77 to 1983–84, average salaries rose 64.1 percent at private institutions, 59.7 percent at church-related institutions, and 55.8 percent at public institutions.

This pattern of change reflects several forces. First, non-public institutions were hard pressed financially in the early part of the period and apparently had to economize by holding down their salary increases. Perhaps as a result of lagging behind public institutions, they energized themselves to compete more effectively for funds and thus made up for past losses and surged ahead of public sector institutions. The latter were handicapped, of course, by the depressed economic conditions that began in 1979 and 1980. This sharply curtailed revenues and thereby limited the extent of faculty salary increases in public institutions. Second, there seem to be cycles in the ability of public and private sectors to offer higher pay to their faculties. In one period private institutions move ahead only to be overtaken later by public institutions. And the cycle repeats itself.

It is common practice to group academic institutions by certain characteristics associated with their purposes so that institutions can compare themselves with similar institutions as they assess their competitive position. The AAUP employs several categories for grouping institutions, based heavily on the scheme employed by the National Center for Educational Statistics. The groups are as follows: Category I for universities offering comprehensive doctoral-level programs; Category IIA for comprehensive institutions offering diverse post-baccalaureate programs but not engaged in significant doctoral-level education; Category IIB for institutions whose primary emphasis is on general undergraduate baccalaureate-level education; Category III for two-year institutions with academic ranks; and Category IV for institutions that do not utilize the standard academic ranks. The categories were revised in 1982–83; this led to creation of a new Category IIC for institutions that emphasize one program area, such as business or engineering; previously, these institutions were spread through Categories I, IIA, and IIB. This revision makes it difficult to present a full set of comparisons for the 1976–77 to 1983–84 period. Hence, we focus heavily on the period 1976–77 to 1981–82.

It appears that the structure of salaries across the various categories has not changed over this period. The cumulative percentage increases in salaries by category for all ranks combined show virtually no difference among Categories I, IIA, and IIB, based on data using the old categories; the same holds true using the new categories for changes from 1981–82 to 1983–84. This stability reflects the fact that the category definitions capture the essential differences among institutions, that few if any institutions change categories, and that we can think of institutions competing with others within their same categories rather than across categories.

One of the most striking characteristics of academic salary structures that first became apparent in the early 1970s was the pronounced disparity in the salaries of men and women faculty members. These differences reflect a multitude of factors arising out of the history of the academic profession which had long been dominated by men. With the growth of the women's movement came a strong push to eliminate unjustified differences in the salaries of men and women by rank. This task could not be completed immediately because so little was known about what accounted for the existing differentials. Nonetheless, colleges and universities, particularly in the public sector, moved quickly to rectify the most glaring discrepancies. But disparities remain.

Did the differences in salaries by gender narrow between 1976–77 and 1983–84? It is not obvious that they did. In 1976–77 the overall salaries of women were about 94 percent of those of men of comparable academic rank ranging from a low of 91.5 percent at the full professor rank to 95.5 percent at the assistant professor rank. By 1983–84 some retrogression appears to have occurred at all three ranks. The salaries of women academics relative to men range from a low of 89.1 percent at the full professor rank to 93.9 percent at the assistant professor rank. This slippage does not necessarily imply, however, that women faculty members are now even further disadvantaged. Of course, that could be possible. More likely, the rapid movement of women into academic positions—up from 31 percent in 1976 to more than 35 percent in 1983— means that by and large women faculty members at all ranks are likely to have less experience relative to their male counterparts and hence are paid less. Unfortunately, our data are such that we cannot control for differences in the duration of experience.

Closer inspection of the data offers no interesting clues about what may be happening. It does appear that women's salaries as a percentage of men's salaries declined somewhat in both the public and private sectors, with no change in church-related institutions. Why this should be so is not clear. However, in Category IIA, we find sharp gains leading to virtual salary parity in public and in church-related institutions; private institutions did least well. Unfortunately, there is no simple way with the available data to develop some understanding as to why the differences persist.

Still another way to examine the salary structure is to compare changes in the distribution of salary levels over time. This is relatively easy to do using the AAUP data published each year that show the distribution of institutions by salary percentile levels. We can compare

1976–77 and 1982–83 using the old classification of institutional categories. We find that the ratios of the 60th to 20th percentiles increased slightly for Categories IIA and IIB; the same ratios declined somewhat for Categories I and III. The same pattern but somewhat more pronounced shows up for comparisons of the 80th and the 20th percentiles. We also observe a slight rise in the ratios of the 80th to the 20th percentile for Categories IIA and IIB. On the other hand, none of these changes are very substantial and hence it is difficult to make a strong case that inequality of salaries within categories is rising to any considerable extent.

By far the most interesting changes in the structure of faculty salaries show up when we examine data by academic discipline. These data are not collected by AAUP. Instead, an annual survey of state universities and land-grant colleges is conducted by the Office of Institutional Studies at Oklahoma State University. The survey was begun in 1974–75, but rapidly changing response rates as the survey became institutionalized make comparisons using data prior to 1976–77 unreliable. A somewhat similar survey has been instituted by Appalachian State University to obtain information for state colleges and universities. Because this latter survey is so new (it was first done on a full scale for 1983–84), it provides no basis for longer period comparisons; moreover, the quality of the data is difficult to assess. Other somewhat similar surveys are undertaken each year, such as that of the Massachusetts Institute of Technology which surveys a select group of public and private universities to gather salary averages by discipline; the results are not published but are circulated informally only to responding institutions. Thus, the most comprehensive and useful data for our purposes are those provided by the Oklahoma State University survey.

Until the late 1970s little attention was paid to salary levels by academic discipline. While some differences always existed, the differences appeared to be reasonably stable and hence of no great import. But with the collection of data on salaries by discipline it has become possible to examine more closely the structure of disciplinary labor markets. As it happens, the period since 1976 has produced some noteworthy changes in salary structure.

Increases in average salaries from 1976–77 to 1983–84, averaged across all disciplines, amounted to 59.7 percent for full professors, 60.2 percent for associate professors, 62.7 percent for assistant professors, and 66.5 percent for new assistant professors. These differences are not out of line with those shown by the AAUP data; the key difference is with the salary for new assistant professors, a figure not obtained through the NCES survey.

This larger increase for new assistant professors reflects several factors. One is the keen competition for the most able new Ph.D.s, and a recognition that high starting salaries are needed to minimize the leapfrogging of salaries that occurs because institutions have less flexibility in raising salaries for people after they are hired. In addition, there is always keen competition to hire the best available people so as to improve the quality of departments. And finally, the buoyancy of non-academic labor markets that generate demand for Ph.D.s has introduced a stronger element of competition between academic and other types of employers.

Just as there have been different percentage increases in overall salaries by rank, the spread of the percentage increases in salaries across the 22 broad disciplines widens with rank. At the full professor rank salary increases ranged from 45.0 to 71.2 percent. The dispersion becomes progressively larger as we move down the ranks, with the widest range of 25.7 to 84.3 percent occurring for new assistant professors.

The nature of the dispersion can be explored in two different ways. In Table 6–10 we show the levels of 1976–77 salaries, the percentage increases over the period, and the 1983–84 salaries for full professors and for new assistant professors; disciplines are listed in descending order by 1983–84 average salaries. This presentation immediately raises the question as to whether greater salary increases over the period went to faculty members in those disciplines with higher salaries in 1976–77. This question is answered by Figure 6–2 which plots for each discipline its salary levels for 1976–77 and 1983–84, with the former on the horizontal axis and the latter on the vertical axis. We observe a strong tendency for the higher paid disciplines in 1976–77 also to be the higher paid disciplines in 1983–84 at both the full professor and new assistant professor levels. The steep slope of the line fitted through the dots indicates in both cases that the higher the 1976–77 salary level, the greater the relative increase in salary from 1976–77 to 1983–84. Thus, the dispersion of salaries widened in relative terms.

We can examine the dispersion in a more detailed way by using information on salary changes by major fields rather than disciplines. The Oklahoma State University survey lists the 22 grouped disciplines already used as well as 204 ungrouped disciplines or fields. Some of the fields are empty for particular ranks, especially at the new assistant professor level. Tables 6–11 and 6–12 show the individual fields whose growth over the 1976–77 to 1983–84 period exceeded or fell short by 15 percentage points of the average percentage increase at the full professor and at the new assistant professor ranks respectively. The 15 percentage point range was picked for simplicity. We find that at the full professor level the number of fields falling outside the range is about evenly divided. This, however, is not the case at the new assistant professor level. Not only were there fewer fields in which new assistant professors were hired in 1983–84, but the number of fields with gains of 15 percentage points above the average gain was somewhat greater than for full professors. Even more surprising is the fact that the number of fields whose growth rate fell 15 percentage points short of the average

Table 6–10. Average Salary Levels By Discipline of Full Professors and New Assistant Professors, 1976–77 and 1983–84, and Percentage Increase in Average Salary, 1976–77 to 1983–84 (Listed in Descending Order of 1983–84 Salary Levels)

Full Professors

Discipline	1976–77 Average Salary	Percent Change	1983–84 Average Salary
Law	30,951	71.2	52,994
Computer Science	27,149	61.9	43,942
Business	25,800	70.0	43,872
Engineering	25,209	70.1	42,875
Physical Science	25,183	61.1	40,563
Mathematics	25,762	53.3	39,488
Psychology	24,959	57.3	39,270
Social Sciences	24,831	57.6	39,173
Biological Science	23,841	63.7	39,008
Public Affairs	25,533	52.6	38,955
Library Science	24,174	58.6	38,342
Interdisciplinary Studies	22,829	37.4	38,229
Architecture	23,202	62.9	37,798
Letters	23,440	58.4	37,120
Foreign Languages	24,266	52.8	37,079
Communications	23,377	56.4	36,563
Area Studies	24,304	48.9	36,189
Agriculture	22,418	60.7	36,025
Education	22,928	56.2	35,819
Home Economics	22,955	55.1	35,607
Fine Arts	22,045	55.1	34,202
Technical and Occupational Studies	22,149	45.0	32,117

New Assistant Professors

Discipline	1976–77 Average Salary	Percent Change	1983–84 Average Salary
Law	20,297	67.6	34,017
Business	16,701	83.0	30,569
Engineering	15,939	84.2	29,356
Computer Science	15,526	84.3	28,266
Library Science	14,602	64.4	24,000
Biological Science	14,564	64.2	23,915
Agriculture	15,030	56.9	23,576
Physical Science	14,050	65.9	23,312
Architecture	15,030	56.9	23,188
Mathematics	14,296	59.2	22,762
Public Affairs	15,561	44.3	22,447
Home Economics	15,169	47.2	22,324
Communications	14,360	52.8	21,949
Social Science	14,017	56.1	21,874
Education	14,283	46.5	20,921
Interdisciplinary Studies	13,635	53.1	20,880
Psychology	13,876	50.2	20,840
Letters	13,321	49.7	19,936
Fine Arts	12,957	52.5	19,758
Foreign Languages	13,263	48.2	19,651
Technical and Occupational Studies	15,517	25.7	19,499

SOURCE: 1976–77 and 1983–84 Faculty Salary Survey By Discipline of Institutions Belonging to the National Association of State Universities and Land Grant Colleges, conducted by Office of Institutional Studies, Oklahoma State University.

106

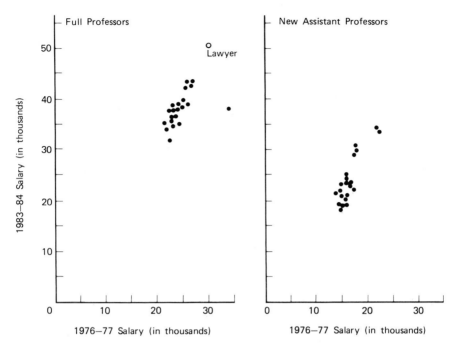

Figure 6–2. Relationship Between 1976–77 and 1983–84 Salary Levels by Discipline for Full Professors and for New Assistant Professors

was three times greater than at the full professor rank. In short, the spread of salaries grew appreciably over the seven-year period.

These data indicate quite conclusively that the academic labor market is neither a single labor market nor is it uninfluenced by what occurs in non-academic labor markets. These observations are not surprising. Everyone knows that engineers and humanists are not interchangeable in the classroom or in research activities. Thus, there are many labor markets structured around major disciplines. Moreover, many new Ph.D.s have always been attracted to non-academic positions, and except perhaps during the tight labor market of the 1960s, more academics leave for non-academic employment than the reverse flow from non-academic to academic employment.

The nature of these many academic labor markets and their links to non-academic labor markets is not yet well understood even though the broad patterns have emerged. The intensive, high level training given to Ph.D.s in different disciplines varies in the extent to which it can be and is utilized in the non-academic sector which includes government (federal, state, and local), other nonprofit organizations, and private firms. In some fields such as chemistry only a minority of new Ph.D.s take academic positions; their training prepares them for work in both the academic and non-academic sectors. In other fields such as the humanities the vast majority of new Ph.D.s take academic positions (or

Table 6–11. Percentage Changes in Average Salary Levels by Major Fields for Full Professors and New Assistant Professors, 1976–77 to 1983–84: Fields with Increases Exceeding Average for Rank by 15 Percentage Points

Full Professors *Average Change in Salary— 59.7 Percent*		*New Assistant Professors* *Average Change in Salary— 66.5 Percent*			
Engineering Materials	96.1	Pathology	145.2	Labor & Industrial Relations	84.8
Petroleum Engineering	85.3	Nuclear Engineering	113.2	Mechanical Engineering	82.6
Genetics	81.9	Social Sciences	105.8	Mining & Metalurgical Engineering	82.6
Social Sciences	81.9	Engineering, other	103.6	Industrial Engineering &	
Environmental & Sanitary Engineering	81.5	Petroleum Engineering	100.0	Management	81.9
Nuclear Engineering	81.1	Bioengineering & Biomedicine	100.0		
Higher Education, general	79.4	Engineering, general	93.9		
Aerospace Engineering	78.3	Home Economics Education	93.4		
Bioengineering and Biomedicine	78.3	Agricultural Engineering	93.0		
Industrial & Management Engineering	78.3	Business and Commerce	93.0		
Letters	75.5	Criminology	92.5		
English Literature	75.3	Business Statistics	90.1		
		Electrical Engineering	89.0		
		Computer Science	86.4		

Table 6-12. Percentage Changes in Average Salary Levels by Major Fields for Full Professors and New Assistant Professors, 1976–77 to 1983–84: Fields with Increases Falling Short of Average for Rank by 15 Percentage Points

Full Professors *Average Change in Salary—* *59.7 Percent*		*New Assistant Professors* *Average Change in Salary—* *66.5 Percent*			
Russian	44.6	Geography	51.3	Social Science, general	43.9
Pathology	44.3	Music	51.3	Spanish	43.8
Home Economics	43.2	Entomology	51.1	Philosophy	43.8
Advertising	43.0	Dramatic Arts	51.0	Education Admin.	42.0
Foreign Languages, other	42.4	Political Science	50.7	Consumer Education	41.9
Education, other	42.1	Psychology, general	50.3	Home Economics, general	41.8
Area Studies	41.8	Microbiology	48.7	Special Education	41.3
Physical Science, general	41.6	Forestry	48.4	Religion Studies	41.0
Public Administration	40.0	Foreign Languages, general	48.2	Education, general	40.6
Bacteriology	38.7	Agricultural Economics	47.0	Food & Nutrition	39.2
Biology	37.4	Food Sciences Technology	46.7	Special Education	37.3
Afro-American Studies	27.9	Speech	46.6	Health Education	36.2
German Literature	23.0	Psychology, other	46.1	Engineering Technology	36.0
Mid East Languages	21.4	Music Education	46.0	Marine Biology	33.2
		Elementary Education	45.9	Applied Design	32.8
		Russian	45.6	Classics	32.8
		Astronomy	45.6	Law	31.8
		Anthropology	45.6	Dance	29.6
		Earth Science	45.4	Public Administration	29.1
		Radio, TV	45.4	Recreation	25.4
		City Planning	44.5	Communications Science	5.5
		French	44.3	Urban Studies	– 2.0

would like to take them), largely because the demand for the skills they possess is so weak outside the academic sector.

The proportions of new Ph.D.s opting for other than academic jobs also vary, depending upon the strength of demand from the non-academic sector. Among the factors affecting the strength of demand is the business cycle, new technological breakthroughs that accelerate the demand for people with certain specialized skills (and perhaps weaken the demand for others with less needed skills), changes in government policy that affect the level of demand for different kinds of specialized skills, shifts in the direction of research activities, and the like. As a result, the flows of new Ph.D.s among sectors and also the flows of past Ph.D.s across sectors are not constant but rather reflect the continuing flux of our economic and social system.

While these data throw but limited light on the operation of academic labor markets, they do illustrate the nature of the zero-sum game. Given the allotment of funds to a college or university for salary increases, the desire of the institution to build quality departments, and differential demand across fields in the demand from the non-academic sector, higher salaries must be offered to retain those people in greatest demand. This leaves less for other faculty not in high demand fields. Thus, we observe a substantial redistribution of salary rewards across disciplines, and there is little that anyone can do about it.

Are Changes in Levels and Structure Related?

The answer to this question is yes, but it needs elaboration. As we have shown, faculty salaries declined relatively during the 1970s and early 1980s while at the same time changes in salary structure were limited to a widening of the averages by disciplines and fields, especially for new Ph.D.s at the assistant professor level. Exactly how are these two developments related?

Because real salaries declined relative to those available in labor markets outside the academic sector, employers in these labor markets whose personnel needs included Ph.D.s in particular disciplines saw the increased possibilities of recruiting academics. At the same time non-academic employment came to look more attractive to academics at least in part because of the salary advantages offered outside academe. Not all disciplines benefited equally from this change—only those in which the skills and knowledge of Ph.D.s in the academic sector were similar to those required in employment elsewhere so that transfer was relatively easy from one sector to another.

As more individuals opted to leave academe for other kinds of employment, and as new Ph.D.s found other types of employment more financially attractive, colleges and universities felt compelled to do what they could to restrain the exodus of people already on their faculties and at the same time to make conditions more attractive for hiring new Ph.D.s.

These efforts were constrained because of limited funds available for faculty salaries. The only way higher salaries could be paid to people in disciplines that were in high demand was to hold down on the increases given to people in other disciplines not in high demand. And thus we witness a widening of the spread of salary averages by field and disciplines. The spreading is most pronounced among new assistant professors all of whom have just been in the market and taken a job. Thus, there is much greater latitude for shifting salary differentials for new assistant professors than for already employed faculty members. This is nothing new; it reflects the way labor markets normally operate.

It is possible for some to argue that a fundamental change has occurred in the way the outside labor market views many academics. Perhaps it is true that the knowledge and skills possessed by people in certain fields are now more valuable than ever before. It is difficult to evaluate this argument without going into specific disciplines and fields. But one should remember that some academics have always been in high demand by other sectors of the economy. To the extent that conditions may have changed somewhat because of dramatic new technological developments, there might have been some widening of disciplinary averages even if faculty salaries had kept pace with inflation or with the salary increases for people performing approximately similar jobs.

One thing is clear. The direction of the linkage between salary levels and structure has been from levels to structure and not the other way around. It is possible to imagine the reverse happening, as in a period of strong economic growth when outside demand for Ph.D.s is strong. This would put upward pressure on the whole salary structure and presumably propel it upward more rapidly. This may have happened in the 1960s. If so, we might have expected faculty salaries to rise more rapidly than they did. Unfortunately, no body of data exists that would enable us to check on the dispersion of average salaries by discipline and field in this earlier period.

We conclude that changes in levels and structure are closely linked. This suggests that the growing undercurrent of dissatisfaction among faculty members that is directed toward their colleagues in fields receiving larger salary increases is misdirected. The fundamental problem is an inability to make any substantial advance that would push up salary levels, across the board, more rapidly.

Conclusions

Sharp declines in real and relative salaries combined with a widening dispersion of salaries across disciplines has depressed and will continue to lower the morale of college and university faculty members.

These declines also weaken the capacity of higher education to serve students and society. The magnitude of these changes puts faculty members into a special category and nurtures their feelings of being ill-treated. Were these changes smaller (say, only half as large) or were they clearly the result of market forces, the current economic position of faculty members would be easier to rationalize for themselves and for others. To the extent that difficult economic conditions forced austerity on government, the declines would be more readily accepted. In fact, however, virtually all other groups of state and local employees have maintained their real and relative salary positions, the only exception being elementary and secondary teachers, who have also fallen sharply behind.

The importance of this demoralization cannot go unmentioned. As a result, faculty members are likely to be less responsive to the varied demands made on them because of feelings they are not adequately compensated. They will be less likely to encourage their best students to pursue academic careers because of the poor salary outlook and their perceptions of slippage in the quality of academic positions. Faculty members with opportunities to move into other types of employment will be increasingly likely to do so because of the financial advantages accruing to them. Faculty members will be more responsive to remunerative outside activities in order to supplement their academic salaries. Current graduate students, witnessing these developments, may respond in similar ways, with more of them opting for better paying non-academic positions; if they do choose academic positions, they may behave more like their senior colleagues.

The tragedy of these developments is that they will undermine a massive and highly effective state and locally operated and financed system of higher education. The implications are serious in several ways. If the quality of the system and its people deteriorate, it will be less able to provide the teaching, research, and public service activities that have long characterized American colleges and universities and that have contributed so much to the development of this nation. Moreover, the system will be less well equipped in the 1990s to respond effectively to the expected retirement of large numbers of faculty members hired during the rapid buildup of higher education in the 1960s.

The only hope of reversing these developments is for faculty members, administrators, responsible public officials, and concerned citizens to redouble their efforts to reorient people's thinking about higher education. The benefits of the educational enterprise must be made more obvious so as to convince people that these benefits are worth the costs of an increased commitment of resources to higher education and to faculty salaries which are a large component of total costs. Unless such efforts are made, the depressed salary levels and twisted structure of salaries will come to be accepted as normal. Once this occurs, any hope of making up for past losses will be forever destroyed.

CHAPTER SEVEN

The Changing Work Environment

Astute observers have long known that the rewards of the academic profession are to an unusual degree *intrinsic*. They are inherent in the work itself (McKeachie in Lewis and Becker, 1979, pp. 3–10). The intrinsic rewards include the satisfactions derived from intellectual curiosity, interest in ideas, exercise of rationality, opportunity for achievement and self-expression, fascination with complexity, ability to solve difficult problems, the pleasure of expertness, and participation in decisions affecting one's life. These rewards are akin to what Thorstein Veblen called the "instinct of workmanship" (1914), and they are related to the ideas of John Ruskin and William Morris on the satisfactions derived from skilled craftsmanship. Professors are in a sense craftsmen of ideas and shapers of people. The intrinsic rewards of the academic profession also have an interpersonal aspect. Ideally, they include membership in a campus community where there is friendship, stimulation of colleagues, the recognition of work well done, and the pleasure of association with promising young people and of being instrumental in their growth and development. They also include participation in professional associations and other groupings representing particular disciplines. These associations—which may be regional, national, or international—provide further opportunities for personal contact with colleagues and for recognition. These intrinsic rewards are distinguished from external rewards which include mainly monetary compensation and social status derived from the profession-at-large or in rare cases from fame that extends beyond the profession.

In the value system of faculty people, the intrinsic rewards are of deep concern and the commitment to work for its own sake is immense.[1] McKeachie (in Lewis & Becker, 1979, p. 7) asserts that "In all

---

1. This may explain (wholly or in part) why faculty compensation even in normal times is generally somewhat lower than that in comparable professions.

studies, intrinsic satisfactions are reported to be much more important than extrinsic rewards." Toombs and Marlier (1981, p. 21) observe: "There is ample evidence . . . that academic life is a deeply socialized existence. The depth of commitment is remarkable" Faculty members, therefore, are bound to be keenly interested in the conditions within and surrounding the work place—conditions that affect the amount and excellence of their work and the quality of their lives as well.

Their incentives, their performance, and their morale are influenced by these conditions. Our own field observations support this conclusion. This is not to say that faculty members are unmindful of monetary rewards. In fact, our campus visits revealed widespread concern among faculty, especially junior faculty members, about their compensation (see Chapter 8). But they also care deeply about the work environment, and this too was manifested in our site visits.

Working conditions in the academic profession have been gradually deteriorating since about 1970. The deterioration has affected most but not all features of the work environment. It has been especially onerous to faculties because it has been accompanied by a slow but substantial decline in monetary compensation (as expressed in real terms after adjustment for inflation). The decline in either monetary compensation or working conditions alone would have been bad enough, but both together have been extraordinarily disquieting. The adverse impact of this double deterioration has been intensified as faculties have compared the present situation with that around 1970, when both compensation and working conditions were probably at an all-time high.

In this chapter we shall review the changes in working conditions over the past 15 to 20 years. Most of these changes are not easily observable and certainly not quantifiable. Trends in working conditions vary greatly among institutions and even among departments and individuals within single institutions. We cannot hope to trace all the variability, but we shall try to note broad changes in the general situation.[2]

Changes in the Work Environment Due to Financial Stringency

For the higher educational community the entire period from about 1970 to 1985 has been one of financial stringency. The stringency has been due partly to the difficulty of keeping income in pace with inflation and partly to changing priorities for public expenditures. It has been expressed in three forms. First, the annual financial

2. In preparing this chapter, we had the benefit of two valuable literature reviews: Sally Loyd's *Faculty Working Conditions and Working Environment* and Roberta Stathis-Ochoa's "*Universities as Workplaces: Changing Conditions*" The two reviews are available on request. For details, see Preface, p. vii.

increments in real revenues, corresponding to proauctivity gains in many goods-producing industries, have disappeared. Second, real revenues per student have declined on the average by about 0.5 percent a year. And third, the period from 1970 has been one of almost constant expectations of financial catastrophe associated with the forecasts of declining enrollments. These factors combined have made money scarce in relation to the institutional standard of living that had been achieved by 1970; they also created an atmosphere of foreboding which led to caution in institutional spending. Under these conditions, most institutions, in order to sustain their basic programs, were forced to economize. Not only did they cut frills, but they substantially reduced the real earnings of their faculty and staff and deferred the maintenance of plant and facilities. Many institutions found it expedient to modify their traditional missions and even to compromise their academic standards in order to attract the students necessary to maintain tuition revenues and state appropriations. Today, colleges and universities everywhere are experiencing the deadening constraints of austerity and diminished expectation, though in most cases their position is far from catastrophic or hopeless. At least not yet. On the whole, colleges and universities have proved to be remarkably resilient. There have been few closures among them and most of these have involved small and obscure private colleges that had never taken root. So far there have been no episodes among them comparable to the gigantic bankruptcies, or near bankruptcies, among major corporations. But the experience of a decade and a half of unremitting financial stringency has left its mark on the faculties. Not only has it produced a slow decline in their real earnings (as shown in Chapter 6), it has also taken its toll on working conditions.

In meeting financial stringency, colleges and universities have often adopted the strategy of maintaining intact their core faculties and staffs. In doing so, they have raised salaries and fringe benefits (in current dollars) as much as finances would allow, and concentrated their cuts on marginal purposes and activities. This strategy has meant that many supportive services and facilities have been curtailed. For example, outlays for secretarial services, laboratory assistance, and paper graders have been reduced. The cutback in services has been only partially offset by the growth of student-aid programs involving "work-study." Ironically, the paring down of support staffs for faculty has been accompanied by a relative growth in number of administrative and general service employees due to increasing effort devoted to student recruitment, fund-raising, public relations, affirmative action, and various requirements of a multitude of public agencies (R. Anderson, 1983, p. 25; H. R. Bowen, 1980b, chaps. 4, 5). The growth of non-academic staff, inevitable though it may have been, has not pleased those faculties whose support services were being curtailed.

The administrative strategy in meeting hard times has led also to inadequacy, obsolescence, and deferred maintenance of offices and laboratories, equipment, library and museum collections, and even inventories of supplies. Deferred maintenance is not new to academe. There never was a time when all the buildings and equipment were completely adequate, up-to-date, and in working order, and when the libraries held all the needed books. It has always been tempting—in good times and bad—to postpone capital maintenance in favor of strengthening instruction, student services, and research and to launch "exciting new programs" rather than to carry out the drab and unspectacular tasks of maintaining plant and equipment. It is difficult to measure trends in leaky roofs, faulty electrical systems, deficits in library collections, and so on, but a few estimates of the growing deferred maintenance have been made. The most comprehensive is that of Harvey H. Kaiser (1984), who estimated the total cost of needed repairs or replacements on the nation's campuses at $40 to $50 billion— an amount that is about two-thirds of a year's total operating budget. This figure is conservative because Kaiser did not include inadequacy of building space, items such as obsolescence of laboratory and educational equipment, and the growing deficit in library acquisitions. He pointed out that the deferred maintenance is rapidly growing worse because many of the structures built during the boom of the 1960s are just now reaching a point when they need a first round of major maintenance and renewal. Richard Anderson, in comparing the condition of higher education in 1968 and 1980 (1983, p. 3), found that the percentage of educational and general expenditures for plant operation and maintenance had not changed appreciably. However, given the growth of the physical plant, expenditures for maintenance and plant operation *per $1000 worth of plant* (at book value) had declined by 43 percent. This decline would have been even greater had the buildings been valued at replacement cost.

A 1984 survey of 43 research universities conducted by the National Science Foundation revealed that half the department heads reported that research equipment was insufficient, 90 percent indicated that lack of scientific equipment "inhibited the conduct of critical research," and the combined opinion was that only 16 percent of present equipment is "state of the art" (*Chronicle of Higher Education*, May 16, 1984, p. 1). At the regional level, comparable estimates have been made. For example, a study for the Texas legislature concluded that it would cost $300 million to repair and rehabilitate the public campuses of the state—exclusive of Texas A & M and the University of Texas at Austin (*Chronicle of Higher Education*, March 14, 1984). A report on the University of California campuses reported that "about $4 billion would have to be spent during the next decade to renovate, construct, and maintain research buildings, libraries, hospitals, and instructional facilities" (ibid.). Moreover, the University of California has issued urgent requests for the updating of their scientific laboratories which have become badly outmoded—especially in comparison with the laboratories of private industry. A similar story is told in a report of the California Postsecondary Education Commission (1983).

Regarding library books, a report of the National Center for Education Statistics (1984, p. 4) showed that between 1978–79 and 1981–82 book acquisitions decreased by 9.1 percent and microform book titles by 6.3 percent.

Financial stringency has had the effect of curtailing funds for professional travel and possibly also for faculty leaves and released time. Financial stringency, however, may not have had a pronounced effect on the *number* of leaves. A study by Andersen and Atelsek (1982) showed that the percentage of humanities faculty being awarded leaves declined only slightly over the period 1977–78 to 1979–80 (p. 18), and that of those institutions reviewing their leave policies, more were thinking in terms of liberalization than restriction (p. 25). However, from scattered information, we are of the opinion that today fewer replacements for those going on leave are provided and as a result the faculty workload is increased.

Financial stringency has undoubtedly led institutions to appoint increasing numbers of more or less temporary, inexperienced, and part-time faculty. As we have emphasized in Chapter 4, there is a place for carefully selected and fully prepared persons who are employed on a temporary or part-time basis. However, a good thing can be overdone. Aside from questions about the ability and preparation of part-time people, they ordinarily cannot carry their share of the load of advising students, serving on committees, recruiting new faculty, selecting new equipment, ordering books for the library, and other departmental or institution-wide responsibilities. When a sample of faculty was presented with the statement, "Higher education faculty have no reason to be concerned over the number of faculty employed part time in higher education," 80 percent disagreed (National Education Association, 1979, p. 13).

Tables 7–1 and 7–2 provide the results of two surveys of faculty opinions regarding working conditions within their departments. Table 7–1 shows that in most departments conditions were stable or were deteriorating. Table 7–2 indicates that a substantial proportion of faculty, in most instances more than half, thought there was room for improvement in various institutional practices affecting working conditions.

Finally, financial hard times have probably been at the root of several other changes affecting faculty. The most important of these is increasing rigor of the requirements for tenure and for promotion in rank and salary. As these requirements have stiffened, faculty members—especially young ones—have felt compelled to pay increasing attention to both research and teaching, undoubtedly at the expense of their private lives. This pressure has been exacerbated by the decline in the voluntary mobility of faculty people among institutions. Mobility is greatly enhanced by successful and visible research and scholarship, and so faculty members in early and mid-career who hope to get ahead are forced, or at least well advised, to pay attention to the length and quality of their publication lists. Faculty time has been pulled in another direction by the demands of accountability and supervision. These are in the form of increasing numbers of reports, red tape, and evaluations which are partly a product of financial stringency. They also derive from the increasing size of institutions and from proliferating layers of supervision and oversight including federal agencies, legislatures, statewide commissions, central offices of multi-campus institutions, and central administrations on local campuses. Finally, the allocation of faculty time has been affected by changes in the participation of faculty in the governance of their institutions. On the one hand, there has been a trend in many institutions toward increasing participation as financial problems have loomed and administrators have sought consensus and support for difficult decisions. In other institutions, there has been a trend toward reduced participation as strong management-oriented administrations have taken emergency action without adequate consultation. Sometimes both trends have occurred in the same institution.

To sum up, the effect of financial stringency on the work environment has been substantial. It has curtailed supportive services and facilities. It has curtailed professional travel. It has led to deferred maintenance and obsolescence of buildings and equipment and has thus increased the difficulty of faculty tasks and reduced the effectiveness of their efforts. Often the failures of maintenance and the inability to increase building space have offended the sensibilities and dignity of faculty members confined to quarters that are inadequate, crowded, ugly, uncomfortable, and lacking in privacy. It has reduced the proportion of the faculty on permanent and full-time appointments and increased their responsibility for the operation and continuity of their institutions and the advising of students. It has changed and increased the rigor of the standards for faculty promotion and mobility, thus diverting their efforts to research and scholarship while at the same time demanding excellent teaching. It has increased the time and effort devoted to accountability.

The Work Environment as Affected by Changes in the Characteristics of Students

A second major factor affecting the work environment has been the profound change in the characteristics of students. These changes began slowly in the 1960s, picked up momentum in the 1970s, and are continuing in the 1980s though at a somewhat slower pace.

Table 7–1. Faculty Opinions of Trends in Their Departments, 1980 to 1982

	Increase	*No Change*	*Decrease*	*No Response*	*Total*
Number of authorized faculty positions	26%	50%	23%	1%	100%
Secretarial and clerical assistance	9	63	26	2	100
Research and related assistance	10	46	33	11	100
Funds for faculty travel	9	34	53	4	100
Funds for faculty leaves with pay	5	53	27	15	100
Number of course offerings	30	47	21	2	100
Library services to faculty members	18	61	15	6	100
Budget for equipment and supplies	21	31	42	6	100
Replacement and renewal of equip- ment	14	41	38	7	100
Maintenance of buildings and facilities	14	50	31	5	100
Faculty control over academic personnel decisions	8	61	22	9	100

SOURCE: John Minter Associates, *Chronicle of Higher Education*, November 23, 1983, p. 20. The data are based on responses from 4,235 faculty members.

Between 1960 and 1985, college and university enrollments mushroomed from 3.9 to 12.2 million students, an increase of 8.3 million. The following figures tell the story (National Center for Education Statistics, *Projections*, 1982, pp. 42, 58, and previous issues):

	Number of students enrolled	*Full-time-equivalents*
1960	3,900,000	3,000,000
1970	8,600,000	6,700,000
1980	12,100,000	8,700,000
1985 est.	12,200,000	8,600,000
Increase from 1960	8,300,000	5,600,000

This growth was fostered by strong encouragement of both youthful and adult learners to enter college and also by low tuition fees and generous student aid. In the 1960s and 1970s one of the major societal objectives was to make of America "a nation of educated people." This goal was on the whole willingly supported by college and university faculties. Indeed, there are few Americans who, in retrospect, would condemn this policy or who would want to revert to an elitist higher

Table 7–2. Percentage of Faculty Reporting that Improvements are Needed in Selected Institutional Practices, 1979

	Percentage indicating improvement needed	
	Four-year institutions	Two-year institutions
Sabbatical leaves	33	35
Travel funds	68	63
Funds for professional growth	70	66
Faculty evaluation	57	57
Definition of faculty merit	76	71
Promotion policies	57	59
Student-faculty ratio	46	48
Teaching materials and equipment	51	45
Faculty load (in hours)	51	54

SOURCE: National Education Association, 1979, p. 15. See also National Education Association, 1972, p. 3.

education. Nevertheless, a substantial portion of the tremendous increase in enrollments consists of persons who differ in gender, social origins, high school preparation, and time commitment to education from the much smaller population of mostly full-time students who attended college before 1970. Table 7–3 presents data on these changes. To these could be added the most numerous of all "new students," namely, white men and women from families and neighborhoods of limited economic and cultural attainments.

Among students from all classes of society many—perhaps a majority—come to college with inadequate high school preparation. Perhaps the most telling documentation of the decline in preparation comes from data collected annually over the years 1966 to 1984 by the Cooperative Institutional Research Program (CIRP).[5] The findings include: sharp declines in the number of students studying subjects requiring mathematical or verbal skills; a fall in the proportion of male students expressing interest in postgraduate careers; a decline of student interest in science; a fall in student scores on standardized tests; and a change in student values toward material things, power, and status and away from altruism and interest in social issues and problems. Along the same line, the Carnegie Foundation for the Advancement of Teaching found, in a 1984 survey of faculty, that 84.4 percent agreed with the statement "Teaching would be easier here if students were better prepared before they were admitted" (*Change*, Nov. 1984, p. 34). Similarly, Minter and Bowen in annual surveys of college administrators and faculties found few respondents reporting improvement and most report-

5. A. W. Astin et al., 1966–84. See also *The Times [London] Higher Education Supplement*, July 16, 1982, p. 6.

ing deterioration or no change in the academic preparation of students in reading, writing, and mathematics.[6]

Experience has proven that the teaching of "new students," and of the many unprepared students, on the basis of traditional methods and standards, is an onerous task that greatly increases the workload of faculties.[7] When done well, it requires special advice, guidance, encouragement, evaluation, and remedial work. On the other hand, if the teaching is carried out on the basis of business as usual, faculties inevitably feel the dissatisfaction and guilt of craftsmen who do sloppy work (Cf. Boberg & Blackburn, ca. 1983, p. 9). At any rate, faculties almost everywhere find their work and their lives complicated by the presence of numbers of underprepared and often undermotivated students, though the severity of the problem varies greatly among institutions. Moreover, the difficulty is compounded in that most of the "new students" are commuters, many part-time, who do not receive the total immersion in college life and academic values that are available to full-time residential students. On the other hand, many of the "new students" bring with them practical experience and seriousness of purpose which were and are often lacking among "traditional" students.

In raising the matter of the changing student populations, we are attempting to be forthright about a real problem. We are mindful of Harold Enarson's warning (1982, p. 48); "In some quarters talk of quality is code language for closing the doors of opportunity for the less well prepared." Our position is that it is the duty of the nation, of the higher educational system, and of the faculties to keep opportunity open by providing excellent education to the widest possible clientele. But this does not alter the fact that the admission to higher education of underprepared students increases the responsibilities and the workload of faculty.

The Work Environment as Affected by Shifts in Enrollments

A third influence affecting the work environment of faculties has been an unprecedented shifting of enrollments among the various academic

6. *Chronicle of Higher Education,* May 26, 1982, pp. 8, 10; *Independent Higher Education,* 1980, pp. 40, 48–50; *Preserving America's Investment in Human Capital,* 1980, pp. 35, 36–38. A fascinating study comparing freshman reading performance over the years from 1928 to 1978 is relevant. In this study, the Minnesota Reading Examination for College Students and the Minnesota Speed Reading Test were administered under comparable conditions to 1,313 freshmen in 1928 and to 1,154 freshmen in 1978—both times at the University of Minnesota. The 1978 freshmen scored significantly lower than their 1928 counterparts on vocabulary, comprehension, and reading rate. The 1978 sample was at least one grade level below the 1928 group on reading ability (Eurich and Kraetsch, 1982).

7. In this regard, it should be noted that calculations of faculty workload ordinarily use as a base full-time-equivalent figures for students rather than headcount. Yet, working with marginal students entails, on the whole, more faculty time and effort than working with adequately prepared students. Thus, the use of FTE rather than headcount as a base tends to understate the increase in faculty workload.

Table 7–3. Changes in the Composition of the American Student Population, 1960–85

	1960	1970	1980	1985 est.
Enrollment by Attendance Status[1]				
Four-year Institutions				
Full-time	71%	73%	71%	68%
Part-time	29	27	29	32
Total	100	100	100	100
Two-year Institutions				
Full-time	55%	52%	39%	37%
Part-time	45	48	61	63
Total	100	100	100	100
Enrollment by Gender[1]				
Four-year Institutions				
Male	63%	59%	51%	51%
Female	37	41	49	49
Total	100	100	100	100
Two-year Institutions				
Male	63%	59%	45%	44%
Female	37	41	55	56
Total	100	100	100	100
Enrollment (full-time equivalent) by Academic Level, Four-year Institutions[1]				
Undergraduate	88%	85%	83%	80%
Graduate	10	12	13	15
First-professional	2	3	4	5
Total	100	100	100	100
Enrollment by Age[2]				
16–21	—	55%	46%	40%
22–29	—	30	31	34
30 and over	—	15	22	26
Total	100	100	100	100
Enrollment by Race[3]				
White	94%	91%	87%	86%
Black and other	6	9	13	14
Total	100	100	100	100
Mean Scholastic Aptitude Test Scores[4]				
Verbal scores	474	455	424	425
Mathematical scores	495	488	466	468

[1] NCES, *Projections*, 1982, pp. 40–41, 58–59; 1970, pp. 23–25.
[2] NCES, *Projections*, 1982, p. 36.
[3] Bureau of the Census, *Statistical Abstract of the United States*, 1984, p. 160.
[4] College-bound high school seniors. Source: National Center for Education Statistics, *The Condition of Education*, 1982, p. 110. Figures for 1985 are estimates of the authors.

disciplines. Whether enrollments are measured by degrees awarded, by total enrollments in various fields of study, or by the plans of entering freshmen, enormous shifting of student choices has occurred, much of it during the 1970s. Among the various disciplines, the gainers, the losers, and those about breaking even were as follows[8]:

Gainers	Losers	Those About Breaking Even
Agriculture	Education	Fine and Applied Arts
Architecture	English	Life Science
Business	Foreign Languages	Economics
Computer Science	History	Physical Sciences
Engineering	Mathematics	Psychology
Health Professions	Political Science	
Home Economics	Sociology	
Law	Theology and related fields	

A glance at this list reveals immediately that the gainers were the strongly professional or practical fields; the losers were the humanistic fields, some of the social studies, and education; and those about breaking even were some of the fields that straddle professional and humanistic aims. These shifts have produced a serious maldistribution of faculties by discipline.[9] Had it been possible immediately to discharge redundant faculty members in the "loser" fields, and transfer the released funds to the employment of new faculty in the "gainer" fields, the maldistribution might have been quickly corrected. But in fact, a prompt shifting of resources has been feasible only to a degree. Some relief has been achieved by terminating non-tenured faculty and also by encouraging and assisting some tenured persons in the loser fields to "retool" for work in the disciplines of high demand. For example, a few mathematicians have become computer scientists. These expedients, however, have yielded only limited results. The institutions, of course, might have ruthlessly disregarded their tenure obligations and discharged large numbers of tenured faculty in the loser fields. They have been restrained from doing so not only because of faithfulness to tenure commitments, but also because they did not wish to decimate the ranks within the humanistic disciplines that had been, and might again become, the mainstay of higher education. Most educators take pride in their institutions as centers of liberal education, and they do not wish to assume the guilt that would be associated with the demise of literature, history, religion, and other humanistic subjects. To make matters

8. U.S. Bureau of the Census, *Statistical Abstract of the United States*, 1981, p. 161; and 1972, p. 133. National Center for Education Statistics, *Digest of Education Statistics*, 1982, pp. 117–21 and previous annual issues; A. W. Astin et al., 1983, pp. 30–31, and previous annual issues; U.S. Department of Commerce, ca. 1982.

9. The shifts in enrollment were in general accompanied also by shifts in funds for research from the humanities and social studies to the hard sciences.

worse, colleges and universities have found great difficulty in recruiting qualified new faculty in the gainer fields. In seeking teachers of business, engineering, computer science, and the like, they have faced the competition not only of other academic institutions, but also of private industry. The situation simply has not favored a massive shifting of resources from humanism to vocationalism. As a result, the gainer fields became understaffed and overworked, and the loser fields overstaffed in relation to enrollments, but not necessarily underworked. Though there have been exceptions in some institutions, most of the loser fields have continued to carry on their programs, the principal changes being somewhat smaller faculties, fewer fresh faces among the faculty, smaller classes, but not necessarily smaller work loads. The overall result of the tremendous shift in demand from humanistic to vocational fields has been a substantial increase in the teaching load of faculties, an increase that has been concentrated in the professional fields.

Another consequence of the massive shift of enrollments has been the increasing disparity in compensation between faculty in the fields that have gained enrollment and those losing it. Traditionally, most colleges and universities have operated with a salary scale that has been roughly similar among the various schools and departments. The main exceptions have been clinical medicine and to some extent law, where compensation rates have been substantially higher than those prevailing in other disciplines. Sometimes, however, market pressure in the other disciplines has brought about differences among them, for example, in the form of initial appointments at a higher step in the salary scale, exceptionally rapid promotion, provision of special funds for equipment and library books, special assistance with housing, etc. But the basic compensation scales have not varied much among different disciplines within institutions except for the special case of medicine. In recent years, however, under the market pressure created by the unprecedented shifts in enrollment, substantial differences among schools and departments in pay scales have become common. This tendency has been accentuated as institutions have reached out for "stars" in their quest for notorious excellence. It is not uncommon today for "distinguished" professors to receive compensation packages—including salary, housing, laboratory equipment, library books, assistants, and secretaries—running well into six digits. No institution would go out of its way to break the traditional salary scale. But in seeking talent for professional fields in competition with industry and private practice, where the opportunities for salary and emoluments vastly exceed those of academe, most institutions believe they have no alternative.

In concluding our discussion of the shift in enrollments, two questions should be considered: Is American higher education experiencing a transitional period when powerful technological and social forces are shaping irrevocably both the academic choices of students and the competitive demands for faculty? Or are our colleges and universities merely going through a temporary phase that could end in the near future and allow a return to traditional relationships between professional and liberal studies?

Clearly, the shift of enrollments toward professional fields can be explained in large part by five important—but not necessarily permanent—social changes. First, the women's movement has brought about rapidly rising college enrollments of women, and at the same time it has led women to prepare for a multitude of careers other than the traditional ones such as homemaking, teaching, and nursing.

Second, the entry into higher education of millions of persons from among the first generation of families to attend college has resulted in a great increase in demand for vocational education. There has been nothing new in this—except in the magnitude of the demand. First-generation students have long tended to elect vocational subjects. The young men have traditionally elected agriculture, business, and engineering; the young women, as noted, home economics, education, and nursing. As a result of their experience in college and in later life, however, these first-generation students have become more sophisticated about higher education and have tended to guide their own children toward a broader liberal education. During the past two decades, higher education has enrolled millions of these first-generation students, and true to form they have elected vocational subjects. But, if history repeats itself, when the time comes for their children to attend, and assuming a reasonably prosperous economy, their choices will probably lean more strongly toward liberal education. The time is not far ahead when the number of second-generation students will increase sharply.

Third, trends in technology may exert an influence on college enrollments. It is commonly asserted that the demands of the new age of "high technology" will increase the amount of technical knowledge required of our people; on the other hand, it is argued, less frequently, that technology in the form of automation will reduce the technical demands on all but a few specialists. Either view, or possibly both, may be correct. It would seem to us, however, that in the prospective state of the world, the need for literature, history, philosophy, pure science, and the arts will increase in proportion as human life is dominated by high technology.

Fourth, it is possible that the demand for professional education will level off or even decline in the near future. Professional fields such as law, medicine, and dentistry are all showing signs of an approaching saturation point.[10] Over the next decade or so, the rate of enrollment in

10. Law school enrollments in the fall of 1983 were down 0.5 percent while entries of first-year students were down about 2 percent (*Chronicle of Higher Education*, Nov. 14, 1984, p. 3).

these and other vocational fields may level off or even decline, and a swing back to liberal education may occur. Those colleges and universities that have kept their liberal arts programs intact during a period of low demand may well be proven wise. A swing back to liberal education would be accentuated if, as appears likely, the demand for secondary teachers expands.[11]

Fifth, many educators and leading citizens as well are concerned about the educational consequences of the shifts of enrollments to vocational and professional fields and are attempting to reverse the recent trends. The many "commission" reports on the reform of higher education which appeared in the early and mid 1980s illustrate these trends. The implications of these reports for faculty are considered on pp. 282–86.

Our conclusion is that a swing back to a better balance between vocational programs and general education would be we believe highly desirable and we believe probable.[12]

Other Changes in the Work Environment

In concluding our account of the work environment, we shall consider several other changes which have influenced the effectiveness and welfare of faculty.

Job Security, Autonomy, and Academic Freedom. Over the years since 1970, faculty members in most institutions have had a sense of increasing anxiety about the continuity of their appointments. Today, this anxiety ranges from slight in the case of well-established professors with tenured positions in strong institutions, to acute for non-tenured persons in unstable institutions. But most faculty members believe—we think rightly—that their job security is at least a little less than it was in more prosperous times. The specific risks they face are falling enrollments, reduced revenues, declining research funds, a weakening of the protection afforded by tenure, and even the demise of their colleges or universities. Especially vulnerable are non-tenured faculty at those numerous institutions which already feel themselves "tenured in." The security of faculty is also impaired because, given the current slackness in the academic labor market, faculty members cannot readily move from one institution to another.[13] It is this lateral immobility that gives

11. However, arts and sciences programs may not be the automatic beneficiaries of changing trends in undergraduate professional education. There are other possible destinations of released funds, for example, "continuing professional education."

12. In a thoughtful analysis, Michael McPherson (1984, p. 24) suggests that the best that can be hoped for is a stabilization of relative demand between technical and humanistic subjects for the next five to eight years.

13. The ability of academic people to move to positions outside academe may be hampered to some extent also by the tendency of outside employers to be less tolerant of individuality, eccentricities, and independent thought than the academic community.

them the sense of "stuckness," a word and concept invented by Rosabeth Kanter. The sense of insecurity has also been accentuated over at least a decade by the constant references in daily campus conversation and in the literature of higher education to an impending decline in enrollments with adverse consequences.[14]

The decline in mobility and security may have impaired academic freedom in subtle ways. We have heard from several sources that faculty members who are fearful for their jobs—particularly younger faculty members hopeful of tenure—feel constrained in their utterances and their writings. For example, in a survey of the National Education Association (1979, p. 13), a quarter of the faculty respondents indicated that there were important constraints on academic freedom, and 80 percent reported that improvements are needed in due process relating to just treatment in salary and other academic matters.[15] Though we have not uncovered any large-scale or major infractions of the spirit or substance of academic freedom, it is plausible that persons with weak market positions might feel inhibited even though there were no intention of actual wrong doing on the part of senior faculty, administrators, trustees, or legislators.[16] We suspect, under conditions of recent decades, that there has been considerable erosion of faculty autonomy and of participation in institutional decision-making; at least, this appears to be the perception of many faculty members. As Richard Anderson found (1983, p. 4): "Faculty perceive their colleges to be less democratically governed. . . . Administrators are making more of the decisions and involving students and faculty less than they did in the early 1970s." Similarly, in a survey of the National Education Association (1979, pp. 14, 16), on most questions pertaining to internal communications and faculty participation in institutional affairs a substantial proportion ranging from a third to a half of the respondents felt that conditions are less than satisfactory or are unsatisfactory. Minter (1982) also reported that 22 percent of faculty felt that faculty control over academic personnel decisions declined during the years 1980–82, whereas 8 percent thought it had increased.

The apparent erosion in faculty participation has been due in part at least to important changes in the governance of institutions

---

14. In a study of the National Education Association (1979, p. 13), 53 percent of the faculty respondents expressed the opinion that declines in the college-age population would adversely affect their institutions.

15. Gerald Reagan (1982) cites five contemporary constraints on academic freedom: student consumerism, vocationalism, program planning, allocation of research funds, and the tenure-granting process. Cf. Hendrickson (1982), who gives five reasons for faculty tension.

16. David Riesman has commented (1983, p. 266) that in the McCarthy era "I came to question the importance of tenure in protecting academic and intellectual freedom against external vigilantes. . . . However, the rise of the vigilante Left in major high-prestige colleges and universities over the last two decades has caused me to reconsider tenure."

stemming primarily from the increasing size of institutions individually and the increasing scope of the entire higher educational establishment. Starting at the top, the control of higher education has been greatly centralized through the increasing role of the federal government, the establishment of state legislative committees on higher education, the creation of statewide coordinating bodies, and the formation of central offices of multi-campus institutions. Much decision-making has been lifted out of the institutions and shifted to higher layers of authority, a process accelerated by faculty collective bargaining. To some extent, because of increasing size and complexity, centralization has also occurred within the individual colleges and universities. This has been especially so as the institutions have become larger and as hard times have struck. Matters once handled intramurally, and on which faculties could be consulted in their unhurried and sometimes cumbersome manner, have tended to gravitate toward the central administrations where they could receive decisive and prompt action. The importation of the "marketing" and "management" mentality into academic administration has probably contributed to the declining influence of faculty in policy-making. This mentality has been an outcome of hard times and it may be that the survival of many institutions can be attributed to "management." There is a clear need, however, for a reconciliation of the values of "management" and the values of faculty participation in academic policy-making. Indeed, the new popular literature on management in non-academic enterprises emphasizes broad participation of employees in decisions affecting them.

Under present conditions, faculty members perceive that their role in academic decision-making has declined, and they are probably right. This change, which has occurred gradually over many years, is in some cases a source of resigned disappointment, in others a cause of serious faculty discontent, and in all a source of poor morale. All of these changes have tended to make academic life more bureaucratic and more rigid. As Clark, Boyer, and Corcoran (1985, p. 23) suggest:

> . . . higher education seems to be undergoing a gradual paradigmatic shift, termed variously from faculty hegemony to student consumerism and from education community to economic industry.

In the same vein, Austin and Gamson (1984, pp. 3, 18), referring to the rising incongruity between the bureaucratic structure of academic administration and the collegial structure of faculties, observed:

> The collegial structure has become so fractured in many institutions that it can do nothing more than provide the backdrop for departmental competition over scarce resources. One result is that decisions normally reserved for the collegial structure are made in the bureaucratic structure. This shift in power away from faculty toward administration is probably the most

important change in higher education that has occurred in recent years. It may move the culture of colleges and universities away from normative to more utilitarian values.

Richard Anderson, who compared conditions in higher education in the late 1960s with those in 1980, detected a significant decline in collegiality. He added (1983, p. 6):

> High levels of democratic governance were especially noticeable in the most effectively managed institutions and were generally absent in the least effective institutions. . . . For an institution to be successful, the faculty must be creative, energetic, and dedicated to their institution. Sustaining these qualities for a prolonged period of time is a monumental task and cannot be achieved through bureaucratic management. . . . The level of institutional financial support and faculty salaries appear to have less effect on faculty morale than the meaningful participation of faculty in governance. Regardless of financial pressures, college and university leaders should maintain their commitment to collegial governance traditions.

Status. Along with the perceived shift of decision-making from faculty to administration to government, there has been a perceived decline in the social status of faculty members. One of the valued rewards of the academic profession has been the comparatively high regard in which it has been held by the general public. This prestige may have reached a high point in the 1960s, when higher education was near the top of the list of social priorities and when faculty members were drawn unprecedentedly into the councils of government and business. Many faculty members feel—whether rightly or wrongly—that they have lost some of the status they enjoyed in the 1960s, and the decline in their real earnings vis-à-vis other occupational groups has reinforced this opinion. We were able to obtain data relevant to trends in the status of faculty from the files of the Roper Center for Public Opinion Research at the University of Connecticut. These data were derived from several surveys of national samples of the population.

Table 7–4 presents responses over the period 1976–83 regarding people's trust in persons of various occupations. "College teachers" come out consistently near the top in this list of occupations. Also, whereas most of the occupations experienced small declines in public trust, college teachers showed a small increase. Table 7–5 reports similar information except that the data refer to institutions rather than occupations, and they cover the longer period from 1966 to 1983 and thus enable one to trace changes during the period of the Viet Nam war and Watergate. In this table higher education is near the top throughout and actually number one at the beginning and at the end of the period, though it drops to number two below "medicine" in some of the intervening years. But the most notable feature of Table 7–5 is that all of the institutions listed lost ground in public confidence—in most instances steep declines—even though their relative positions remained roughly the same. Thus, faculty persons are probably right when they perceive a decline in their absolute status even though their relative position has remained high.

Other interesting information surfaced in our study of public attitudes toward the professoriate. In a Roper organization survey covering the years 1973 to 1977,[17] people were asked whether various institutions had too much or too little power or influence. "Educators" were near the low end of the scale; they were infrequently deemed to have "too much power." Correspondingly, educators were perceived near the high end relative to "Don't have enough influence." We would interpret these results to connote trust in educators. In another Roper survey conducted in 1977 and 1980, people were asked about their opinions (ranging from poor to high) of eleven different occupations.[18] "Educators" were fourth from the top of the list but only slightly behind the leaders, namely, television and news reporters, and bankers, and ahead of businessmen, lawyers, and U. S. Senators.

In a 1980 survey conducted by the Connecticut Mutual Life Insurance Company, the respondents were asked, "Do you think there is an untapped reservoir of leadership in this country?" To this question, 87 percent responded "yes." Another question followed: "If so, which of the following groups do you think may provide the greatest untapped leadership potential?" "Educators" were third on the list behind "top management of large corporations" and "top managers of small companies."[19] Finally in an imaginative survey conducted by Yankelovich, Skelly, and White in 1979, people were asked whether or not it would be good for the country to have as President someone who was a member of various specified groups. The responses, as shown in Table 7–6, indicate that a substantial number of people believe it would not be preposterous for a college professor to follow in the steps of Woodrow Wilson and become President of the United States.

Insofar as public opinion surveys are valid indicators of status, faculty members occupy a *relative* place of considerable prestige and respect in American society. As is well known, however, the trauma connected with the Viet Nam war and Watergate led to a catastrophic decline in public confidence toward most social institutions and most categories of leadership, and the nation has not yet fully recovered. Public attitudes toward faculty and toward the colleges and universities they serve are still affected by these distressing events of nearly twenty years ago. Yet, one can probably say with some confidence that "higher education" and "college professors" are among the most highly

17. As reported by the Roper Center for Public Opinion Research.
18. As reported by the Roper Center for Public Opinion Research.
19. As reported by the Roper Center for Public Opinion Research.

Table 7–4. Responses to the Question: How would you rate the honesty and ethical standards of people in these different fields? Percentage Responding "Very High" or "High."

	1976	1979	1981	1983
Clergymen	—	61%	63%	64%
Druggists, Pharmacists	—	—	59	61
Medical Doctors	56	51	50	52
Dentists	—	—	52	51
COLLEGE TEACHERS	44	42	45	47
Engineers	49	46	48	45
Policemen	—	37	44	41
Bankers	—	39	39	38
TV Reporters, Commentators	—	—	36	33
Funeral Directors	—	26	30	29
Journalists	33	32	32	28
Newspaper Reporters	—	—	30	26
Lawyers	25	26	25	24
Stockbrokers	—	—	21	19
Business Executives	20	19	19	18
Senators	19	19	20	16
Building Contractors	23	18	19	18
Local Political Office Holders	—	14	14	16
Congressmen	14	16	15	14
Realtors	—	13	14	13
State Political Office Holders	—	11	12	13
Insurance Salesmen	—	15	11	13
Labor Union leaders	12	13	14	12
Advertising Practitioners	11	10	9	9
Car Salesmen	—	8	6	6

SOURCE: Gallup organization as reported by the Roper Center for Public Opinion Research.

regarded groups of our nation. In concluding this section, it is perhaps appropriate to observe that concern for status on the part of faculties is not merely a matter of vanity; status ordinarily translates into financial support and influence (H. R. Bowen, 1968).

Needless Aggravations. During the many years of financial stringency when administrators in many institutions have been hard-pressed to keep their institutions solvent, faculty members have been subjected to numerous indignities and petty aggravations. For example, they are often embarrassed as well as handicapped by obsolete equipment, shabby quarters, lack of privacy, and inadequacy of assistance. They are offended by the many irresponsible and tasteless criticisms of the tenure system and by the not very subtle hints that their departure or early retirement would be welcomed. On some campuses, faculty members are resentful of the relatively luxurious facilities and staff available to administrators as compared with those available to faculty

Table 7–5. Responses to the question: As far as people in charge of running some of the institutions in this country are concerned, would you say you have a great deal of confidence, only some confidence, or hardly any confidence at all in them? Percentage responding "Great deal of confidence."

	1966	1971	1973	1975	1977	1979	1981	1983
HIGHER EDUCATION	61%	37%	44%	36%	37%	41%	34%	36%
Military	62	27	40	24	27	29	28	35
U.S. Supreme Court	50	—	33	28	29	28	29	33
Medicine	73	61	57	43	43	30	37	30
Television News	25	—	41	35	28	37	24	24
White House	—	—	18	18	31	15	28	23
Executive Branch, Federal Government	41	23	19	13	23	17	24	—
Organized Religion	41	27	36	32	29	20	22	23
Congress	42	19	29	13	17	18	16	20
The Press	29	18	30	26	18	28	16	19
Major Companies	55	27	29	19	20	18	16	18
State Government	—	—	24	—	18	15	—	18
Local Government	—	—	28	—	18	—	—	18
Law Firms	—	—	24	16	14	10	16	12
Organized Labor	22	17	20	14	14	10	12	10
Wall Street	—	—	—	—	—	—	12	—
Advertising Agencies	21	—	11	7	7	11[1]	—	—

[1] 1978.

SOURCE: Louis Harris and Associates as reported by the Roper Center for Public Opinion Research.

("Too Many Administrators?" *Chronicle of Higher Education*, Oct. 5, 1983, p. 21). In some institutions they are embarrassed also by clumsy and redundant attempts to evaluate their work.[20]

Responsible evaluation of faculty is essential. But the academic profession may be one of the most frequently evaluated groups in our society. Every faculty member regularly subjects himself to inspection by one group or another as he meets his classes, attends committee meetings, gives outside lectures, writes books and articles, and consults. In addition, most institutions carry out regular student evaluation of faculty, and frequent appraisal of faculty is associated with accreditation, proposals for foundation and governmental grants, proposals for sabbatical leaves, and recommendations for salary increments and advancement in rank.

These are perhaps small things but they speak loudly about the

---

20. There is an enormous literature on evaluation of teaching with many conflicting points of view. We find the most useful and balanced discussion in McKeachie (1983a, pp. 37–39). See also Kellman (1982, p. 29); Kearl (1983, pp. 8a–9a); and Larsen (1983, pp. 10a–11a).

Table 7–6. Responses to the question: Considering the state of the country and the world today, do you think it would be good for the country or *not* good for the country to have as President someone who was a member of the designated groups?

	Good for the Country	Not Good for the Country	Not Sure
Business Executives	60%	25%	15%
Women	60	23	17
Black	56	22	23
Jew	46	24	31
College Professor	40	43	17
Labor Leader	31	54	16
Atheist	19	60	21
Priest	14	69	17

SOURCE: Yankelovich, Skelly, and White, 1979, as reported by Roper Center for Public Opinion Research.

respect accorded faculty members and about the very concept of a professor. They do not help in the recruitment of new and capable persons to the profession or in retaining those already there. Most petty aggravations have probably been unintentional, and perhaps inevitable under conditions of financial stringency. Nevertheless, some of them might have been corrected, and all of them might have been more bearable, if relationships could have been conducted more tactfully and with more concern for the sensitivities of a great profession in a time of adversity. Loyd (1985, p. 101) has summarized the matter:

> The importance of avoiding needless aggravations cannot be overemphasized. Especially for institutions which cannot dedicate resources to improving the situation of faculties, perhaps the most important course of action is to emphasize continuity where it can be emphasized and to avoid, more than ever, threats to academic freedom, threats to tenure, interference with faculty governance, impediments to collegiality, unnecessary red tape, and other forms of interference.

It should be emphasized, however, that most institutions have, in fact if not in appearance, been conducted with deep concern for commitments to tenure and with humaneness toward non-tenured persons. In an important study, Kenneth Mortimer found that over the period 1978 to 1983, only about 1,200 tenured persons had been discharged for financial reasons from four-year colleges and universities. This was from a tenured population of several hundred thousand (Scully, 1983b, p. 21).

Perquisites. Academic employers have traditionally relied on faculty members' sense of vocation, love of academic life, and need for security—rather than munificent compensation—for motivating and retaining them. In so doing, academic employers have traditionally tried to create a favorable work environment, to provide various non-monetary benefits and amenities, and to form meaningful communities that would make the institution a good place for faculty to live and work. Being a faculty member has often been more like belonging to a large and caring family than working for a corporation. The community concept has probably survived more significantly in small institutions located in small cities than in huge institutions in big cities. But the community concept survives, to some extent at least, in many colleges and universities. Nearly everywhere, faculties and their families enjoy certain privileges and benefits that add to their real income or welfare. These include admission to campus events, use of sport and recreational facilities, use of libraries and museums, attendance at social events, tuition remission for faculty children to attend college, use of institutional purchasing departments, subsidized housing, and even, in a few cases, cemeteries. These are important benefits that colleges and universities can bestow and for some of which the marginal costs tend to be low, assuming that the services and facilities involved are provided anyway for students and other groups. Furthermore, these benefits for faculties are considered to yield a return to the institutions through encouraging faculty to fraternize with one another and with students, and thus promote collegiality, and also to help in the recruitment and retention of faculty. We have found that the social and cultural ambience of colleges and universities as enriched by these perquisites is highly valued and is one of the features of campus life that attracts people to the academic profession and holds them in it. The pleasures of campus life are often mentioned when faculty members say that they are willing to stay even though real income and other working conditions may be deteriorating. So far as we can tell, these perquisites have not been curtailed very much, if at all, during the recent period of financial stringency—though some of them are under assault by the Internal Revenue Service (Farrell, 1983, p. 7).

Faculty Development. Another set of benefits for faculty members more closely related to the academic program than the perquisites above is "faculty development" and "renewal." The traditional method has been sabbatical or other leaves and released time for research or other purposes. As of 1981, about three-fourths of all colleges and universities were awarding sabbatical leaves, some automatically to all faculty and some only to those chosen on a competitive basis (Andersen & Atelsek, 1982, p. v). The time away was usually one-half year on full pay or a full year on half pay. Many other leave arrangements were common, often with the salary paid by an outside agency.

Another kind of "faculty development" which has enjoyed something of a boom in the past fifteen years enables or sometimes compels faculty members to undergo periods of education or on-the-job training intended to update skills, to "retool" for new work assignments or new careers, or to improve teaching, counseling, or research. Faculty reactions have been mixed (Oldham & Kulik, 1983). On the whole, the many efforts toward "faculty development" seem to us to have had a modestly favorable impact on overall trends in working conditions, though in some cases they have been spectacularly successful.

Conclusions

Given the steady erosion in real earnings for American faculty over the past decade and a half, the question of whether the faculties' work environment similarly has deteriorated, or has held steady, assumes great significance. If compensation, already widely perceived to be inadequate, is reinforced by perceptions of a work environment that is growing less attractive, the consequences could be severe. We suggest that any profession afflicted by both declining real income and a perceived degeneration of the work milieu, compounded by an outlook for "more of the same," would soon have difficulty in attracting, and even retaining, highly talented persons.

From our study of working conditions including our field visits, we conclude with considerable confidence that in many respects the conditions of faculty work have been slowly deteriorating over the period since 1970. The deterioration has of course been less in some institutions than others, but few have gone unscathed. However, an overwhelming majority of institutions have adjusted to financial stringency. They have managed not only to stay afloat but to maintain their stability. With exceptions, conditions in academe as of 1985 are not desperate. Yet, with the deterioration of working conditions combined with the slow erosion of faculty pay and with the uncertainty of the future, faculty morale has sagged. But it has not collapsed.[21] Even given the steady barrage of pessimistic forecasts about the state of higher education, conditions and attitudes have not worsened to the point that multitudes of faculty members are actually abandoning the profession.[22] And given the continuing large flow of young people from the graduate schools, there is as yet no acute shortage of acceptable candidates for academic positions (except in a few disciplines). When faculty members are asked whether they like their work and whether they would choose academic careers if they could start over, the answer is strongly in the affirmative, a finding confirmed in our campus visits.

21. In responding to questions about their morale, 72 percent of faculty members indicated that their morale was very or fairly high. Only 5 percent reported that it was very low (National Education Association, 1979, p. 12).

22. From a recent survey, one learns, however, that 11.5 percent of a large sample of faculty members have seriously considered leaving to take a job outside academe and that another 5.3 percent have considered early retirement (John Minter Associates, 1983, p. 20).

Yet, looking ahead after fifteen years of austerity the situation is still perilous. The long-predicted decline in the youthful population is still ahead. The political and financial conditions of the country are still not favorable to large increases in appropriations for higher education. The gap in compensation and working conditions between the academic profession and comparable employments is still ominously wide. Perhaps most important, the quality of education being provided by our colleges and universities is below par. Many of our institutions have become huge, impersonal, and layered with administrative tiers; the high school preparation of our students is inadequate; the adjustment of academic programs and methods to the needs of new clienteles is still incomplete; and the excessive concentration on job-related studies at the expense of liberal education compounds the problems. The conditions of work in academe are far from ideal and could get worse. We believe, however, that a nation as dedicated to education as the United States is would not allow the kind of slow retrogression now clearly under way to become more acute. We believe that faculties feel a cautious optimism and that is why they are not defecting in large numbers. The situation is different, however, for those gifted young people who might be candidates for future academic careers. They have not yet been socialized to the academic profession and have made few or no commitments to it, and, as we shall show in Chapter 11, they are staying away in droves.

In interpreting these conclusions, it would be well to remember that the base year for judging trends in working conditions was the prosperous late 1960s and early 1970s. An earlier base year might have produced a different set of conclusions. A recent informal survey of faculty persons who have been members of the AAUP for fifty years or more, produced interesting responses. To the broad question, "Has the American professor's quality of life improved in the past fifty years?" 90 percent replied "Yes" (*Academe*, Nov.–Dec. 1983, p. 9). But they were responding in terms of the exceptional conditions of the Great Depression, World War II, and the post-GI period. These eras are hardly suitable as a base for judging American higher education in the late twentieth century.

CHAPTER EIGHT

The Faculty Condition:
Observations from the Field

Between November 1983 and May 1984, a team of six researchers visited a highly diverse group of 38 campuses and conducted a total of 532 interviews with faculty and administrators.[1] The objective of these campus visits was to further our understanding of how faculty members viewed their circumstances. We particularly wanted to know whether they thought the quality of their professional lives had improved or declined over the past decade and a half. Accordingly, we sought to assess possible changes in faculty morale and collegiality and to determine whether faculty autonomy had been compromised. We also asked faculty members about their compensation and their job security. Finally, we canvassed faculty and administrators for their views of the quality of persons currently being attracted to academic careers.[2]

We found that campus moods ranged from buoyance to deep depression, that changes in faculty compensation ranged from substantial increases to sharp declines in real earnings, and that perceptions of the work environment ranged from highly positive to wretchedly negative. At one extreme, faculty characterized campus administrators as autocratic adversaries; at the other, as effective, vigorous, supportive leaders. Given our efforts to ensure that our sample of campuses adequately reflected the diversity of American higher education, such bold contrasts should not be surprising.

---

1. The six campus visitors were E. Howard Brooks, Claremont McKenna College; Martin J. Finkelstein, Seton Hall University; Wilbert J. McKeachie, University of Michigan-Ann Arbor; Howard R. Bowen and Jack H. Schuster, Claremont Graduate School; and Patricia J. Foster, Loma Linda University, who also served as site-visit coordinator.

Campus visits ordinarily lasted two or three days. At four campuses, two visitors each spent two days.

A note on the methods used for site visits and a list of campuses in the sample are found in Appendix A.

2. The issue of faculty quality is discussed in Chapters 2 and 11.

Before we discuss our observations from the field, a prefatory word on "organizational culture" may help to place our research strategy in clearer perspective. Every organization—colleges and universities are certainly no exception—embodies two co-existing but sometimes quite divergent realities. One reality is "objective," and many of its properties are readily susceptible to measurement. "Hard" facts and trend lines describe this entity and thereby attest to the objective condition of the campus and its faculty. To illustrate, one can determine student credit hours generated per full-time-equivalent faculty at a given college; these data can, in turn, be broken out by department or by full- and part-time faculty. The quantifiable dimensions of the organization are endless: faculty travel allocations by academic unit; endowment yields for the past five fiscal years; distribution by age of the physics faculty (and, accordingly, anticipated retirement dates); tuition revenues generated per full-time-equivalent faculty in history; freshman Scholastic Aptitude Test scores over the past ten years; and comparative salaries of male and female associate professors (or philosophers versus economists) by time in rank or highest degree earned. Taken together, such characteristics define the objective reality of the organization.

The second organizational reality is grounded in perceptions. Organizational theorists remind us of the importance of intra-organizational myths, sagas, and perceptions. The people who make up the organization may experience it in ways that are not easy to square with the "objective" reality of the printouts. It is the faculty's *perceptions* of their circumstances, however inconsonant these perceptions may be with the "facts," which drive their responses to their environment, which determine how they behave and how they transmit the organization's culture to their colleagues, to their students, and to others. Accordingly,

> We need to be able to perceive and understand the complex nature of organizational phenomena, both micro and macro, organizational and individual, conservative and dynamic. We need to understand organizations in multiple ways, as having "machine-like" aspects, "organism-like" aspects, and others yet to be identified. [Jelinck, Smircich, & Hirsch, 1983, p. 331]

This observation underscores a crucial point. The condition of the American professoriate is only in part discernible from the measurable, tangible aspects of campus life. The faculty condition cannot be understood apart from the faculty's own perceptions of its condition.

Four overarching themes emerged from our campus visits: the faculty dispirited, the faculty fragmented, the faculty devalued, and the faculty dedicated.

The Faculty Dispirited

To take the pulse of the faculty at the 38 campuses we visited, we asked about morale, collegiality, autonomy, and attitudes toward the campus administration. Although their meanings may be imprecise, these terms served as cue words that helped to evoke faculty attitudes.

Faculty Morale. Perhaps the most elusive dimension of the faculty condition is morale. At the outset, two considerations must be acknowledged. First, any given informant is subject to the human frailty of assuming that his or her own morale extends to a wider circle of colleagues. As one vice president at a research university warned us, "Faculty morale is overrated and volatile; people project their morale on everyone else."

Second, because faculty morale is a function not only of campus-specific conditions but also of general developments in higher education, it is difficult at best to apportion responsibility for morale between local and generic factors.

Given these caveats, we found that faculty morale varied considerably, ranging from excellent at one liberal arts college to very poor at nine campuses. Between these extreme cases, faculty morale at twelve campuses seemed to be good; at nine campuses, fair; and at the remaining seven, poor (Table 8–1).

We asked faculty and administrators about changes in faculty morale over the past five or ten years. On the basis of their responses, we concluded that faculty morale had strongly improved at only two campuses and had substantially deteriorated at five. It had improved slightly at ten campuses, had lost some ground at nine, and had remained relatively stable at eleven (Table 8–2). Our overall impression was of some volatility, but little dramatic change, in faculty morale.[3]

Table 8–1. Assessment of the Condition of Faculty Morale, by Type of Campus

Campus Classification	Level of Faculty Morale				
	Excellent	Good	Fair	Poor	Very Poor
Research universities	0	3	5	0	1
Doctorate-granting institutions	0	0	2	2	2
Comprehensives	0	3	0	4	3
Liberal arts colleges	1	4	1	0	1
Community colleges	0	2	1	1	2
	1	12	9	7	9

SOURCE: Site Visits, 1983–84

3. "Change" can be misleading. To illustrate, faculty morale at Swarthmore was judged to be "excellent" and to have remained at that same high level in recent years; thus, Swarthmore ranked at the midpoint ("about the same") on the scale of change. There simply was little room for upward movement there. By contrast, at one community college, where morale was still very low, we found some evidence of movement in the right direction and tallied that campus as showing "some improvement."

Table 8–2. Assessment of Change in the Condition of Faculty Morale over Past 5–10 Years, by Type of Campus

| | CHANGE IN FACULTY MORALE | | | | |
| | Improvement | | About the | Deterioration | |
Campus Classification	Substantial	Some	Same	Some	Substantial
Research universities	1	5	1	1	1
Doctorate-granting institutions	0	0	4	1	1
Comprehensives	0	2	3	4	1
Liberal arts colleges	0	2	2	2	1
Community colleges	1	1	2	1	1
	2	10	12	9	5

SOURCE: Site Visits, 1983–84.

As Table 8–3 shows, about a third of the faculty members and administrators asked about changes in faculty morale felt that it had improved over the past five or ten years, whereas 44 percent perceived a decline. Faculty held a somewhat dimmer view of trends in morale than did department chairs or campus administrators: 28 percent of faculty members, but only 17 percent of administrators, thought that morale had decreased substantially.

Patterns by type of institution are difficult to discern. As Tables 8–1 and 8–2 show, each category includes campuses judged to exhibit high and low, improving and declining, faculty morale.[4] The research universities in our sample appeared to be doing fairly well; only one was hurting badly. Similarly, faculty morale at the four more selective liberal arts colleges (classified Liberal Arts I in the Carnegie scheme), already strong, seemed to be getting even stronger. By contrast, faculty morale at the majority of comprehensive colleges and universities appeared to be waning. Similarly, declining faculty morale characterized the less selective liberal arts colleges (the Liberal Arts II category) we visited, though the decline was less dramatic.

Faculty morale is, of course, the product of numerous factors. Seven factors appeared to be particularly influential.

The first such factor is campus leadership—or, more accurately, the faculty's perceptions of campus leadership. As is noted in a later section, faculty members were quick to criticize campus administrators for any perceived malfeasance. Contrary to the common assumption that faculty prefer the administration to keep far away from academic matters, however, we frequently found faculty members either voicing gratitude for strong presidential leadership or hoping it would materialize. This phenomenon was most visible at middling institutions uncer-

4. For campus classifications, see Appendix A and Carnegie Council on Policy Studies in Higher Education, 1976.

Table 8–3. Responses to a Question About Change in Faculty Morale[1]

PERCEIVED CHANGE IN FACULTY MORALE

Respondent Group	Number	Improvement		No Change	Deterioration		Don't Know
		Substantial	Slight		Slight	Substantial	
Administration	133	14%	24%	23%	21%	17%	2%	100%
Department chairs	96	14	22	21	21	22	1	100
Faculty	130	14	13	22	22	28	1	100
	359	14	19	22	21	23	1	100

[1] The question asked: "Has faculty morale changed significantly over the past 5 to 10 years?
SOURCE: Site Visits, 1983–84.

tain about how best to mobilize for the future. Although the faculty's natural ambivalence about forceful administrative leadership was evident, leadership which provided institutional direction in tough times was clearly appreciated by the faculty. Conversely, those leaders who responded to events idiosyncratically, who failed to establish a clear course, contributed significantly to faculty malaise.

Second, faculty morale was strongly influenced by the perceived role of external agencies. At public campuses, signals from a supportive governor or, conversely, from a stringent, accountability-minded legislature were readily understood. At almost every public campus, the faculty (and administrators, too) were sharply critical of a public policy process that was seen as providing far fewer resources than were deemed necessary to carry out the campus mission.

A third, and very powerful, determinant of morale was the level of faculty compensation—especially the magnitude of the most recent salary increase. This "bread and butter" factor is discussed more thoroughly later in this chapter, but it is worth noting here that some presidents were raising unprecedented sums of money (John Brademas at New York University and Eamon Kelly at Tulane University come particularly to mind). Few conditions inspire faculty more than newfound campus affluence, especially when part of the largesse is passed on to faculty in the form of significant salary increases.

The fourth factor shaping faculty morale was the adequacy or inadequacy of working conditions. This issue is examind at greater length in a later section.

A fifth factor—one that struck us as having a very positive impact on the faculty's outlook—might be called the "phoenix factor." Several of the campuses in our sample had experienced some very rough financial times over the past decade; a few had nearly succumbed. At these places (New York University, Southern Connecticut State University, Borough of Manhattan Community College, and Fordham University are examples) the crisis was still vivid in the memory of almost all the people we interviewed. Those campuses had since backed away from the brink and were now enjoying decent, even solid, financial conditions. The joy of survival—of having been removed from the critical list—had obviously boosted morale. In the apt words of a scholar at one such campus, "Nothing makes your feet feel better than taking off a pair of tight shoes."

The sixth factor was a general sense that shifting values and increasing demands have changed the academic enterprise for the worse. Time and again, older faculty members reported a dramatic transformation in their world of work. They said in various ways that the "life of the mind" has been badly compromised by the rapid pace and relentless pressures of contemporary academic life. One university professor, who was about to take early retirement, expressed his disappointment as follows:

> Somehow the sheer joy of being a scholar has been eroded, and it's not just because of the money. They have taken the fun out of teaching. You feel so pressured; it's hard to find time to sit and think. We have gotten into a publish-or-perish-type syndrome where we publish trivia and we're not reflecting on what we're doing. . . . We have lost our sense of scholarliness.

The seventh and final factor—and one that seemed to weigh heavily with almost all ambitious faculty members—was the self-perception of "stuckness." This frustration over career immobility was voiced succinctly by the faculty council chairperson at an urban university: "Morale is worse because you can't move to another job." A senior faculty member who was opting for early retirement complained, "There are no more discussions about how *big* we want to get." At that campus, as at several others, faculty members were preoccupied simply with survival. "If we can just manage to tread water for the next ten years," the history chairperson at the same university said wistfully, "we can get by."

Our findings suggest neither a uniformly bleak outlook nor any sense of dramatic change. "Middling" morale was the norm, although many faculty members were uneasy about the future. Even those who said that their present situation was acceptable or good often expressed anxiety about enrollment projections and fear that the condition of their campus could deteriorate rapidly. In sum, many faculty were nervous; apprehension was their dominant mood.

Collegiality. The term "collegiality" has many meanings. It can refer to the quality of relations among colleagues within an academic department or among faculty members in different academic departments or, at a complex campus, across schools or colleges within a university. The term may also be applied to the relationship between the faculty and the administration. We told our respondents that, for the purposes of

Table 8–4. Assessment of the Condition of Faculty Collegiality, by Type of Campus

Campus Classification	Level of Faculty Collegiality				
	Excellent	*Good*	*Fair*	*Poor*	*Very Poor*
Research Universities	0	5	4	0	0
Doctorate-granting institutions	0	3	1	2	0
Comprehensives	1	1	5	1	2
Liberal arts colleges	1	3	2	1	0
Community colleges	0	1	1	4	0
	2	13	13	8	2

SOURCE: Site Visits, 1983–84

this study, "collegiality" meant the relationship *among faculty members* both within and among departments.[5]

Collegiality has held up reasonably well in recent years. As Table 8–4 shows, we judged the state of collegiality to be excellent at two campuses and good at thirteen more. On the other hand, collegiality was consistently reported to be poor at eight campuses and very poor at two. At the remaining thirteen campuses, the condition of collegiality was assessed as fair. We found no extreme changes in recent years; at none of the campuses we visited did most of the respondents report that collegiality had either improved or deteriorated substantially. But, as Table 8–5 shows, some slight deterioration in collegiality was reported at fourteen of the thirty-eight campuses, and some improvement in collegiality was reported at six campuses.

In the aggregate, faculty were divided on the question of whether collegial relations had improved or deteriorated (Table 8–6). Of 370 individual respondents, 25 percent reported some improvement in collegiality on their campuses, while 31 percent perceived recent declines.

Faculty comments made clear that collegiality has held up reasonably well within departments. However, as a seasoned academic vice president cautioned, wide variations may exist on a given campus: "Collegiality is fairly high at the departmental level," he noted, "except in a few departments which are civil wars about to happen." In any case, a high level of collegiality among faculty members did not necessarily mean that the faculty situation is good. Rather, it may represent a response to growing campus problems. At campuses experiencing tensions and pressures, faculty members tend to bond more closely together. "When you have a common 'enemy'," observed the faculty

5. Our interview question was: "How would you describe the strength of collegiality on your campus (in your department)? In other words, is there a sense of community and shared purposes and interests here?" We then asked, "Has collegiality changed significantly over the past 5 to 10 years?" and then coded these responses on a five-point scale from "substantial improvement" to "substantial deterioration," making allowances for "don't know" responses.

Table 8–5. Assessment of Change in the Condition of Faculty Collegiality over Past 5–10 Years, by Type of Campus

| | CHANGE IN FACULTY COLLEGIALITY | | | | |
| | *Improvement* | | *About the* | *Deterioration* | |
Campus Classification	*Strong*	*Some*	*Same*	*Some*	*Sharp*
Research universities	0	2	6	1	0
Doctorate-granting institutions	0	1	3	2	0
Comprehensives	0	1	4	5	0
Liberal arts colleges	0	1	3	3	0
Community colleges	0	1	2	3	0
	0	6	16	14	0

SOURCE: Site Visits, 1983–84.

senate chair of a university that was undergoing difficult times, "collegiality blossoms." A campus senate president of a growing campus put it another way: "It's a different kind of collegiality now; it had been based on friendship, but now it is based on common goals."

Even where collegial relations were reported to be good, collegiality was almost invariably much weaker among departments and, within more complex institutions, among schools. A visitor to a research university described the typical situation: "Collegiality is fairly strong within departments, modest within the colleges as a whole, but weak among the colleges." The increasing size and complexity of many of our compuses erodes the capacity for collegial relations. One senior faculty member, yearning for bygone days, said sadly, "We used to be a 'band of brothers,' but now. . . . " His voice trailed off. Another bemoaned the disappearance of faculty colloquia and picnics, and still another reminisced nostalgically about the old pool table around which faculty used to congregate in the late afternoon. This spirit, it seems, has been lost. A frequent complaint was that faculty are meeting their responsibilities to a minimal degree, choosing instead to flee campus in order to pursue other interests. One associate professor probably spoke for many frustrated faculty at various campuses when he said: "This becomes a kind of ghost town after 1:30 or 2:00 o'clock. Most students leave for their part-time jobs, and faculty, too, are on their way!"

Collegiality has been further undermined by tensions between junior and senior faculty, and by perceived salary inequities. Both issues are discussed later.

Faculty Autonomy and the Administration. Faculty members at most campuses were satisfied that their autonomy had been reasonably well preserved.[6]

6. Our interview question was: "Do you feel that faculty here have as much autonomy as in the past?"

Table 8–6. Responses to a Question About Change in Collegiality[1]

PERCEIVED CHANGE IN FACULTY MORALE

		Improvement		No	Deterioration		Don't
Respondent Group	Number	Substantial	Slight	Change	Slight	Substantial	Know
Administration	130	10%	21%	35%	24%	8%	2% 100%
Department chairs	95	2	20	44	23	9	1 100
Faculty	145	7	15	45	21	8	4 100
	370	7	18	41	23	8	3 100

[1] The question asked: "Has faculty morale changed significantly over the past 5 to 10 years?"
SOURCE: Site Visits, 1983–84.

No one suggested that the faculty member's traditional freedom in the classroom had been infringed upon in any direct way. There the professor remains the master of his domain. Thus, academic freedom does not appear to have been weakened. But some faculty members were concerned about the subtle threats to academic freedom that result from the vulnerability of non-tenured faculty in a tight academic labor market; several seasoned observers commented that junior faculty were less willing to be bold in their teaching and writing. While such assertions are plausible, we came across no hard evidence to substantiate such claims.

Faculty at most campuses reported that they enjoyed a respectable degree of autonomy in the governance process. At several campuses, however, the administration was generally perceived as autocratic or meddlesome. By way of example, one campus visitor to a community college found:

> there is some clear sense that professional autonomy is declining as reflected in (1) less of a decisive role in faculty appointments and evaluations (the recently instituted faculty evaluation package was entirely designed by administrators); (2) the tendency of administrators to check up increasingly on faculty; and (3) the dean's initiating curricular revisions or changes without faculty consultation.

Fortunately, such conditions were not common. At several public institutions, faculty often expressed bitter resentment that the state political apparatus was interfering unduly with campus affairs. At still other campuses, they were very upset with their own governing boards.

As might be expected, faculty members were quick to criticize the institution's president for neglect, that is, for disregarding the primacy of faculty claims on available campus resources. Thus, the presidents of various campuses were roundly castigated for a long list of sins: spending money on new buildings rather than raising faculty salaries, giving pay raises to their administrative colleagues when faculty salaries were inadequate, siding too closely with students in campus disputes, and not taking the faculty senate seriously, to name several typical transgressions. A university dean may have identified the pivotal issue when he outlined the faculty's high expectations of the administration on his campus: The faculty wants no less than "a signal from the administration that it believes the faculty role to be the most important in the University, that the most important thing in the University is what the faculty do." Such a signal may be a long time coming, there and elsewhere.

A senior faculty member at a private university summarized the faculty's concern about its dwindling influence:

> It's not that being a faculty member is not prestigious. It's still that. But it's *only* that. We've lost control over the kind of students we encounter in our classrooms and over the direction of our campus.

All in all, the faculty mood was glum. Making allowances for the varied circumstances of the thirty-eight campuses in our sample, we found that faculty members tended to be apprehensive and discontent. The common view was that faculty life had once upon a time been better and that conditions could very well get worse, maybe a lot worse, in the foreseeable future. Our overall sense was that faculty were frustrated and dispirited.

The Faculty Fragmented

While aware that the historical homogeneity of the faculty has been disintegrating over the last century, we were nonetheless surprised and disturbed at the extent of faculty fragmentation revealed by our interviews. Three megatrends have contributed to this fragmentation. The first is academic specialization: Speciality has begat sub-specialty in a process that continues unabated. At many higher educational institutions, parochial, campus-bound loyalties have long since disappeared as the professoriate becomes more cosmopolitan and more discipline-defined. The second megatrend is the growing heterogeneity of the faculty in terms of religious background, race, and gender, as noted in Chapter 4. The third megatrend playing a role in faculty fragmentation is the increasing diversity of the higher education system itself; faculty are now diffused throughout an incredibly wide array of colleges and universities that differ drastically in missions and clienteles.

Survey data have repeatedly demonstrated that faculty attitudes and values differ along many dimensions, including type of institution, discipline, and age. To some extent, such differences are natural and healthy, attesting to the vitality and dynamism of the academic enterprise. But the movement toward differentiation of the faculty, as described below, has gone beyond healthy diversity.

We found considerable tension among various segments of the faculty, a tension attributable in some measure to the heavy emphasis on research. On campus after campus (the community colleges in our sample were an exception) the clarion call to scholarly research rang loud and clear. This was especially true at the research universities, where the ethos of scholarly productivity has long obtained; in those settings, the commitment to research was a reinforcement of the long-standing faculty reward structure. But the publishing obsession was also evident at a number of institutions where in the past scholarly productivity was rare and effective teaching was the paramount criterion by which faculty were hired and promoted. Now, however, solid teaching no longer sufficed for promotion at these campuses. As a result of the research imperative, both junior and senior faculty saw themselves at risk, though for very different reasons. In this section, we discuss not only the junior faculty, mid-career faculty, and senior faculty, but also some "special cases": part-timers, "nomads," and minority faculty.

Junior Faculty. That non-tenured faculty on the tenure track were under great stress was evident not only from their own testimony but also from the comments of many senior faculty and administrators who observed their plight. The junior faculty at many of the campuses we visited had become, in a sense, "privatized"; that is, the overwhelming pressure to produce and publish had isolated them. According to a number of the deans and department chairs whom we interviewed, the junior faculty had ceased to function as fully participating members of their campus communities, even within their own departments, as they "burrowed toward tenure."

"I like the University," said an assistant professor of English who has an impressive publishing record by the standards of his struggling university. "I feel that I'm doing all the right things—but it may not help." And he is right on both counts. An assistant professor of history with a lengthy list of publications to his credit, including a book with the Harvard University Press, mused, "That would get me tenure at Wesleyan or Brown, but maybe not here." "Here" is a master's-granting institution that is waging an uphill battle to keep its liberal arts program intact. He added:

> I'm a stone realist. I see my predicament as a function of larger social forces. I understand that I was born too late. If I had been born ten years earlier . . . I'd probably be an associate or full professor at a more prestigious institution.

And he is probably right.

The pressure to obtain tenure is immense. At a few of the places we visited, such as the University of Pennsylvania, the odds were long against obtaining tenure. ("Negative tenure prospects" was the term used.) At many campuses that until recently had only infrequently denied tenure, the lives of countless assistant professors were filled with dread or resignation, as the result of a confluence of factors. First, vacancies were scarce; many departments—especially in the humanities—were already heavily tenured-in, and those departments that still had one or two assistant professors were understandably ambivalent about becoming 100 percent tenured. Second, demographic trends and declining student interest in the liberal arts made many campuses apprehensive about the future. Third, because of the academic labor market prevailing in most fields, assistant professors could be "churned" and readily replaced (or so it was assumed) with equally promising or better new hires.

A senior professor of English in a well-regarded department worried about the injustice of dramatically shifting standards:

> There are a considerable number of associate professors here who over their whole careers have not produced as much as it now takes to get promoted to associate professor.

At some of the campuses we visited, the tenure barrier was widely perceived to be virtually insurmountable. We asked in our interviews, "Has it become more difficult to obtain tenure on your campus in recent years?" As Table 8–7 shows, a sizable majority of respondents (73 percent) thought that tenure had become more elusive, fewer than one per cent thought tenure was now easier to obtain, and the remaining 22 percent saw no change. Interestingly, this perception of increased difficulty applied across all types of institutions, ranging from 84 percent of the respondents at the research universities to 50 percent at the community colleges.

At a few institutions, obtaining tenure was still regarded as fairly automatic, or "axiomatic," as one observer put it. But we more commonly encountered campuses where exceedingly able people were being denied tenure because the governing board or administration

Table 8–7. Responses to a Question About Difficulty in Obtaining Tenure[1]

		Perceived Change in Difficulty of Obtaining Tenure					
Respondent Group	*Number*	*Much More*	*More*	*About the Same*	*Less*	*Much Less*	*Don't Know*
Administration	116	41%	34%	22%	0%	1%	3%	100%
Department chairs	84	33	39	24	1	0	2	100
Faculty	113	35	38	21	0	1	5	100
	313	36	37	22	*	1	4	100

* Less than 0.5 percent
[1] The question asked: "Has it become more difficult to obtain tenure on your campus in recent years?"
SOURCE: Site Visits, 1983–84.

simply would not make the "thirty-five-year, million dollar plus" commitment in a field where enrollments were sagging. One president of a master's-granting college anguished over the dilemma he faced in his managerial role: while fully aware that some of his junior faculty were by far the best his institution had ever employed, he nonetheless felt an overriding responsibility to avoid any personnel commitments that would saddle the institution with expensive long-term financial obligations.

Perceptions of the difficulty of obtaining tenure did not always square with reality: that is, with actual tenure decisions. For instance, at one respected eastern university, where almost everyone said that tenure had become practically beyond reach, the statistics demonstrated just the opposite: During the eight-year period from 1972 to 1979, 55 percent of those considered for tenure had been successful. But during the 1980–83 period, the success rate had actually improved to 80 percent. Of course, figures on positive and negative decisions fail to take into account those instances where term contracts are not renewed or where faculty members are "counseled out."

Whatever the underlying facts, we found repeatedly that the pressures bearing on junior faculty members were formidable. Coupled with low salaries and high living costs in urban areas, the probationary period for young faculty members was often a grueling and lonely ordeal.

Mid-Career Faculty. The tenured faculty, especially associate professors, were also threatened by the new emphasis on research. At the research universities, where promotion to full professor has long depended on a respectable publications record, nothing much has changed. But at those institutions where the criteria for promotion are shifting, where effective teaching or even mere longevity were no longer sufficient for promotion to full professor, mid-career faculty were feeling the pinch. And judging from our sample, such campuses are legion. The mid-careerists were caught in particularly awkward circumstances. Unlike junior faculty, they had achieved tenure and were more or less secure ("financial exigencies" and the like aside). But, looking over their shoulders, they saw the new breed of well-trained young faculty fixated on scholarship and performing—albeit out of dire necessity—at levels heretofore rarely seen on campus. And looking ahead, they saw the senior faculty, anointed under a different reward structure and now safely ensconced in the highest-paying grades. As presidents and deans sang praises to the institution's redirected mission, many mid-careerists contemplated their marginal prospects for promotion with a deep sense of inequity.

Senior Faculty. Interviews with senior faculty members produced some highly disconcerting findings. We encountered many senior faculty who had played loyally by the "old rules," not just for years but for decades, but were now faced with a shifting reward structure. Some were angry, embittered, alienated from the new order. Not infrequently they expressed outrage about market-driven compensation, differential pay policies which they considered unjust and even humiliating. From the viewpoint of some of them, such compensation policies tended to favor not only the new breed of self-centered young faculty but also the high-demand academic fields. Many members of the senior arts and sciences professoriate, the "old guard," felt they had been relegated to subordinate status. Often, they were resentful toward the ungrateful administration and some were suspicious of the upstart assistant professors.

What forces fuel the renewed emphasis on scholarship? Two can be easily identified. The first is the timeless impulse to emulate the most prestigious universities. As far as we know, no one has tallied the number of institutions claiming to be the Harvard of the South or the Princeton of the West or the Yale of the Midwest, but the total number of such claimants is probably large. The institutional pecking order has long been in place, and while some elegant teaching institutions deserve their fine reputations, it is the first-rank research universities that continue to dominate the pantheon of American higher education.

The second factor is the cyclical nature of the academic labor market, which at present heavily favors the buyer. As is noted in Chapter 11, campuses gleefully report that, in most fields, they can now hire young faculty members whose credentials are far better than those of previous newcomers to academe. Holding doctorates from the most respected universities, these young scholars are already deeply imbued with the research ethos.

It is not surprising, then, that campus aspirations are spiraling upward and that many a doctorate-granting university which still falls far short of top rank now explicitly espouses the goal of becoming "a leading research university." Such aspirations are not new, but they are now being pursued with single-minded determination. In the groves of academe, to question the importance of research approaches heresy. Still, we cannot help but wonder whether the stampede toward scholarship—or what passes for scholarship—serves the nation's needs or the longer-run interests of those campuses which historically have been strongly committed to excellent teaching. We fear that the essential balance between teaching and scholarship has been lost, that the scales are tipping too far toward the former at many institutions.

Part-Time Instructional Staff. Much has been written about the proliferation of part-time faculty: Apparently, administrators are hiring more and more part-timers, both to preserve some measure of flexibility in staffing and to save money. We did not interview part-time faculty members; they were not included in our research design. But we did ask full-time faculty and academic administrators for their perceptions of the use of part-timers. The responses were revealing. For one thing, we found the use of part-timers to be very extensive, though more so on some types of campuses than others. For instance, we visited one community college where part-timers constituted 79 percent of the faculty by headcount. Other campuses had comparable ratios of part-time to full-time faculty. In no instance was it claimed that part-time faculty were integrated into the academic departments to which they were assigned or into campus life. Occasionally, a chair or dean would mention that a few of the part-timers, apparently eager to establish themselves as legitimate candidates for potential full-time openings, attended faculty meetings, made themselves readily available to students, and generally "behaved like regular faculty members." Nonetheless, the overall conclusion was inescapable: Typically, the part-timer is viewed as an expedient, and the part-timer's understandable response is a minimal commitment to the institution.

The "Nomads." On each campus we visited, we sought to meet with junior faculty members who, in pursuing their academic careers, had been unable to secure a tenure-track appointment. Most of these itinerant scholars were employed by the better institutions, those that wanted well-trained doctorate-holders with solid teaching experience and hired part-timers only as a last resort. Unlike part-timers, these "nomads" or "gypsies" occupied full-time positions and were more or less compensated as such. But they were on terminal contracts for a semester or a year, maybe two. Usually they were filling in for faculty members absent on leave. Their circumstances did not provide them with much solace. One nomad put it this way: "The title 'Visiting Assistant Professor' is one of the greatest euphemisms of modern academic life." We could not disagree. Those we interviewed were invariably spending much of their energy trying to find new jobs to go to when their current appointments expired.

These nomads were, it seemed to us, in a class by themselves. Some had pursued their doctorates for a decade and, as far as we could discern from their curriculum vitae, had done well as students at first-rate institutions; many had had extensive experience as teaching and research assistants, and some had taught as part-timers along the way. But their numbers now outstripped the number of available "regular" positions, and their attitudes reflected despair and dogged determination rather than optimism. Some said they had tried every plausible strategy to land a "real" job, had done all the "right things," but without result. One, a Berkeley-educated classicist filling in for a year at a southern university, said in a tone of resignation: "As far as I know, I've been iced. I thought I could succeed if I did it professionally." Another expressed his frustration concisely: "Being an outsider is a malignancy." But not all their tales were so wrenching. For instance, a Soviet Jewish mathematician reported that, after spending several years at Berkeley in a marginal academic position, he had just learned of his appointment as an assistant professor at an eastern university. Nonetheless, to us the interviewers, the overall experience was depressing.

The laws of supply and demand seem too immutable to permit any happy ending for these nomadic scholars.

To summarize: various segments of the faculty today find themselves in widely differing circumstances in terms of professional pressures, stature, and security. Perhaps this has always been the case, perhaps there has never been a single, cohesive faculty—not at Hutchins's Chicago or Butler's Columbia or Eliot's Harvard or Tappan's Michigan or, for that matter, at seventeenth-century Oxford or fourteenth-century Bologna. Today, however, the economic and demographic forces pressing so hard upon our institutions of higher education appear to have divided the faculty even more sharply than in previous years. The faculty seems to be growing more segmented than usual, and the end of this movement toward fragmentation is not in sight.

A Note on Minority Faculty

No aspect of our campus visits was more alarming than the situation we found with respect to minority faculty members. Over the past two decades, higher education has made considerable progress in opening the faculty to ethnic minorities, but that movement seems to have ground to a halt and may even be in reverse gear. Almost everywhere we went, we were struck by the scarcity of minority faculty. More unsettling still, a decline in their numbers is anticipated for the near future. We will not attempt here to summarize previous findings on the topic. Our comments derive primarily from our campus visits (which included two historically black colleges) and our conversations with knowledgeable professionals who have dealt with these matters for years.[7] The fundamental problem lies in the difficulties involved in attracting able minority-group members into doctoral programs and then into academic careers.

Two points bear mentioning at the outset. First, we are talking here about *ethnic* minorities, not women. Although women have long been an under-represented minority within the professorial ranks, the challenge of attracting talented women to academic careers involves somewhat different issues. While women in the professoriate still face significant problems, they have made noticeable progress. For instance, at the thirty-eight campuses in our sample, the total number of ladder-rank male faculty members had remained almost exactly the same over the last five years, whereas the number of women in comparable

---

7. These include J. Herman Blake and John U. Monro, Tougaloo College, Mississippi; Cleveland L. Dennard, formerly Atlanta University Center; Donald Deskins, University of Michigan; and Norman C. Francis, Xavier University of Louisiana.

appointments had increased by about 11 percent.[8] To put it another way, women accounted for virtually all of the growth in full-time faculty members.

The second caveat is that our discussion focuses on black faculty members. We did not address issues related specifically to other under-represented ethnic minorities, whose situations within the academic profession are at least as problematic as that of blacks.

Though specific experiences varied, at every one of the thirty-six predominantly white campuses we visited, administrators expressed concern about the difficulty of attracting black faculty members. One respected private university in the South has been losing ground in its efforts to retain its few black professors. Another had lost its last black arts and sciences professor. Berkeley, on the other hand, reported a modest increase in ladder-rank black faculty, from 20 in 1973–74 (1.3 percent of all ladder-rank positions) to 33 in 1982–83 (2.1 percent), despite having made few new hires of any faculty.[9] The University of Michigan had fared slightly better: about 3 percent of its full-time faculty were black—reportedly the highest percentage among the Big Ten universities (Mackay-Smith, 1984, p. 37; see also Farrell, 1984, pp. 1, 20).

We believe that the crux of the problem is a shortage of adequately trained young black scholars. Here we differ with advocates of affirm-ative action who insist that the problem lies in the historical and con-tinuing reluctance of white-dominated institutions to conduct their appointment and promotion processes in accordance with the prin-ciples of affirmative action. At most—indeed, nearly all—of the cam-puses we visited, faculty and administrators acknowledged that the current number of minority faculty members was embarrassingly small, and the figures certainly support the observation. But most campus leaders insisted that they were trying hard to remedy the problem, though they felt some despair about their ability to do so. To be sure, affirmative action may have lost much of its appeal, given the strong buyers' market. Top-rate talent is available in most fields, and virtually all institutions have an unprecedented opportunity to hire exception-ally able non-minority faculty. Nor have we any reason to assume that institutions are in fact doing all they might or should be doing to attract minority scholars. Nonetheless, we do credit the frequent assertions of campus administrators that a serious supply problem exists in most fields.[10]

8. Derived from campus figures for full-time professional ranks, by gender (*Academe,* Sept. 1979, pp. 319–367 (for academic year 1978–79) and July–Aug. 1984, pp. 2–63 (for 1983–84)).

9. The total number of minority faculty members at Berkeley rose from 72 (4.9 percent) to 141 (9.0 percent) over the same period.

10. As recently as 1981, 3.8 percent of all doctorates were awarded to blacks. But 49 per-cent of those doctorates were earned in a single field, education.

But that is only the beginning of the problem. The evidence we see points to a rapid drying up of the *potential* black professoriate. Data on graduate school enrollments are available elsewhere (e.g., Deskins, 1984), and we need not recite them here. We turn instead to anecdotal evidence.

Consider the comment of a highly regarded senior professor at Morehouse College, who shook his head sadly when he said:

> It's a shock these days to find a *good* student who wants to go on to a teaching career. Teaching was always a great profession, but now a kid goes for a summer job at, say, Equitable or Metropolitan Life or Chase Manhattan or Merrill Lynch. They don't think much about teaching after that.

Similarly, the chairman of the political science department at a historically black college commented that he saw few, if any, good students being attracted to teaching careers:

> The best students I see are basically interested in law schools. Whether any good students enter academic careers is in part a function of the opportunity structure in society at any given time. Faculty role models can't be impecunious if we want to attract talented students into teaching.

The comments we heard from historically black and majority white campuses alike were uniformly depressing. Poor compensation is the most frequently cited explanation. According to a black biology chairman:

> It is tougher to get good students to go to graduate school because of perceived better pay in other careers like medicine. . . . There are more options out there, but not for teaching positions.

Some, to be sure, resist the attraction of higher pay. A young black physicist with a Ph.D. from MIT said he had based his decision to teach on "traditional" academic grounds: "I received offers from Xerox and Corning Glass that were 40 percent to 50 percent higher than what I could get teaching, but industry looked to me more like factory work." And a mid-career historian, trying hard to make financial ends meet, insisted that he loved teaching but exclaimed:

> I have a younger brother—he's 31 which is 10 years younger than I am— and he went to work for a Fortune 500 company. Would you believe he's making *five times* my salary? His *bonus* this past year was larger than my salary!

The relatively poor salaries offered in academe were just part of the problem. The extended "pledgeship" at a subsistence level of income during graduate school, as well as the problematic prospects of attaining tenure after an arduous probationary period, make academic careers even less attractive to talented, mobile minorities. We were left with the strong impression that academe will not be in a position to compete for highly able minorities unless some extraordinary interventions occur.

The Faculty Devalued

It may be axiomatic that faculty members rarely feel adequately compensated. Our analysis of faculty compensation (see Chapter 6) attests to an economic situation that has adversely affected most faculty in most settings over the past decade and a half. The evidence derived from our thirty-eight campus visits confirms that perceived inadequacies in compensation and in working conditions are deeply troubling to both junior and senior faculty in most disciplines and at all types of institutions. Several themes emerged from these visits.

Variability. Conditions varied considerably from campus to campus. At a handful of the campuses in our sample, compensation was *not* an issue that rankled faculty. One dean of faculty at a community college conceded:

> Senior, tenured faculty earn over $50,000 to begin with. Then by teaching a one-course overload each semester plus summer school, they wind up with over $60,000. We call this "multiple positioning."

A few faculty members were candid enough to report being somewhat embarrassed at their high level of pay, especially when they took into account the money they made through overload or summer teaching. At most campuses, however, the opposite viewpoint dominated, even though the faculty at our sample campuses tended to be somewhat better compensated than faculty at similar institutions.[11]

11. Using ratings of faculty compensation for 1983–84 published by the American Association of University Professors (*Academe*, 1984, pp. 2–63), we determined that faculty salary and compensation at the campuses we visited were slightly higher than that for other campuses in the same AAUP classifications. Some of the campuses we visited were near the very top of their compensation groups, among them Fordham, Jersey City State College, New York University, St. Mary's College, Southern Connecticut State College, the University of Michigan (even after its hard treatment at the hands of the legislature), the University of Pennsylvania, and, at the top, with all faculty ranks compensated at the 95th percentile or better, Swarthmore College. Similarly, four of our six community colleges—Cypress College, Joliet Junior College, Montgomery College-Rockville Campus, and Borough of Manhattan Community College—were above mid-range in compensation among two-year campuses. On the other hand, some campuses we visited were in the lower compensation ranges, including Alma College, the Universities of Alabama-Birmingham, Denver, Louisville, and Nevada-Reno, Washington State University, and (hovering near the bottom of its classification) the University of New Mexico.

Viewed in terms of changes in faculty compensation over the past five years (1978–79 to 1983–84), the distribution of campuses was quite even; by our calculation, about as many campuses outperformed as underperformed their peer groups in faculty compensation over this past half-decade.

Salary Differentiation. An issue that may be just as disturbing to faculty as the relatively low compensation scales in academe, vis-à-vis other professions, is the issue of perceived compensation inequities within academe. Of course, differences in compensation have always existed among fields. To cite the most obvious example, medical and law faculty generally receive "off-scale" salaries. Within the liberal arts, faculty compensation at many institutions has historically been responsive to market conditions, though only to a limited extent. Now, however, market-responsive pay differentials appear to have been unleashed at many institutions. At roughly one-third of the campuses in our sample, including the most prestigious research universities and several of the most respected liberal arts colleges, the traditional inhibition against marked pay differentials had gone or was disappearing. We found that differential pay policies were causing a lot of distress, especially among arts and science faculties within universities and, as noted previously, among the more senior professors. As one senior historian at a private research university exclaimed, "I'm not sure we want to know what an assistant professor of economics is making. I suppose we're more afraid of the humiliation." The dilemma is not lost on administrators. One chief executive officer fretted:

> Consider the different marketplaces for, say, business and liberal arts faculties. It's a pill. A professor of philosophy is making $28,000 and has been here forever, then we hire in an assistant professor of management at $34,000. How can we both love and pay our humanities and social science faculty? My concern is how to maintain a sense of community in the face of this dysfunction.

The Workplace in Decline. At almost every campus we visited, faculty expressed strong misgivings about their working conditions. Usually, the problem was not any increase in formal workload: neither the number of courses nor the number of students to be taught seems to have increased much in recent years. Rather, faculty complained about having too little office and laboratory space, woefully obsolete equipment, inadequate supplies, and far too little clerical support. At many campuses, travel funds were cited as being preposterously low. On the other hand, sabbatical leave policies seem generally to have remained intact, despite the pressure of tightening budgets. Moreover, administrations were often credited with having provided at least a modicum of support for research (especially stipends for summer research projects) and for attempting to launch faculty development programs. While some faculty voiced their unhappiness with the caliber of the undergraduates they were obliged to teach, quite a few thought that the decline in student quality had been arrested. And we were impressed with the commitment to teaching expressed by faculty at institutions which admitted large numbers of marginal students. As one community college professor, a novelist with several books to her credit, said of her colleagues in English, "People who come here to teach are quite realistic. They know they'll be teaching writing skills. Very few here are highly frustrated."

Coping with Adversity. Despite their widespread unhappiness over compensation, many faculty members, even at those campuses not near the top of the pay scale, were not as badly off financially as one might assume. Not surprisingly, our interviews revealed that faculty families very frequently have two incomes. A very high proportion of young faculty families seem to have adopted this means of coping. Moreover, many faculty members were earning more, sometimes a lot more, by working at outside jobs. Some of these jobs were academic in nature: for example, teaching overloads at the home campus or lecturing part time elsewhere. Others had no purpose other than making financial ends meet. We interviewed an array of entrepreneurial faculty, from the chair of a biology department who sells weight loss products to faculty members engaged in real estate sales or wedding photography. Although available evidence is short, we suspect that faculty efforts to supplement their earnings have increased in recent years. As Table 8–8 shows, a third of the 276 interviewees who were asked whether outside earnings had increased significantly in recent years said they did not know, 36 percent were not aware of any change, and 30 percent thought that such income had increased. More instructive, only 2 percent thought that supplemental income had decreased. A professor of art at a state college said it best: "Everyone I know is doing something else. I think they'd like to have a job that they could give everything to, but people find they can't afford it."

We interviewed several business and engineering school deans who felt that faculty consulting was getting out of hand. But they were exceptions. Most department chairs, deans, and chief academic officers did not believe that faculty efforts to supplement their incomes interfered with their ability to carry out their campus responsibilities.

Table 8–8. Responses to a Question About Faculty Outside Earnings[1]

Respondent Group	Number	*Perceived Change in Faculty Outside Earnings*				
		Increased	*About the Same*	*Decreased*	*Don't Know*	
Administration	90	36%	34%	0%	30%	100%
Department chairs	83	24	45	2	29	100
Faculty	103	29	29	2	40	100
	276	30	36	2	33	100

[1] The question asked: "Have faculty outside earnings increased significantly in recent years?"

NOTE: These percentages may underestimate the extent of outside earnings. The department chairpersons interviewed were predominently from departments of biology, history, physics, and political science; none were from professional schools, where such earnings presumably are more commonplace.

SOURCE: Site Visits, 1983–84.

The high cost of housing was a frequent target of faculty complaints, not only at urban campuses but in communities where one
would least expect such a problem. Housing costs seemed to present
particular difficulties to young faculty families, who were already under
a great deal of career-related pressure and who were also having a hard
time finding suitable places to live.

The Faculty Dedicated

In the final analysis, despite the many frustrations of academic life,
most faculty members choose to accept the bargain. No single finding
stands out so consistently across all thirty-eight campuses than the
unwillingness of faculty to abandon their academic careers. They like
their work, whatever its shortcomings. Indeed, many of them *love* their
work. The vast majority of faculty, around 90 percent, indicated that
they would choose the academic profession if they were starting over
again (Table 8–9).

We interviewed some faculty members who had elected to retire
early, without benefit of special early retirement plans, and some who
had embraced the opportunity afforded by newly devised or expanded
early retirement plans. The so-called golden handshake has attracted
many more faculty members than was anticipated, as happened at the
nineteen-campus California State University system, including the CSU
campus in our sample, California State University, Los Angeles. More
disturbing, it seemed to us that an increasing proportion of early retir-
ees had been pressured, not so subtly, into hanging up their caps and
gowns. Campus administrators and department chairs occasionally
mentioned the need to press for "selective" early retirements because of
the presumed recruiting opportunities that consequently would
become available.

Some faculty members in high-demand fields had succumbed to
the temptation of higher—often much higher—pay in the private sector. We found a small but steady stream of faculty who had opted for

Table 8–9. Responses to a Question About Satisfaction with Having Chosen an
Academic Career[1]

Respondent Group	Number	Yes	No	Don't Know	
Administration	56	93%	5%	1%	100%
Department chairs	59	95	3	2	100
Faculty	146	90	5	5	100
	261	92	5	4	100

[1] The question asked: "Are you glad you chose the academic profession? In other words, would you do
it again if you were starting over?"
SOURCE: Site Visits, 1983–84

different workplaces as engineers, computer scientists, economic and management consultants, accountants, finance specialists, and clinical psychologists. As mentioned previously, the number of faculty who continue to teach full time while working part time at outside jobs seems to be increasing. Occasionally, such a division of the faculty member's time may create problems. As one university academic vice president said bluntly, "There's a real difference between faculty who are *professionally engaged* here and those who are merely *employed* here."

Overall, genuine voluntary attrition—that is, among tenured faculty—is almost invisible, though little legends pop up here and there. Almost invariably the observers left behind interpret such defections as being motivated by the prospect of better pay, coupled with dissatisfaction over campus working conditions. By way of illustration, we found a tenured university sociologist who had abandoned academe for a career in real estate, a handful of economists who had chosen the International Monetary Fund or some similar agency, a logician who had left his tenured position in the philosophy department to work for a computer company, and (perhaps most unexpected of all) a promising medieval historian who had opted for a marketing research career. Two sociologists at different campuses became restaurateurs. A few faculty members, Ph.D. in hand, decided to return to school to prepare for more lucrative careers: an assistant professor of chemistry, said to be destined for tenure, decided instead to enter medical school; an anthropological linguist, likewise untenured but not deemed to be at risk, chose to earn a degree in computer science. Some faculty sought an environment where they could pursue their research interests free of burdensome teaching loads, as in the dual case of a senior and a junior chemist who left a small, undernourished university to work for the National Bureau of Standards.

Despite this sprinkling of exceptions, the vast majority of faculty we encountered planned to stay put. At all but one campus, the exodus of faculty members was perceived to be negligible. As one business school dean observed, "No faculty . . . are leaving voluntarily. It's a lifestyle decision. The psychic income is considerable." The one exception was a small urban campus where compensation was extremely low; here, quite a few faculty were said to be leaving, primarily to relocate to other campuses.

Sometimes the faculty member's enthusiasm for his or her work was boundless. One professor of psychology preparing for his retirement suggested:

> When historians look back on the period 1950 to 1990 and ask what occupation was the best, they'll conclude, "Faculty members had it best of all!"

A liberal arts college historian said:

> I love what I'm doing. Where else could I have as much freedom to do what I love to do?

A department chair at a research university exclaimed:

> This is the greatest life and job that one can have; it is the epitome of what one can do with one's life.

A young economist at a growing master's-granting college probably spoke for many faculty members when he expressed his appreciation for his academic legacy:

> In the lower levels of academe, like my institution, we've inherited perquisites from the higher echelon without the responsibilities of pure scholars. Thus our legacy includes twelve-hour teaching loads, the tenure system and academic freedom. Not a bad deal!

At first glance, it would appear that the faculty, despite occasional complaints, remain *en masse* fiercely loyal to academe. But profound dissatisfactions only partly attributable to constrained campus resources appear to lie just beneath the surface. The indisputable attractions of campus life are being offset more and more, it would seem, by a host of petty annoyances and serious frustrations. We found it disconcerting, therefore, that nearly half (47 percent) of the 122 faculty members, including department chairs, who gave an unequivocal answer to the question, "Do you ever consider leaving academe altogether?", said, "Yes."

A Summing Up

The conclusion we reach, on the basis of our several hundred interviews, is that professorial life is still, on balance, attractive to its practitioners. Sometimes their commitment is marginal, as seems to be the case with the young art historian at an eastern university who conceded, "It's the least objectionable way I know of to make a living." But much more often, our interviewees testified to a genuine, deep-seated love of teaching and research that was inspiring to us.

Perhaps the best overview of our findings is provided by our assessments of the overall quality of faculty life and of changes in the quality of faculty life at the campuses we visited. These assessments are based in part on faculty responses to specific questions and in part on our interpretation of all the evidence observed in our site visits: evidence on morale, compensation, attitudes about administrative leadership, faculty attrition, and so forth. Seeking a global appraisal, we concluded that the quality of faculty life at two-thirds of the campuses—twenty-five of the thirty-eight—ranged from fair to excellent. At three of

Table 8–10. Quality of Faculty Life, by Type of Campus

Campus Classification	Excellent	Good	Fair	Poor	Very Poor
Research universities	0	4	4	0	1
Doctorate-granting institutions	0	1	3	1	1
Comprehensives	0	3	0	5	2
Liberal arts colleges	3	3	0	0	1
Community colleges	0	2	2	2	0
	3	13	9	8	5

SOURCE: Site Visits, 1983–84.

the campuses in this group—all of them liberal arts colleges—the quality of faculty life was judged to be excellent; at another twelve, good. But at a third of the sample—thirteen campuses—we concluded that the quality of faculty life was poor or very poor (Table 8–10). Comparing types of institutions, one finds that faculty at liberal arts colleges seemed to be faring best (despite recent pressures on liberal arts studies), with one glaring exception. On the other hand, seven of the ten comprehensive colleges and universities in our sample were judged to be in difficult straits, at least with respect to quality of faculty life.

In our assessment of recent changes in the quality of faculty life, we concluded that the overall situation had improved at fifteen campuses, had deteriorated at sixteen campuses, and had remained about the same at ten campuses (Table 8–11). Faculty at the research universities and, again, at the liberal arts colleges, appeared to be not only maintaining but improving their circumstances. While there were no instances of strong improvement, at three campuses the faculty condition was worsening sharply.

In summary, the condition of American faculty strikes us as being neither bleak nor bountiful. Most faculty members see themselves as occupying a tenuous middle ground. Many of the people we interviewed were committed, sometimes passionately so, to academic life

Table 8–11. Change in Quality of Faculty Life, by Type of Campus

Campus Classification	Improvement		About the Same	Deterioration	
	Strong	Some	Same	Some	Sharp
Research universities	0	4	2	2	1
Doctorate-granting institutions	0	2	3	1	0
Comprehensives	0	3	1	5	1
Liberal arts colleges	0	3	3	0	1
Community colleges	0	1	1	4	0
	0	13	10	12	3

SOURCE: Site Visits, 1983–84.

and to the freedoms and satisfactions it affords. But few were unreservedly enthusiastic. Almost all criticized some aspect of the faculty condition; quite a few shared the view of an eminent professor of English at a major university, who characterized the faculty condition as "profoundly troubled."

The satisfactions and frustrations of faculty life appear to hang in uneasy balance.

Part Three

THE ACADEMIC LABOR MARKET

In this section we set forth the results of our studies of the demand for and supply of faculty, the colleges and universities being the source of demand and appropriately educated people the source of supply. This subject leads us into a discussion of faculty attrition, the changing market for Ph.D.s, and projections of new faculty appointments over the years from 1985 to 2010. Finally, we consider the flow to academe of persons at various stages in their careers, including especially persons of exceptional talent who are of critical importance even though their numbers are small.

CHAPTER NINE

The Flow of Faculty to and from Academe

The American faculty is a corps of nearly 700,000 educated persons. Each year, even when higher education is depressed, thousands of people enter the academic profession, and even in prosperous years thousands leave it. The flow of these people into and out of the profession largely determines the caliber of the professoriate as defined by such qualities as intelligence, breadth and depth of learning, creativity, motivation, social responsibility, concern for students, and personal integrity. In this chapter, our purpose is to trace these flows with special attention to the sources from which academic people are drawn and to the destinations of those who leave. This subject is of great importance because, over the next twenty-five years, our colleges and universities will probably require nearly as many faculty appointments (full-time-equivalent) as there are members of the professoriate today.

The Academic Labor Market

The movement of faculty into and out of academe, both as to total number and composition, varies from year to year depending on conditions in the academic labor market. In this market, colleges and universities are the buyers, or the source of demand; educated individuals are the sellers, or the source of supply. The transactions in this market consist of appointments to faculty positions by colleges and universities and the relinquishment of these positions by faculty members. A large portion of the transactions, however, involve mere transfers from one institution to another as faculty members resign from one college or university and simultaneously accept appointments in others. From the standpoint of the present discussion, which pertains to the professoriate as a whole, such transfers are merely internal moves in a game of musical chairs. This internal mobility does not affect the number or the characteristics or the caliber of faculties in the aggregate but only the institutional location of particular individuals. The movement of faculty from one institution to another may improve efficiency by helping to place faculty members where they can be most happy and productive, but it does not change the numbers or the characteristics of the professoriate as a whole. In this and the following chapter, we shall be concerned only with the entry of individuals into academic positions *from other pursuits*, and the departure of individuals from the academic profession. We shall be looking at changes in the faculty personnel of the whole higher education system, not at the game of musical chairs within that system.

The number of faculty persons serving the nation's colleges and universities at any given time is determined primarily by the aggregate workload of the institutions, that is, by student enrollments and by research and public service responsibilities. Growth in the workload will usually lead to increases in the numbers of faculty persons; decline will usually lead to decreases. The number of faculty persons hired in any academic year will also be affected by the number of replacements for persons departing from academe through retirement, death, and resignation. The total "hires" then will equal the number of replacements plus or minus the net change in size of the faculty.

The supply of people to fill these positions depends partly on the attractiveness of the academic profession—relative to other occupations—in the matters of compensation, working conditions, and status. The supply also depends on less obvious influences such as the opportunity afforded by the profession (relative to other occupations) to serve one's fellow human beings and to engage in stimulating work.

The Supply of Faculty

The persons to fill the positions that may open up in the next twenty-five years will be drawn in part from people who will be attending graduate school with a view to entering the academic profession directly. It will be drawn also from the several million persons who will have engaged seriously in advanced study but will be currently working outside academe in a wide range of scholarly, scientific, and professional fields. Some of these people will at some time in their lives have had actual academic experience—either full time or part time. Some of them may under some conditions be receptive to offers of positions in the academic profession.

Discussions of the supply of faculty are often based on three assumptions: (1) that the inflow into the profession will consist almost entirely of young persons averaging perhaps thirty years of age who enter directly from graduate school; (2) that the outflow will consist mainly of persons retiring at a predetermined fixed age plus a few persons who become ill or die before retirement; and (3) that those persons who become academics may change institutions from time to time, but that they will remain somewhere in the higher educational system throughout their careers. It is true that many faculty members do enter the profession directly from graduate school and that many stay in the profession to retirement. Indeed, the practice of employing persons just out of graduate school is encouraged by the financial reality that in the short run such persons are less costly to the employing institutions than those who have had experience in non-academic employments. Overall, however, there are many paths from college, through graduate school, to academic appointments, and to retirement. At one extreme, a small minority of people go the whole distance without interruption. At the other extreme, a large number who attend graduate school accept positions outside academe and never become members of the professoriate. iate. Between the extremes are many whose careers are interrupted for longer or shorter periods by military service, work outside academe, family demands, travel, and other activities, but who enter the profession for a part (or several parts) of their careers.[1] Given all these possibilities it is clear that there are numerous people who hold M.A.s, Ph.D.s, and other advanced degrees and certificates who are not employed in academe, and who might under some conditions be persuaded to enter or re-enter the professoriate. These people are a latent supply which constantly hovers over the market and which may become an actual supply as conditions change. Among these people are:

1. Students of graduate and professional schools who, though planning on different careers, might be lured into academe.

2. Professional employees of government, business, and other non-academic organizations.

3. Self-employed persons who have professional training and experience.

4. Professionally trained persons not employed and not in school:
   (a) Homemakers
   (b) Persons who have dropped out of the labor force for travel, for further education, to be near a spouse, for other personal reasons, or because of discouragement.
   (c) Unemployed persons actively seeking jobs.

5. Academic administrators or other academic employees who may be qualified for faculty positions.

6. Part-time faculty who may transfer to full-time status.

7. Immigrants who are professionally trained.

---

1. Toombs (1979, p. 10) found in a survey of a sample of persons listed in *Who's Who in America* that 59 percent of academic persons had had non-academic experience during their careers and that about one-fifth of non-academic persons had had academic experience. He referred to this phenomenon as the "permeability" of the profession.

These people include practising professionals, for example, physicians, lawyers, engineers, scientists, economists, psychologists, journalists, accountants, artists, musicians, and many other persons who are learned in academic disciplines and are in other ways suitable for academic work. Indeed, there are many people in non-academic positions in government, business, and other organizations, and many self-employed persons as well, who have already had successful full-time experience as faculty members and could easily resume academic work if they so chose. And there are also thousands of persons who are serving successfully as college teachers in part-time positions and who might become full-time faculty members.

We are not suggesting that all these professional or specialized persons are, or could easily become, qualified for full-time academic appointments. Probably only a minority, but a sizable minority, might be so qualified. We certainly do not recommend that the nation rely solely or even primarily on these people for its future professoriate. But we do contend that in shaping the future of the professoriate it would be a mistake to overlook the many potential sources, to rely solely on those young men and women who are in graduate school at the time and are specifically planning academic careers.

We emphasize the diversity of potential sources for two reasons. First, during the current and possibly future depressed period in higher education, the caliber of young persons who are specifically preparing for the academic profession may not be equal to the caliber of those who joined the profession in the late 1960s and 1970s. There is considerable evidence, as we shall show in Chapter 11, that some of the best talent has been, and is being, attracted to medicine, law, business, engineering, science, and other fields and away from academic careers. The higher education community, therefore, may be well advised to extend its recruiting beyond the traditional sources. It may be able to attract people who bypassed jobs in higher education in their youth but still have a hankering to be members of the professoriate. The "lost generation" need not be lost forever. It may be able also to attract people who are seeking opportunities for career change in mid-life. Moreover, sad to say, the youthful candidates for faculty positions in the near future will include the generation that was characterized by declining test scores, inadequate high school preparation in the basic subjects, and college education with underemphasis on general education. If a wider net is cast, academic institutions will be able to overcome gaps in the age distribution of the faculty, to achieve a wider selection of women and minority groups, and to attract persons of particularly rich experience. However, we are not advocating that those young people who have chosen to prepare for the academic profession and who will be emerging from graduate school should be passed over. Rather, we are suggesting that the faculties of the future need not be drawn exclusively therefrom.

Because of the vast numbers of persons in the society who have had professional training and experience, there is not likely ever to be any absolute shortage of persons to fill all the academic positions. Few institutions, or even departments, are ever forced to retrench or shut down for absolute lack of staff. Temporarily, they may be compelled to fill in with part-time or superannuated people, or to employ persons less than ideally qualified. But there is seldom an absolute shortage. With suitable inducements, people can be drawn into academe on relatively short notice—as illustrated in the GI period and in the rapid buildup of the 1960s. It is also true for the long run when it is possible to add almost any number of people to the professoriate partly through the training of young people and partly through recruiting people of all ages from other pursuits. The question is not: Can all the positions be filled? Rather, it is: What *caliber* of people can be attracted to these positions?

The answer to this question depends on the prospective reward structure of the academic profession *compared with that of other occupations.* The reward structure consists primarily of compensation, working conditions, and status. It will affect young people who are deciding on which careers to prepare for and will also affect those already in the labor force who are qualified, or could become qualified, for academic appointments. In the next twenty-five years when there will be so many academic jobs to fill, it will behoove American society to provide the finances, and the higher educational community to provide the other conditions, that will make academic employment competitive with other comparable professions.

Faculty Attrition

The supply of faculty is not a one-way street. Faculty members enter academe, but they also leave, and leaving is not always via retirement (Waggaman, 1983, pp. 6–16). A full treatment of supply, in this case negative supply, requires a consideration of exit from the profession. Exit may occur in the following ways:

1. Retirement
2. Death or illness
3. Voluntary departure to accept a position in government, business, or other non-academic organizations or to become self-employed
4. Involuntary separation: non-renewal of contract, denial of tenure
5. Dropping out of the labor force
   (a) for personal reasons, for example, to care for a family or to follow a spouse who has been transferred
   (b) for travel and further education
   (c) because of boredom, discouragement, burnout

6. Transfer to administrative or other non-faculty positions in higher education

7. Transfer from full-time to part-time faculty position

8. Emigration

All of the events in the list above involve departures of faculty. Collectively they are usually referred to as *faculty attrition*. They include all departures except those involving only transfer from one academic institution to another.[2]

Annual faculty attrition measures the number of replacements that must be appointed each year to maintain the faculty at its current level. If growth of faculty is desired, the number of "hires" must exceed the amount of attrition; and if faculty retrenchment is desired, the number of "fires" may be reduced by the amount of attrition. Attrition is thus a critical factor in estimating the number of "hires" or "fires" that may be necessary or possible in any given year—even though attrition may not always conveniently involve the most dispensable people or the disciplines with redundant faculty.

Attrition varies from time to time depending upon several factors. The age composition of the faculties affects the number of retirements, departures due to illness or death, and the number dropping out from burn-out or "mid-life crisis." The ratio of women to men has, in the past at least, affected the number dropping out for personal reasons such as care of family or following spouses who are moving to jobs at new locations. Factors that influence age of retirement affect the number retiring in any given year. Institutional policies designed to push people out—policies relating to probation and tenure, dismissal for cause or financial exigency, and early retirement—will have an impact on the number of departures. Institutional practices regarding the transfer of faculty members to administrative posts will make a difference. Finally, and perhaps most important, the attractiveness of academe as a place of employment—relative to employment in business, government, and other industries—will affect the number of voluntary separations. Clearly, attrition rates are not static but vary from year to year as a resultant of all these factors. However, barring cataclysmic events such as war or deep depression in higher education, the combined attrition rate from all causes tends to be confined to fairly narrow limits and to change rather slowly. Our review of the literature on attrition as summarized in Appendix C leads us to two conclusions. First, a conservative estimate of average faculty attrition for the period 1985–2010 is about 4 percent a year. This breaks down to roughly 1.3 percent for retirement and death, and 2.7 percent for departures for all

2. See Lovett, 1984, for a fascinating account of "300 men and women who had held full-time academic positions but who had left those positions for other employment between 1976 and 1980." See also Zey-Ferrell, 1982, and Stecklein & Willie, 1982.

other reasons. Second, the future rate of attrition over the years ahead is likely to rise if the faculties grow older on the average, if the ratio of women to men increases, if institutions become more adept at forcing or enticing redundant faculty members to leave or retire, or if the gap widens between compensation and working conditions within academe and those outside. On the other hand, the rate of attrition might fall if some of these developments were reversed. Our guess is that in the years ahead—especially after 1995—the factors tending to bring about a rise in attrition will outweigh those tending to make it decline.

In the following chapter (10), we shall attempt to project the number of job openings or "hires" for faculty over the years 1985 to 2010, using varying assumptions. The future rate of faculty attrition is the dominant factor determining the likely number of job openings. In preparing these projections, we have adopted three basic assumptions regarding attrition, a low one of 3 percent per annum, a middle one of 4 percent per annum, and a moderately high assumption, which we believe to be the most plausible of the three. This high assumption includes a rising rate of attrition beginning at 4 percent in 1985, gradually increasing to 6 percent in 2000 as retirements increase substantially, and leveling off at 6 percent thereafter.

These are important figures because they dominate the estimates of the number of new faculty appointments that might be made assuming no change in the overall size of the faculty, or the number of positions that might be eliminated without involuntary separations. For example, if attrition were at the rate of 4 percent a year, the faculties could be cut by about 32 percent in ten years without any firings. The 32 percent is appreciably more than the prospective percentage decline in enrollments over the next decade. Attrition is of course not a cure-all for adjusting faculty members to declining enrollments. It does not always occur at the "right" time and in the "right" disciplines and does not always affect the highest paid members of the faculty. Yet the normal rate of departure of faculty could go a long way toward meeting the need for retrenchment.[3]

The assumed rates of 3, 4, and 4 to 6 percent refer of course to all higher education. The rates will probably differ among various types of colleges and universities, among institutions individually, and among academic disciplines. We suspect that the rates will tend to vary inversely with affluence and distinction, and be lower at major private

---

3. It should be noted that the attrition rates we are considering refer to persons leaving the profession, and do not include those moving from one institution to another. From the point of view of each institution, the game of musical chairs has a place in adjusting faculty numbers to changes in demand. For any one institution, a resignation is a resignation regardless of what happens to the individual concerned. Moreover, the game of musical chairs helps distribute the faculties among institutions where they can be most effective and opens up fresh opportunities for them, even if it does not remove individuals from the academic profession.

universities, at public flagship universities, and at well-known liberal arts colleges, and to be higher at other types of institutions. But overall, we believe that attrition rates will be higher, and the opportunities for the appointment of new faculty members greater, than is generally realized. Many of the studies of faculty attrition refer only to full-time or tenured faculty and ignore the substantial attrition among junior and part-time faculty. Indeed this attrition is under the control of the employing institutions and not wholly subject to the independent decisions of the incumbents. A vacant position created by the voluntary or enforced resignation of a full-time junior person or of two or three part-time persons creates an opening for a new full-time appointee exactly as the resignation of a tenured person creates such an opening. In such cases the ability to fill a position with a first-class, full-time appointee depends on finances, not on the availability of a position. The problem of faculty staffing is not so much an inadequate number of job openings as it is the finances and the perspicacity to fill each opening with a person of excellent capability.

The Process of Career Choice, Professional Training, and Induction

We have referred to the varied sources of supply of faculty. However, the one feature that is common to all potential faculty people is college education and advanced study. The undergraduate college, however, often plays a decisive role in students' choice of professional careers and in their decision to proceed to advanced study. Many young men and women who enter graduate (or professional) school have already acquired a taste for serious learning and have been encouraged by their undergraduate professors to "go on" to graduate school. Many colleges take pride in the percentage of their students who attend graduate school. Thus, for many, the undergraduate college is the place where at least tentative decisions about advanced study and academic careers are made. Another frequent pattern is that students on graduation from college accept jobs in business, government, the armed forces, etc., but later discover that they miss the pleasures of learning and of the campus style of life and decide to prepare for academic careers by entering graduate school. Still others enter graduate school with the intent of becoming practicing professional persons but eventually decide on academic careers. Whatever the decision-making process, the graduate and professional schools have an important role in the induction of people into the academic profession. This role is to prepare them in terms of substantive learning as evidenced by advanced degrees and to socialize them with respect to academic values, rewards, and style of life.

At the time of first entry into an academic position, the age of faculty members is seldom below twenty-five, often as high as forty or

Table 9–1. Time Required for Doctoral Studies, 1981

	Median Lapse of Time from Baccalaureate to Doctorate		
	Years formally registered for study	*Total time overall*	*Median Age at Doctorate*
Engineering, Mathematics, and Physical Science	5.7 years	7.2 years	29.6 years
Life Sciences	5.9	7.3	30.1
Social Sciences	6.5	9.0	32.0
Humanities	7.7	10.8	33.5
Professional fields offering Ph.D.	6.6	11.1	34.2
Education	7.0	13.5	37.3

SOURCE: National Research Council, *Summary Report*, 1981, pp. 32–33.

more, and probably averages around thirty-three. As shown in Table 9–1, the median length of time for which doctoral candidates are registered for graduate study ranges from six to eight years. Scientists are at the low end of the range and social scientists and humanists at the high end. Because many students do not go through graduate school without interruption, or attend part time, the total elapsed period from the baccalaureate to the doctorate ranges from seven to thirteen years. The median age of the candidates at the time the doctorate is awarded varies from thirty to thirty-seven years. These data reveal the long period of gestation involved in the education of potential members of the academy, and suggest that a lead time of roughly ten years from the baccalaureate degree on the average is needed to prepare young people for the academic profession. However, the time needed for acquiring a Ph.D. could be shortened through an acceleration scheme. For example, a financial program that would enable graduate students to attend full time (instead of part time as is now so common) would both improve graduate education and shorten the elapsed time required for the degree. Also, the recruitment of doctoral students before completion of their dissertations—as was common in the 1950s and 1960s—would reduce the time.

Generally, after recruitment by colleges or universities, doctoral recipients who enter academe face a probationary period of several years. At the end of the probationary period, they may be dropped and forced to find employment elsewhere, either within or outside academe, or they may be granted lifetime tenure as career faculty members in the employing institutions.

This process of preparation, socialization, recruitment, and probation is quite rigorous. It is not guaranteed, however, to produce a highly motivated and well-educated professoriate. This will happen only if two conditions are met: (1) that the graduate schools are able to attract people who are on the whole potentially well qualified for the profession—including an ample supply of the rare, brilliant ones who are needed to provide the intellectual leadership of the nation; and (2) that the colleges and universities are able to recruit and retain a generous share of these qualified people. If either of these two conditions cannot be met, the caliber of the academic community will sooner or later decline.

These two conditions were largely met during the Great Depression of the 1930s when opportunities in higher education were promising relative to openings in other industries and occupations. They were also met during the 1960s and early 1970s when a vast expansion of higher education occurred and much talent was attracted. In view of the long lag between choice of academic career and actual entry into the profession, the people who emerged from the graduate schools and entered academe during the 1970s and early 1980s were still of high caliber as is almost universally testified by presidents and deans throughout the nation. Indeed, a large majority continue to report that on the whole their faculties are the best ever (Minter & Bowen, 1982).

The high quality of present faculties is explained also by the ability of the employing institutions in recent years to be selective. There were relatively few academic appointments to be made and relatively large numbers of persons completing graduate programs. But as of 1985 there are early warning signs that the caliber of faculties may be beginning to deteriorate. People who would once have planned to enter academe are seeking, or have been forced to seek, other outlets for their talents and energies. Meanwhile, increasing numbers of people already in academe are beginning to think of alternative options—though as yet there is no stampede out of the profession. The faculties of 1985 may be the best prepared in our history, but after years of declining compensation and deteriorating working conditions they are not the most highly motivated in our history and do not have the highest morale. Under these conditions, the current crop of young people, as they contemplate embarking on the long journey to academic careers, may be having second thoughts. Ten years from now the caliber of persons available to the profession may have diminished. These matters will be considered in some depth in Chapter 11.

The Amazing Expansion of Advanced Study

Over the past several decades, especially since 1950, there has been a phenomenal growth in the number of persons seeking advanced degrees and in the facilities and opportunities for advanced study. Since the late 1970s, however, the growth has leveled off. For doctors' degrees, it would have declined sharply except for the entry of thousands of women into doctoral study (see Table 4–2). The growth in number of first professional degrees has continued into the 1980s. This growth has been fueled in part by the entry of thousands of women into fields such as law, medicine, dentistry, veterinary medicine, and theology. The trends in degrees awarded are shown in Table 9–2 and Figure 9–1. As a result of the growth, the population as of 1985 contains more than six million persons who have received masters' degrees, well over a million who have received first professional degrees, and 750,000 who hold earned Ph.D.s.[4] Allowing for duplication, perhaps seven to eight million persons hold one or more advanced degrees of some kind. Perhaps as many as 90 percent of these people are employed outside academe and have little interest in joining the academy. They nevertheless make up the population from which almost all academic people are drawn. It is because this population is so large that an absolute shortage of academic people is highly unlikely. Indeed, the size and diversity of this pool suggest that higher education, which requires only about 700,000 faculty members (full time and part time), could employ whatever quantity of faculty it might need—but the quality of the people it would be able to attract from the pool would depend on the compensation and working conditions it could offer and the effectiveness of its recruitment efforts.

Traditionally a majority of new faculty appointees have been younger persons who have recently completed or have been about to complete their advanced study and have chosen to seek academic employment. It is of special interest, therefore, to compare trends in the number of persons appointed to academic positions with trends in graduate enrollments and advanced degrees awarded. These comparisons are made in Table 9–2. As shown in this table, graduate enrollments, advanced degrees awarded, and faculty appointments all traced a similar pattern of growth during the period from 1950 to 1969. After 1969, graduate enrollments continued to grow, though at a diminishing rate, and so did masters and first professional degrees awarded. Ph.D.s awarded leveled off, but did not decline significantly. Meanwhile, faculty appointments took a downward trend. Clearly, academic employment became a diminishing outlet for persons with newly minted degrees. See Figure 9–1.

The comparisons may be clarified by comparing the number of Ph.D.s awarded (Table 9–2, column 5) with the number of faculty appointments (column 8). During the five years 1950–54, about 47,000 persons were recruited into the higher educational faculties. This was a relatively small number of appointments, compared with the numbers for later quinquennia, because higher education was then in the

---

4. These figures are estimates derived from the cumulation of historical data on the number of earned degrees awarded (with allowance for mortality). National Center for Education Statistics, *Digest of Higher Education Statistics*, 1983, p. 132; National Center for Education Statistics, *Projections of Education Statistics to 1990–91*, 1982, p. 70.

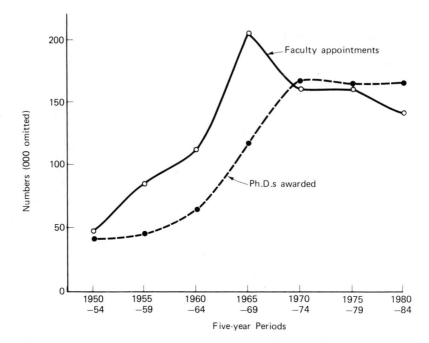

Figure 9–1. Ph.D.s Awarded and Faculty Appointments, by Five-Year Periods, 1950–84

doldrums. Yet the number of appointments exceeded the number of Ph.D.s awarded and provided a moderately brisk academic market for new Ph.D.s, especially so considering that not all of the new Ph.D.s were headed for academe. In the 1960s as the nation began to gear up for the great expansion of higher education that was triggered by Sputnik and fueled by the postwar baby boom, the number of new faculty appointments increased rapidly and steadily. During the period from 1960 to 1970, the number of new faculty appointments exceeded the number of Ph.D.s awarded by a wide margin. But in the early 1970s the tide turned. New faculty appointments declined fairly sharply while the production of new doctorates stabilized.

Non-Academic Employment of Ph.D.s

As the output of new Ph.D.s continued at around 167,000 each five-year period, or 33,000 a year (Table 9–2), exceeding slightly the number of appointments to academe, a relatively new development occurred, namely, large-scale employment of Ph.D.s outside academe. Tables 9–3 and 9–4 provide estimates of the numbers involved. The growth in

Table 9–2. Graduate Enrollment, Advanced Degrees Awarded, and Faculty Appointments, by Five-Year Periods, 1950–54 to 1980–84 (000 omitted)

1	2	3	4	5	6	7	8
							Estimated Number of Persons Appointed to Faculty Positions from Outside Academe: Over Five-Year Periods
	Graduate Enrollment:	Number of Advanced Degrees Awarded[2]: Over Five-Year Periods					Total Appointments:
Five-Year Period	Average for Period[1]	Masters	First Professional	Ph.D.s	Increase in Number of Faculty FTE[3]	Replacements[4]	Increase plus Replacements[5]
1950–54	242	298	n.a.	41	22	25	47
1955–59	336	329	n.a.	46	54	32	86
1960–64	493	488	133	65	72	41	113
1965–69	831	877	165	118	139	66	205
1970–74	1,084	1,315	241	168	77	85	162
1975–79	1,308	1,540	332	165	60	102	162
1980–84 (est.)	1,418	1,529[5]	369[5]	167[5]	32[5]	110	142

[1] National Center for Education Statistics, *Projections of Education Statistics*, 1964, p. 8; 1967, p. 19; 1976, p. 34; 1982, p. 52. American Council on Education, *Fact Book*, 1980, p. 57. Council of Graduate Schools, Jan. 1983, p. 5. Estimates of the authors by interpolation. Data do not include first professional degrees.
[2] National Center for Education Statistics, *Digest of Education Statistics*, 1982, p. 130; National Center for Education Statistics, *Projections of Education Statistics*, 1964, p. 12; 1968, p. 31; 1978, pp. 38–39; 1982, p. 70.
[3] National Center for Education Statistics, *Digest of Higher Education Statistics*, 1982, pp. 105, 107; 1968, p. 56; 1964, p. 24.
[4] Computed assuming attrition at 4 percent of average number of faculty.
[5] National Center for Education Statistics, *Projections of Education Statistics*, 1982, p. 70.

employment of Ph.D.s in business, government, hospitals, private consulting, and other occupations outpaced the growth in their employment by colleges and universities. By 1985 an estimated 319,000 Ph.D.s, or 43 percent of all those holding the degree, were working outside academe.[5] This major development had the fortunate effect of preventing serious unemployment among Ph.D.s. In fact, the rates of unemployment have been surprisingly low considering that a substantial increase in the number of Ph.D.s occurred precisely when the rate of Ph.D. appointments in higher education declined sharply (as shown in Table 9–2).

5. About a half of all science Ph.D.s were employed outside academe, one-fourth of those in the social sciences, and one-sixth of those in the humanities (National Research Council, 1982, pp. 20, 50). Data are not available on the employment of Ph.D.s earned in professional fields such as education, business, engineering, computer science, clinical psychology, and communications. However, it is evident that substantial percentages of these people are employed outside academe.

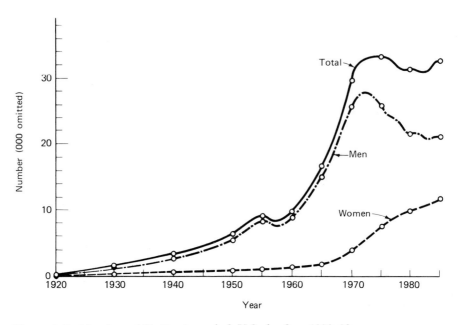

Figure 9–2. Number of Ph.D.s Awarded, U.S., by Sex, 1920–83

One may ask why doctoral production continued at a high rate after 1975 despite the worsening academic market. One reason is the long lag between entry into Ph.D. programs and receipt of the degree. Many young people undoubtedly had chosen the academic profession as undergraduates in the late 1960s or early 1970s when higher education was prosperous and then continued despite the onset of the depression in higher education. Another reason, given the long period between the baccalaureate and the doctorate, is that some candidates are entering graduate study in the 1980s planning to be ready for the academic market in the 1990s, when it is expected to pick up (Scully, 1983). Still another factor is that the women's movement encouraged and emboldened many young women to seek Ph.D.s. Figures on the number of doctorates awarded by sex tell the story (Figure 9–2). Men responded to the changing market conditions by a 18 percent reduction in number receiving the Ph.D. over the period 1975 to 1982.[6] Simultaneously, women were entering the market in unprecedented numbers. The combined effect of the declining number of men and the increasing number of women was to stabilize the total. Had women claimed only their traditional share of the Ph.D.s of 10 to 15 percent, the total would have been less by several thousand. But the most compelling reason for the continuation of doctoral production at a high rate was that opportunities outside academe did open up on a large scale. They opened up

6. See also Table 4–2.

Table 9–3. Estimated Employment Trends for Persons with Ph.D.s

	1	2	3	4	5
	Number of Active Ph.D.s in the U.S. Population[1] (000 omitted)	Estimated Percentage of Faculty with Ph.D.s[2]	Number of Faculty[3] (000 omitted)	Estimated Number of Ph.D.s Serving as Faculty Members[4] (000 omitted)	Estimated Number of Ph.D.s Employed Outside Academe[5] (000 omitted)
1950–51	101	37%	165	61	40
1955–56	134	39	198	77	57
1960–61	204	41	236	97	107
1965–66	273	49	340	166	107
1970–71	398	55	474	260	138
1975–76	522	58	628	364	158
1980–81	643	60	678	407	236
1984–85 (est.)	750	62	695	431	319

[1] Estimated from data covering Science, Engineering, Social Science, and Humanities in National Research Council, 1982, pp. 20, 50; and by cumulating doctorates awarded over successive 40-year periods reduced by 15 percent to allow for withdrawal from the labor force, retirement, net migration, and mortality.
[2] Estimated by the authors from many sources, especially American Council on Education, *Fact Book*, 1980, p. 118; National Education Association, 1979, p. 19.
[3] National Center for Education Statistics, *Digest of Education Statistics*, 1983, p. 103; 1964, p. 24. Includes full-time and part-time faculty.
[4] Column 2 multiplied by column 3.
[5] Column 1 minus column 4.

partly because of expansion of the market for scientists, engineers, economists, managers, and other professional persons. They also opened up simply because those with Ph.D.s are on the whole among the brightest, most energetic, and most versatile individuals in our society and tend to be attractive to employers. As Radner and Miller (1975, p. 349) commented, "The doctorate will in the future serve as a base for a widening variety of career employments."

Unemployment or Underemployment among Ph.D.s

The question of unemployment or underemployment among Ph.D.s has received a great deaal of discussion in recent years (Zumeta, 1981). Excellent data on this subject have been gathered by the National Research Council and are presented in Table 9–5. These data refer only to scientists, engineers, social scientists, and humanists and do not include persons with Ph.D.s in professional fields such as theology, business, journalism, and education. The figures are shown separately for the entire population of Ph.D.s in the included fields and for persons who have received their degrees recently. These figures leave the impression that unemployment or underemployment has decidedly not been rampant among Ph.D.s either in the scientific or humanistic fields

Table 9–4. Employment of Ph.D.s by Type of Employer and Year of Doctorate

	Year Doctorate Granted[3]				
	1960–64	*1965–68*	*1969–72*	*1973–76*	*1977–80*
Physical Sciences and Engineering[1]					
— Colleges and Universities	54.4%	47.5%	38.2%	36.1%	34.2%
— Business & Industry	34.5	38.2	46.1	50.1	52.8
— Government	7.8	10.5	10.7	9.3	7.7
— Other	3.3	3.8	4.8	4.4	5.2
Total	100.0	100.0	100.0	100.0	100.0
Life Sciences[2]					
— Colleges and Universities	67.3%	63.8%	62.6%	58.9%	62.1%
— Business & Industry	14.1	16.6	15.4	20.1	19.2
— Government	12.1	13.6	14.3	13.3	11.5
— Other	5.9	6.0	7.6	7.6	7.0
Total	100.0	100.0	100.0	100.0	100.0
Social Sciences					
— Colleges and Universities	71.2%	65.3%	62.1%	57.7%	54.0%
— Business & Industry	12.4	17.4	17.4	15.9	16.2
— Government	7.4	7.4	8.1	10.5	11.3
— Other	8.9	9.7	12.2	15.4	18.3
Total	100.0	100.0	100.0	100.0	100.0
Humanities					
— Colleges and Universities	92.9%	86.1%	85.8%	77.0%	75.0%
— Business & Industry	1.8	5.8	5.5	9.8	10.1
— Government	0.3	2.0	2.4	5.8	5.0
— Other	4.1	4.4	5.8	7.0	9.8
Total	100.0	100.0	100.0	100.0	100.0

[1] Includes mathematics, computer science, chemistry, physics, astronomy, and earth and environmental sciences.
[2] Includes agricultural, medical, and biological sciences.
[3] Because of rounding, totals do not add precisely to 100.

SOURCE: National Research Council as reported in *Chronicle of High Education*, Oct. 5, 1983.

(cf. Zumeta, 1982, pp. 335–38). The hard core of unemployment (defined as not employed and seeking work) is of the order of 1 or 2 percent,[7] an almost irreducible minimum given a less than perfect labor market. However, this is too rigid a definition of unemployment or underem-

7. A survey of Danforth scholars who had completed their work for the Ph.D. revealed unemployment rates varying by class from 0 to 3.8 percent. *Danforth News & Notes*, March 1979, p. 4.

ployment. Some of those working part time would prefer to work full time; some of those with postdoctoral appointments are in a temporary holding pattern and would prefer to have permanent jobs[8]; some of those not employed and not seeking employment would prefer to have jobs had they not been discouraged from seeking work by a slack labor market; and some of those not reporting may have been avoiding acknowledgment of their unemployment. We would guess that the true unemployment among different groups of Ph.D.s would be roughly in the range of 3 to 6 percent—still below unemployment rates for the general labor force. These figures do not spell catastrophe. They seem remarkably low when one takes into account both the tremendous outpouring of Ph.D.s in recent years and the unemployment rates in the general labor force which have varied from 7 to 11 percent in the early 1980s. These figures also convey a sense of the high ability and versatility of persons who have earned the Ph.D. They are by no means confined to academic work.

Table 9–6 presents data on rates of unemployment or underemployment as of 1981 among various academic disciplines. Though, as would be expected, the rates were higher in the humanities than in the sciences, the differences among fields were surprisingly small.

A careful study of the employment situation among humanists was conducted in 1984 by Laure Sharp under the auspices of the National Endowment for the Humanities (Sharp, 1984). Her work was based on a restudy of the excellent data collected by the National Research Council and already cited in Tables 9–5 and 9–6. Sharp summarized her results (p. vii) by saying that "This investigation leads to a more positive assessment of the overall employment picture [for Ph.D. humanists] than was suggested by press reports and anecdotal evidence about the plight of unemployed Ph.D.s . . . periodic surveys never confirmed the existence of high unemployment rates for humanists; (2.9% in 1977, 2.2% in 1979, and 1.5 percent in 1981)." More specifically, her study revealed (p. vii) that in 1981, 89 percent of the 76,000 surveyed humanists were in the labor force and that among these 2 percent were unemployed, another 2 percent were working part time and looking for full-time work, and 3 percent were working part time by choice. She noted that in no humanistic discipline were fewer than 70 percent working full time in their chosen fields; moreover, unemployment rates were not appreciably affected by the prestige of the institutions from which humanists received their degrees. Unemployment rates, she found, were not very different among different categories of Ph.D. recipients

8. Zumeta (1982a, pp. 18–19, 34–35), in a detailed study of postdoctoral research appointments, concluded that these positions, though temporary, cannot be interpreted as merely a way out for young Ph.D.s who fail to find regular employment either inside or outside academe. Rather, he suggests that for a variety of reasons, the postdoctoral appointment has become a more or less regular part of the education and training of scientists and scholars.

Table 9–5. Employment Status of Doctoral Scientists, Engineers, Social Scientists, and Humanists, 1981

	Scientists, Engineers, and Social Scientists		Humanists		Social Scientists[2]	
	All Graduates (1938–80)	Recent Graduates (1975–80)	All Graduates (1938–80)	Recent Graduates (1975–80)	All Graduates (1938–80)	Recent Graduates (1975–80)
Employed Full Time	88.5%	84.5%	83.6%	84.0%	90.0%	92.5%
Employed Part Time	—	2.7	—	8.2	—	3.1
Employed Part Time, Seeking Full Time Work	0.5	—	2.0	—	0.7	—
Employed Part Time, Not Seeking Full Time Work	2.4	—	4.0	—	2.2	—
Post Doctoral Appointment	2.9	9.7	0.7	0.9	0.7	1.2
Not Employed, Seeking Employment	0.7	1.2	1.4	2.5	0.7	1.4
Not Employed, Retired and Other	3.6	—	6.1	0.6	4.1	—
No Report	0.4	0.3	0.6	0.6	0.5	0.1
	100.0[2]	100.0[2]	100.0[2]	100.0[2]	100.0[2]	100.0[2]

SOURCE: National Research Council, *Science, Engineering, and Humanities Doctorates in the United States, 1981 Profile*, 1982, pp. 16, 30, 32, 47, 58, 61. These data include all Ph.D.s except those in Professional Fields as follows: Theology, Business Administration, Home Economics, Journalism, Speech and Hearing Science, Law and Jurisprudence, Social Work, Library and Archival Science, Education, and a few other fields.

[1] Social scientists also included in "Scientists, Engineers, and Social Scientists."

[2] Because of rounding, totals do not add to precisely 100.

Table 9–6. Employment Status of Doctoral Scientists, Engineers, Social Scientists, and Humanists Who Received Their Degrees Between 1938 and 1980, by Selected Disciplines, 1981

	Percentage Employed Part Time and Seeking Full-Time Work	*Percentage Not Employed and Seeking Employment*
Sciences:		
Mathematics	0.2	0.5
Computer Science	1.0	0.2
Physics/Astronomy	0.1	0.6
Chemistry	0.3	0.7
Earth & Environmental Sciences	0.8	0.8
Engineering	0.2	0.1
Agricultural Science	0.2	0.4
Medical Science	0.3	0.7
Biology	0.8	1.5
Psychology	1.1	1.1
Other Social Science	0.8	0.8
Humanities		
History	2.2	1.2
Art History	2.7	4.2
Music	3.1	2.2
Speech and Theater	1.6	0.8
Philosophy	2.6	0.6
Language and Literature		
English–American	1.5	1.2
Classical	2.6	1.9
Modern Foreign	2.8	2.1

SOURCE: National Research Council, 1982, pp. 32, 61. *Science, Engineering, & Humanities Doctorates in the United States: Profile 1981*, pp. 32, 61. These data differ from those of Table 9–5. In that table the rates are computed as percentages of the total Ph.D. population, whereas in this table the rates are computed as percentages of the relevant Ph.D. labor force, omitting retired persons, persons not seeking work, and those not reporting employment status.

though they were somewhat higher for women than for men. Sharp indicated, however, that the situation had been greatly eased by the steep decline in the number of Ph.D.s awarded in the humanistic fields—from a peak of 5,170 in 1974 to 3,560 in 1982 (p. 4).

Further enlightenment on the situation in the humanities is provided by a study conducted by May and Blaney (1981). In answering the question What happens to Ph.D.s who cannot find jobs as teachers?, their answer was (p. 83):

> They have gotten and will get other types of jobs. Most will enjoy their work as much as they would have enjoyed teaching. Very few face unemployment . . . the proportion believing themselves underemployed will be similar to that among the Ph.D.s who did find teaching jobs [p. 85]. [May and Blaney continue [p. 111] Our contention is simply that the

number of people who equip themselves to be scholars should not be a
function of actual or anticipated fluctuations in the academic job market. . . . It will not be a waste if large numbers continue to earn doctorates
and go on to posts in insurance companies or government agencies. In
fact the cultural life of the United States will be significantly richer if the
training of scholars can be divorced from the preparation of teachers and
if it becomes no more extraordinary for a corporate vice-president to have
a Ph.D. in philosophy than a law degree and no more remarkable for
someone in business or civil service to publish a scholarly book or article
than to win an amateur golf tournament or to be elected to local office. The
humanities could become more integral to American life.

A revealing data set has been compiled by the National Research
Council on career plans of new doctorate recipients at the time the
degree is awarded. These data show the percentage of recipients having
made definite career arrangements and the percentage still seeking positions. The data are especially interesting because they are stratified by
field of study and are available annually over a long period, 1958 to 1982.
Table 9–7 presents these data for selected years in terms of the percentage of new doctorates still seeking positions, that is, not placed, at the
time the degree is awarded. As the table shows, the percentage not
placed was fairly stable in 1958 and 1964, was appreciably higher in
1970, and was still higher in 1976 and 1982. Clearly, new Ph.D.s were
not being snapped up instantly in the 1970s and 1980s, as they had been

Table 9–7. Percentage of New Doctorate Recipients Still Seeking Positions at
the Time Degree was Conferred, by Field of Study, Selected Years, 1958–1982

	1958	1964	1970	1976	1982
Physics and Astronomy	13.3%	17.8%	22.8%	28.5%	24.3%
Chemistry	9.3	9.5	18.1	21.2	16.3
Mathematics	15.1	15.8	20.6	28.1	23.6
Engineering	15.4	16.5	24.7	27.3	25.9
Biochemistry	11.4	9.8	13.4	17.2	19.7
Other Biosciences	16.0	13.7	19.7	22.4	22.9
Medical Sciences	12.6	12.9	17.8	20.4	22.7
Psychology	18.6	14.9	20.6	28.2	28.5
Anthropology	15.1	12.0	12.4	29.4	40.5
Economics	14.8	9.1	10.9	17.7	15.7
History	15.8	9.6	13.6	34.4	34.1
English and American Literature	9.9	11.0	13.8	37.3	32.4
Foreign Language and Literature	14.8	9.3	18.8	39.1	32.5
Education	15.2	13.1	22.6	24.5	24.6
All Fields Combined	14.3	13.0	19.4	25.7	24.7

SOURCE: National Research Council, *Doctorate Recipients from United States Universities: Summary
Report, 1982*, pp. 16–21.

in the 1960s. More time and effort—and presumably worry—were required in getting placed than had been common in the 1950s and 1960s. These figures also imply the necessity on the part of many aspirants of settling for second or third or tenth choices in their acceptance of job offers. Yet, as shown in Tables 9–5 and 9–6, unemployment rates have been surprisingly low. We interpret these data to mean that job-seeking among new Ph.D.s has become more difficult in recent years than it was in the halcyon 1960s, but that jobs are eventually found.

Many new doctorates have been forced to find positions outside the traditional labor markets, and more time and ingenuity have been required to seek out such positions. Thus, the change in the labor market for Ph.D.s has shown up more plainly in data on the time required for placement rather than on rates of unemployment. Table 9–7 also shows that the percentage of new doctorates still seeking positions is on the whole higher in the humanities and social studies than in the sciences, and this is no surprise.

Solmon and Zumeta (1981) in observing the increasing employment of Ph.D.s outside academe remarked (p. 22), "If Ph.D. holders have successfully demonstrated their unique productivity in nontraditional work settings during a period of their relative abundance, they may have created permanent new sources of demand for their services." This thought has an ominous ring, because it suggests that academe is likely to face increasingly intense competition with industry and other employers for the available Ph.D. talent.

Concluding Comments

This chapter has been a wide-ranging analysis of the academic labor market. It has been provocative in the sense that the conclusions reached are not wholly congruent with anecdotal accounts or journalistic reports about conditions in this market. In preparing this chapter, we have been plagued by the shortage of reliable statistics available for our purposes. We have made use of several excellent data sets of the National Center for Education Statistics and the National Research Council, but have also found it necessary to improvise by developing our own rough estimates of some of the needed information. Our main conclusions are:

First, through various forms of advanced study, the nation is creating a great and growing pool of highly educated people who engage in a multitude of professional (but non-academic) activities for which advanced study is required. We suspect that in the future there will be increasing two-way traffic between this pool and the academic community. The pool will be an increasing source of new faculty and at the same time it will confront academe with stiff competition for both new Ph.D.s and established members of the professoriate.

Second, the number of openings for new appointments to the nation's faculties is likely to be greater than is usually assumed. Our studies lead to the conclusion that attrition might average 4 percent in the years ahead to the late 1990s and might even reach 6 percent by the beginning of the 21st century. The actual attrition rate is likely, however, to be volatile because it is a function of two independent variables: conditions within academe and conditions in the economy outside academe. The worse the inside conditions, the greater the attrition; and the better the conditions outside, the greater the attrition. At the annual rate of 4 percent, over ten years about 32 percent of all faculty positions would become vacant through attrition, and over twenty-five years 70 percent would become vacant. We see the problem of faculty replacements as not so much a shortage of vacancies as a shortage of funds necessary to make solid appointments as distinguished from non-tenure-track and part-time appointments. It should be noted, however, that success in the appointment of full-time tenured faculty would ultimately result in a reduction of the attrition rate.

Third, the average lead time for preparing Ph.D.s is of the order of seven to eleven years from the date of the baccalaureate degree. Since many students make their career choices during their undergraduate years, the average elapsed time between decision to enter advanced study and completion of Ph.D. programs may easily be ten years or more. These facts suggest that the latter half of the 1980s is none too early to encourage promising young people to consider academic careers.

Fourth, the number of Ph.D.s awarded during the 1970s would probably have fallen quite sharply had the women's movement not propelled many women into Ph.D. programs. The net result of the decline in number of degrees awarded to men and the increase of number awarded to women has been a fairly stable production of Ph.D.s at just above 30,000 a year.

Fifth, despite much opinion to the contrary, unemployment rates among Ph.D.s have remained remarkably low. Though there have been differences among the various fields of study in the time and trouble involved in achieving placement, the market has not been glutted with unemployed Ph.D.s. This has been due in part to an unexpectedly resilient academic labor market and also to the increasing flow of Ph.D.s to non-academic employment. It is something of a miracle that the market has been able to absorb most of the increasing number of new Ph.D.s.

Sixth, the data suggest that people with advanced degrees have proven their versatility and employability and have successfully penetrated new markets.

One of the principal themes of this book is the need for higher education to be strongly competitive for the best talent in the labor market wherever it can be found—in the graduate schools, in the pool of professional talent, among the unemployed or underemployed. To seek the best talent is always expensive. But we would argue that in helping higher education to get through a difficult period, funding agencies should recognize the need to make faculty salaries and working conditions competitive to assure the recruitment and retention of genuine talent. In the same vein, it would be healthy for the profession and for the institutions if new job openings could be created, for example, by converting part-time positions into full-time positions, by employing more Ph.D.s on community college faculties, and through tactful forms of early retirement.

CHAPTER TEN

Projections of New Faculty Appointments, 1985–2010

Many observers see that the future of American higher education as somewhere between dismal and catastrophic. The assumptions underlying this pessimism are, first, that the number of job openings for new, younger faculty members will be minuscule; and second, that the faculty now in place will grow steadily older on the average and gradually become decrepit, demoralized, and spent. These assumptions do have an element of truth in them. Given well-known demographic trends, higher education is not likely to be a buoyant growth industry in the decade 1985–95. On the other hand, it does not follow that the faculty now in place will remain exactly where they are, or that there will be scant job openings, or that an aging faculty will inevitably become an ineffectual one. The assumptions that dominate thinking about the future are too limited. They do not take into account the broad range of possibilities and opportunities for stability or even growth in enrollments and for substantial turnover of faculty that would create thousands of job openings for "new blood." We acknowledge that a pessimistic scenario could materialize, but we believe that it is by no means inevitable. To show the range of possibilities before the higher education community, we have attempted in this chapter to project the number of new faculty appointments that may be possible, under varying assumptions, over the twenty-five years from 1985 to 2009. Our results give cause for guarded optimism but offer no guarantee that the prophets of gloom are wrong.

The Projections

The number of possible faculty appointments is determined mainly by two variables. One is change in the number of faculty members. An increase in numbers, other things equal, will give rise to new appointments; a decrease will extinguish existing appointments. The other variable is faculty attrition, that is, the rate at which faculty members leave academe for any reason. Thus, the number of faculty appointments available in any period will equal the change in the number of faculty (which may be positive or negative) plus attrition. In projecting future job openings for the period from 1985 to 2009, we estimated the rates of attrition and the number of faculty on the basis of various assumptions as described in detail in Appendices C and D (pp. 294–300). Into these basic projections we then wove various other assumptions pertaining to enrollments, full-time faculty vs. part-time faculty, funds for research and public service, student-faculty ratios, and rates of attrition. In the end, we created nine different scenarios or models intended to describe the range of possibilities that lie ahead. In formulating our projections, however, we did not take into account the prospect of such overwhelming and destabilizing events as large-scale war, persistent deep depression, a technological revolution in instructional delivery systems, or such potentially pervasive public policy initiatives as national public service or dramatic revisions in immigration policies. With these qualifications, we turn to the nine projections which are presented in Table 10–1 and Figure 10–1 and are discussed in the remainder of this chapter.[1]

Projection I (the steady state model) assumes that the higher educational system will continue about as it was in 1980–81 with no change in total enrollment or rates of faculty attrition. This assumption is probably unrealistic because substantial changes in the number of persons in the youthful age group are certain. However, it should not be dogmatically rejected. As shown in a 1980 study, if the college attendance of adult learners 35 and over continued to increase at the actual rate of growth of the recent past, total enrollments in higher education (expressed in full-time-equivalents) would be approximately stable for the rest of the twentieth century[2] (H.R. Bowen, 1980a, pp. 26–27). It should be noted also that, as of 1984–85, the nation was five years into the decline in number of college-age persons but enrollments had nevertheless remained remarkably steady.

Projection II (traditional student cohort) is at the opposite end of the range of possibilities. It is based on the assumption that the number of faculty will vary with the population of ages 18 to 21, and that faculty attrition will be at the rate of 4 percent of the total faculty. This model implies a very low level of faculty appointments over the first three five-year periods. It is doubtful, however, that enrollment and faculty requirements would follow the fluctuations in number of persons 18 to 21 years of age. Young people of these ages supply only about a half of

1. We are indebted to several predecessors in the task of projecting the demand for faculty: Radner & Miller (1975); Cartter (1976); and W. G. Bowen (1981).

2. Each age cohort within this group was weighted according to actual attendance in higher education in 1980.

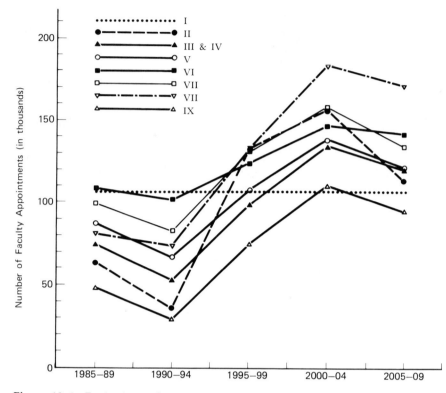

Figure 10–1. Projections of Number of Faculty Appointments, Full-Time-Equivalent 1980–81 to 2009–10, with Varying Assumptions

college enrollments. College students are drawn from a wide range of age groups. Thus, Projection II probably should be rejected.

Projection III (contemporary student cohort) seems to us more plausible than either I or II. It is based on the assumptions: (a) that college enrollments, and therefore the number of faculty members, will vary with the number of persons in the population of ages 16 to 34; and (b) that faculty attrition will be at the rate of 4 percent. These assumptions imply substantial numbers of faculty appointments throughout the next twenty-five years. The remaining models are variants of Projection III.

Projection IV (reduced part-time faculty) is like III except that it adds the assumption of a shift from part-time to full-time faculty, thus increasing the relative number of full-time faculty. This shift would not affect the total number of projected full-time-equivalents. It would, however, strengthen the faculties by reversing the current trend toward increasing employment of part-time people. It would bring the ratio of part-time to full-time faculty back to where it stood in 1970. With these assumptions, the number of part-time faculty would gradually decline from 212,000 in 1980 to 127,000 in 2010, thus creating about 28,000 additional full-time positions. This change would not and should not eliminate all part-timers. Many part-timers are capable people who make substantial contributions. Their role is to complement the full-time faculty, to facilitate prompt adjustments upward or downward in the overall size of faculty, and perhaps to effect economies in a temporary period of hard times.

Projection V (reduced student-faculty ratio) is like III except that it deals with the effect of lowering the ratio of students to faculty. The assumption is that the student-faculty ratio would be reduced by about 10 percent from 14.3 to 12.9 over the period 1985 to 1994 and held constant thereafter. This change would bring about a substantial increase in the number of faculty positions associated with a given enrollment. We think, on grounds of educational quality, that the professoriate should be enlarged relative to the student population—especially in view of the kinds of educational tasks that lie ahead. But we stop short of predicting that such expansion will occur.

Projection VI (increase in older adult learners) is also like Projection III except that it explores the effect of a possible increase in the percentage of older persons attending college. Specifically, it assumes that the percentage of all students in the age groups over 35 will increase from 7.7 percent in 1985 to 12 percent in 1995 and later years. This change would have the effect of raising enrollment and thus increasing the required number of faculty. We think this projection is not unrealistic in that the population beyond the usual college age is so huge that only small percentage increases in college attendance among this group would be necessary to offset a decline in number of younger students (H. R. Bowen, 1980a, pp. 26–27).

Projection VII (increased research and public service) is like Projection III except that it assumes steadily increasing support for research and public service over the years 1985 to 1994 and leveling off thereafter. The assumed increase by 1995 would be sufficient to increase employment of full-time faculty by 10 percent, or 46,000 persons.

Projection VIII (high faculty attrition) investigates the effect of an increase in the rate of attrition from 4 to 6 percent. We believe that the lower rate may be appropriate in the near future but that it will be too low as the faculties become older and retirements and deaths increase. Six percent may then turn out to be a reasonable, even conservative, projection. The assumed increase from 4 to 6 percent produces a substantial increase in the number of faculty appointments. Indeed, Projection VIII yields the largest number of faculty openings of any of the nine projections.

Projection IX (low faculty attrition) is like III except that it is based on the assumption of a level 3 percent (instead of a 4 percent) rate of attrition throughout the years from 1985 to 2009. This projection yields a smaller number of new appointments than any of the others. We believe that an attrition rate averaging as low as 3 percent is possible but unlikely.

Table 10–1. Projections of Number of Faculty Appointments, Full-Time-Equivalent, 1985–89 to 2005–09, with Varying Assumptions[1] (in thousands)

Projection I (Steady State)
Assumes: No change from the situation in 1985.

5-Year Periods[2]	Change in Number of Faculty[3]	Attrition[4]	Total Appointments[5]
1985–89	0	107	107
1990–94	0	107	107
1995–99	0	107	107
2000–04	0	107	107
2005–09	0	107	107
1985–2009	0	535	535

Projection II (Traditional Student Cohort)
Assumes: Number of faculty will vary with population of ages 18 to 21; rate of attrition at 4%.

5-Year Periods[2]	Change in Number of Faculty[3]	Attrition[4]	Total Appointments[5]
1985–89	−31	95	64
1990–94	−50	87	37*
1995–99	+49	87	136
2000–04	+60	98	158
2005–09	+9	105	114
1985–2009	+37	472	509

Projection III (Contemporary Student Cohort)
Assumes: Number of faculty will vary with population of ages 16 to 34; rate of attrition at 4%.

5-Year Periods[2]	Change in Number of Faculty[3]	Attrition[4]	Total Appointments[5]
1985–89	−28	102	74
1990–94	−42	95	53*
1995–99	+7	92	99
2000–04	+39	96	135
2005–09	+18	102	120
1985–2009	−6	487	481

Projection IV (Reduced Part-Time Faculty)
Assumes: Number of faculty will vary with population of ages 16 to 34; rate of attrition at 4%; reduction in ratio of part-time to full-time faculty.

5-Year Periods[2]	Change in Number of Faculty[3]	Attrition[4]	Total Appointments[5,6]
1985–89	−28	102	74
1990–94	−42	95	53*
1995–99	+7	92	99
2000–04	+39	96	135
2005–09	+18	102	120
1985–2009	−6	487	481

Projection V (Reduced Student-Faculty Ratio)
Assumes: Number of faculty will vary with population of ages 16 to 34; rate of attrition at 4%; reduction in student-faculty ratio.

5-Year Periods[2]	Change in Number of Faculty[3]	Attrition[4]	Total Appointments[5]
1985–89	−3	91	88
1990–94	−22	89	67*
1995–99	+8	90	98
2000–04	+43	96	139
2005–09	+20	101	121
1985–2009	+46	467	513

Projection VI (Increase in Older Adult Learners)
Assumes: Number of faculty will vary with population of ages 16 to 34 but with increase in number of students in age group over 35; rate of attrition at 4%.

5-Year Periods[2]	Change in Number of Faculty[3]	Attrition[4]	Total Appointments[5]
1985–89	+4	105	109
1990–94	−3	105	102*
1995–99	+17	107	124
2000–04	+35	112	147
2005–09	+24	118	142
1985–2009	+77	547	624

Projection VII (Increased Research and Public Service)

Assumes: Number of faculty will vary with population of ages 16 to 34; rate of attrition at 4%; increased financial support for research and public service.

5-Year Periods[2]	Change in Number of Faculty[3]	Attrition[4]	Total Appointments[5]
1985–89	– 5	104	99
1990–94	–19	102	83*
1995–99	+30	103	133
2000–04	+39	110	149
2005–09	+18	116	134
1985–2009	+63	535	598

Projection VIII (High Faculty Attrition)

Assumes: Number of faculty will vary with population of ages 16 to 34; rate of attrition increases from 4 to 6% over the period 1985 to 1999 and remains steady at 6% thereafter.

5-Year Periods[2]	Change in Number of Faculty[3]	Attrition[7]	Total Appointments[5]
1985–89	–28	110	82
1990–94	–42	116	74*
1995–99	+ 7	128	135
2000–04	+39	144	183
2005–09	+18	153	171
1985–2009	– 6	651	645

Projection IX (Low Faculty Attrition)

Assumes: Number of faculty will vary with population of ages 16 to 34; rate of attrition at 3%.

5-Year Periods[2]	Change in Number of Faculty[3]	Attrition[8]	Total Appointments[5]
1985–89	–28	77	49
1990–94	–42	71	29*
1995–99	+ 7	69	76
2000–04	+39	72	111
2005–09	+18	76	94
1985–2009	– 6	365	359

* Low point

[1] See accompanying text and Appendix D for definitions, data, and methods used in deriving this table.
[2] Academic years, i.e., 1985–86 to 1989–90.
[3] Computed from the number of faculty in full-time-equivalents as shown in Appendix D.
[4] Assumes annual attrition rate of 4% of average number of faculty during each five-year period.
[5] Sum of change in number of faculty plus attrition.
[6] Total appointments would be the same as in Projection III, but the number of full-time appointments would be increased and the number of part-time appointments reduced.
[7] Assumes annual attrition rate rising steadily from 4% to 6% from 1985 to 1999, and thereafter constant at 6%.
[8] Assumes level annual attrition rate of 3%.

In presenting these nine projections, we do not pretend to forecast faculty appointments precisely. Indeed, we are in full agreement with Radner and Miller (1975, p. 347) who wrote:

> Manpower forecasting is a notoriously inaccurate business. It is quite possible that all the current work on both the supply and the demand sides . . . will prove in due course to be wrong. For one thing, if the projections now being made and debated are taken seriously, actions will be taken that may invalidate the projections; and, indeed, the projections are partly for the purpose of encouraging the reexamination of policies.

Our purpose is to suggest that there is a wide range of plausible possibilities most of which do not require deep or sustained cuts in the overall size of the faculty or in the number of appointments. The estimated number of faculty appointments for the several projections are summarized as follows:

	Projection Model	Low Period 1990–94	Total Appointments 1985–2009
I	Steady State	107,000	535,000
II	Traditional Student Cohort	37,000	509,000
III	Contemporary Student Cohort	53,000	481,000
IV	Reduced Part-Time Faculty	53,000	481,000
V	Reduced Student-Faculty Ratio	67,000	513,000
VI	Increase in Older Adult Learners	102,000	624,000
VII	Increased Research and Public Service	83,000	598,000
VIII	High Faculty Attrition	74,000	645,000
IX	Low Faculty Attrition	29,000	359,000

The opportunity also exists to combine several of the changes and thus to intensify their effect. Imagine, for example, that the ratio of students to faculty could be lowered, the ratio of full-time to part-time faculty could be raised, and the funding of research and public service could grow—all simultaneously. In that event the effect on the size of the professoriate would be tremendous.

Whatever decline in faculty actually materializes, if any, will likely occur over a span of 10 years (from 1985 to 1994) rather than suddenly. The rate of change is likely to average no more than 1 to 1.5 percent a year over 10 years—much less than any likely rate of faculty attrition. Whatever retrenchment is necessary, therefore, may be far less traumatic than is usually assumed. Indeed, the problem facing higher education may not be primarily the impending decline in enrollment and faculty. Rather, the more critical problem may be the inadequacy of available public and private revenues. In the past fifteen years, the financial squeeze on higher education has been due partly to intermittent economic depression mixed with inflation, partly to deliberately imposed limitations on the taxing power of the state, and partly to the attempt to shift public spending from social purposes to national defense. To revitalize higher education, an extended period of moderately rising financial support is needed.

Disaggregation and Mobility

The projections of new faculty appointments presented in this chapter are aggregative. They refer to the combined academic market of the nation, not to subdivisions of that market, and they are based primarily on demographic data. We have not attempted to break the totals down into components such as geographic regions, types of institutions,[3] and academic disciplines. We are aware that projections classified in these ways would be interesting and valuable but we doubt their feasibility. For one thing, necessary data about past changes in these components are not available and we do not have the resources to gather them. More important, we doubt (partly on the basis of the well-known unreliability of long-run manpower projections) that forecasts of faculty appointments for subdivisions of the academic market would prove fruitful. Instead, we believe that it is pertinent to consider the *process* by which the academic market adjusts to shifts in demand by geographic regions, types of institutions, and academic disciplines.

This adjustment is achieved through the *mobility* of faculty. There are two types of mobility, short-run and long-run, and they usually overlap in time. Short-run mobility occurs when faculty members, *while staying within academe*, move from one region to another, or from one type of institution to another, or from one discipline to another. Long-run mobility occurs through the normal process of faculty turnover, that is, departure of faculty from academe for various reasons and entry of new faculty from outside academe.[4]

Mobility of any kind tends to be sluggish. People do not respond instantly to every change in demand. However, the adjustments required of faculty tend to be easiest and quickest in the case of regional shifts, almost as easy in the case of shifts in type of institution, but more difficult and slower in the case of shifts in academic disciplines. For example, if the nation's population and its college enrollments are declining in the "frost belt" and expanding in the "sun belt," newcomers will head for the sun belt, and many older academics will move there as well. Physical moving is very common among academics, and few spend their lives in a single location. From the geographic

---

3. For example, public-private, four-year–two-year, large-small, university-college, urban-rural.

4. Persons who enter from graduate study, or service as teaching assistants, or nonacademic employment within colleges and universities are defined as outside academe.

standpoint the academic labor market is national in scope and reasonably homogeneous. Similarly, when adjustment to differential trends in
type of institutions is required, for example, during the rapid growth of
regional state universities, newcomers to the profession will flock to the
new openings, and many established people will move from traditional
types of institutions to newer types.

The adjustment problem is considerably more difficult—but not
overwhelmingly so—in the case of changes in demand among academic
disciplines. Over the past 15 years, we have witnessed substantial
changes in demand for rapidly growing fields such as business administration, engineering, and computer science and shrinking fields such as
philosophy, anthropology, and history. In earlier decades differences of
the same magnitude occurred but were masked by the overall expansion
of enrollment which allowed almost all fields to grow—though at different rates. But with the leveling off of total enrollment, the changes in
demand forced some fields to contract while others grew and the differences became glaringly evident. Short-run mobility of faculty persons
among academic fields is sometimes possible as when mathematicians
become computer scientists or English literature professors turn to
remedial writing. It is sometimes possible for historians to become professors of business, or physicists to become engineers. Many colleges
have deliberately sponsored such switches, some have instituted
retraining programs in order to display their faculties more efficiently
and to avoid layoffs, and some faculty union contracts mandate retraining opportunities for faculty about to be dismissed as redundant. But
switches among distantly related disciplines are difficult. It is hard for
anthropologists to become accountants, or art historians to take up
mechanical engineering. In general the adjustment among academic
disciplines is a long-run process involving the withdrawal from academe of persons in fields with surpluses and the entry of other persons
into fields with shortages. The withdrawals may take the form of normal
turnover through retirement, illness, death, and voluntary departures
for a host of reasons. They may also happen through the release of non-
tenured people and enforced resignations of tenured people. The surprising fact is that through normal turnover tolerable adjustments have
been achieved over surprisingly short periods of time and generally
without abrogation of tenure. The adjustments are never perfect. There
are always pockets of redundant faculty in some fields, and shortages in
others, but as of 1985 few disciplines are so lacking in personnel that
legitimate students must be turned away, and few are so burdened with
redundant personnel as to precipitate a crisis. The secret of this success
is that changes in demand usually occur slowly. Only a small fraction of
the faculty corps in declining fields must leave each year, and only a
small number of additions are needed in growing fields each year to
keep pace with the rate of changes in demand. It must be recognized,
however, that the adjustment process is more difficult in a period of

steady or declining total enrollments than in a period of growth. Growth is the great lubricator of the adjustment process.

As we consider the state of higher education in 1985, we expect changes in the demand for faculty by region, by type of institution, and by discipline and changes also in the allocation of faculty to accommodate these changes. We expect that through the decisions of academic administrators and faculty members a lagged response to these changes will maintain a reasonable though never perfect equilibrium. We expect also, if there is a period of growth—as expected after 1995—the pain of further adjustments will be alleviated. We also guess that the relative growth of those academic disciplines which have expanded in recent years will taper off or decline in the next decade. We refer here to medicine, law, business, dentistry, engineering, and computer science, all of which seem headed for saturation (see Chapter 7, pp. 125–26). But we do not believe that it is in our power to estimate the distribution of the faculty by region, by type of institution, or by discipline over the next twenty-five years.

Conclusions

The main conclusion from these assumptions and calculations is that it is not beyond possibility for the higher educational system to have openings for new faculty sufficient to allow a steady infusion of "new blood." Undoubtedly there will be periods of slack opportunities for recruitment, especially in the years 1990–94. The magnitude of the decline in enrollment over the next decade is likely to be of the order of 15 percent, about comparable to the decline in enrollment in the early 1950s after the departure of the GIs. But there need be no extended period in the next twenty-five years when there will have to be net layoffs of faculty. Even during the trough of 1990–94 at least 30,000 new faculty (full-time-equivalents), and conceivably as many as 100,000, could be drawn from outside academe into the higher educational system. And on the average from 1985 to 2009, as many as 70,000 to 130,000 new full-time faculty might be appointed during each five-year period (Table 10–1). These figures, it should be noted, refer only to new (full-time-equivalent) persons recruited to the higher educational system; they do not include faculty who move from one college or university to another. Though these figures fall far short of the roughly 200,000 appointments from outside that were made each five-year period from 1965 to 1979, they are not inconsiderable.[5] They are certainly large enough to provide a meaningful infusion of young and energetic people.

5. Given 4 percent attrition and a faculty in 1984–85 of about 520,000 full-time equivalents, only 22,000 replacements a year, or 110,000 each quinquennium, would be needed to maintain a "steady state."

Over the next twenty-five years (1985 to 2009), the number of appointments will probably equal two-thirds or more of the entire faculty as of 1985.[6] Thus, the caliber of the faculty as a whole will be largely determined by the appointments made between now and 2010. It will behoove academic leaders to fill such positions as they can offer with vigorous recruitment and intense concern for quality. And for the very reason that financial rewards may be wanting, institutions should do their best to offer intrinsic rewards that will be attractive to brilliant and imaginative teachers and scholars. Faculty personnel policies are in great need of attention and stabilization to the end that colleges and universities can offer solid career commitments. In this connection, the recent and acrimonious debates over academic tenure have been extremely counter-productive.

Over recent years, there has been much discussion within higher education of the need for faculty turnover. It has taken the form of contriving ways to shed older faculty, to dismiss younger aspirants for tenure, and to encourage the departure of less competent persons in mid-career. It has become received doctrine that the greater the faculty turnover, and therefore the larger the number of positions to be filled with "new blood," the better. We agree that turnover is healthy provided it is managed tactfully in ways that do not convey to the rank and file of the faculty a sense of insecurity and inadequate appreciation and provided it does not alienate young people from the profession. But just as low rates of turnover may be unhealthy, so high rates of turnover may be even more unhealthy. They are likely to occur at times when the general labor market for professional people is booming and the academic labor market is in the doldrums, a condition to which higher education is likely to be vulnerable at times over the next twenty-five years. The problem will then not be getting rid of the deadwood or of the young ones who did not make tenure, but retaining the core of the faculty. At such times, institutional loyalty, attachment to the profession, continuity, collegiality, and security of the faculties will be invaluable assets. These assets are not cultivated by blatant talk of getting rid of the deadwood, eliminating tenure, advocating legislation to single out faculty for early retirement, etc. Moreover, efforts to increase turnover suffer from the implied assumption that new appointees will all be more competent than those departing, an assumption that is never wholly valid and that may be less so in the years ahead when academe may be relatively unattractive to prospective candidates.

---

6. A study of prospective retirements in the public universities of Michigan yielded the conclusion that by the year 2000 these institutions "will have retired (at age 70) between 20 and 30% of all current faculty. If deaths and resignations are included or if incentives for early retirement are implemented, the retirement rate approaches 40% of current faculty" (Prince, 1984, p. 2).

If conditions in the future should produce a "flight" from the academic profession, there would be thousands of vacant positions. Many of them would probably be filled from the latent supply of educated people who are employed outside academe, are unemployed, or underemployed. On the average they would probably be filled with persons of lesser qualifications than those being replaced. Thus, the ultimate effect of the exodus might be a deterioration in faculty qualifications, not a decline in numbers. It might involve an increase in part-time persons, temporary appointees, junior faculty, etc. The conditions giving rise to an exodus would also be a warning signal to those contemplating entry into the academic profession. Such a warning would be likely to influence the more capable persons. These are the ones who are most sensitive to deteriorating conditions, and also the ones with the most career options. Thus, in two ways, the capability of persons in the profession would be adversely affected. We do not expect a mass exodus because we believe the nation would step in to save the profession before conditions became that extreme. However, given continuing deterioration of compensation and working conditions in higher education and prosperity in the outside economy, a considerable defection would be possible with disastrous consequences for the quality of the academic profession. This defection would include not only those currently employed but also those contemplating academic careers.

On the question of "new blood," it should be noted that we have considered in this chapter only the flow of people into faculty positions from other pursuits. We have not considered the quite considerable shifting of faculty members from one college or university to another. This process has important value. It often enables faculty members to be promoted when they are faced with no available openings in their present situations; it sometimes enables them to be placed in positions closer to their interests; it also enables them to gain refreshment and stimulation from new colleagues and new experiences; and it helps to combat declining morale. Often the learning curve and the morale of faculty members rises when they move from one academic environment to another. The game of musical chairs is an important feature of academe and will surely continue to be played during the years ahead.[7] Indeed, if inter-institutional moves are counted, it is likely that the turnover rate will be at least double the 3 to 6 percent rates we have assumed. Through this process, persons who may be "old hat" at one institutions can become new blood at another (*Times* [London] *Higher Education Supplement*, Sept. 24, 1982). Another possibility for maintain-

7. An innovative organization known as "National Faculty Exchange" is serving as catalyst and broker in arranging temporary exchanges of faculty between or among institutions. The purpose is to gain some of the advantages of mobility at a time when new positions are scarce. The address of the organization is Indiana University-Purdue University at Fort Wayne, 2101 Coliseum Boulevard East, Fort Wayne, Indiana 46805.

ing or increasing the vitality of academic people is through the many kinds of programs with the generic name of "faculty development."

In the end, the prospective financial situation for higher education cannot be ignored. To attract the caliber of people needed, colleges and universities must have the resources to keep faculty compensation and working conditions reasonably competitive. In their recruiting they must have the means to compete with business, law, medicine, engineering, and other fields which have recently drawn a disproportionate share of talent. In Chapter 13 of this book, we consider the matter of faculty recruitment at some length.

CHAPTER ELEVEN

The Flow of Exceptional Talent to Academe

As earlier chapters demonstrate, the American professoriate is a complex mosaic, and its present condition has both positive and negative aspects. Although the data do not suggest a profession presently in grave trouble, there have been some disturbing signs: declining compensation and a deteriorating work environment. Since these factors, among others, shape faculty morale, it is not surprising that morale is shaky on many campuses. Whatever frustrations are voiced, however, voluntary exodus from faculty ranks has not risen appreciably above normal levels.

That so remarkably few have abandoned academic careers makes it very hard to assess the extent to which faculty members have become psychologically disengaged, retaining their jobs but reducing their commitment to their academic responsibilities. Nonetheless, one cannot reasonably maintain that the academic profession is in serious jeopardy as of 1985. On the contrary, the evidence indicates that the full-time professoriate still appreciates the manifest advantages of academic life.

These considerations dictate an exploration of a question central to understanding the condition of the professoriate: What do we know about the quality of persons now being attracted to academic careers? The answer to this question is crucial since not only the future condition of the professoriate but also the national welfare depend upon the talent of those young people now being recruited to academic careers and of those who may be attracted to such careers in the foreseeable future.

Given that the compensation picture is growing steadily less attractive and that even exceptionally able young scholars face formidable obstacles in the struggle to gain tenure—if, indeed, they can get a full-time appointment to begin with—it has become commonplace to mourn "the loss of a generation of scholars." Perhaps every academic personally knows of well-trained—sometimes superbly prepared and strongly committed—young men and women who, in another age, would be launched on promising academic careers but who, in today's labor market, have been squeezed into marginal academic roles or have simply given up and pursued other options. No one knows how many such cases there are, but surely they number in the tens of thousands. And it is virtually impossible to estimate the number of highly talented young people who may be potentially outstanding scholars but who never even contemplate academic careers. Thus, whatever the strengths of the full-time professoriate now in place, we cannot afford to ignore the question of whether academe will be able to compete for exceptional talent during the trying decades ahead.

In this chapter, we attempt to "x-ray" the metaphorical pipeline leading to academic careers. The following sections examine the career plans of undergraduates; the quality of graduate students contemplating academic careers; the quality of junior faculty, especially those employed in top-rated departments; and the career choices, over time, of highly able populations.

Decisions Prior to Graduate School

The size and quality of tomorrow's professoriate will be determined by decisions that are being made today. Those people who ultimately seek to enter academic careers take the first steps in that direction well in advance of their entry to graduate school. This section looks at the career plans of two groups: college freshmen and graduating seniors.

Career Proclivities of College Freshmen. For nearly two decades, the Cooperative Institutional Research Program (CIRP)—conducted by the American Council on Education in conjunction with UCLA's Higher Education Research Institute—has annually surveyed the entering freshman classes of some 300–400 diverse higher educational institutions.[1] One item on the freshman survey asks respondents to indicate their probable "career occupation," choosing from among more than forty options that include both "college teacher" and "scientific researcher."

Since 1966, when the CIRP was launched, the proportion of freshmen planning careers in college teaching or scientific research has plummeted (Table 11–1). Thus, in 1966, 1.8 percent of the freshmen said they planned to become college teachers; by 1979, the proportion had

---

1. The findings from the CIRP survey are published in an "annual report of national normative data on the characteristics of students entering college as first-time, full-time freshmen" (A. W. Astin et al., 1966–84, 1983 report, p. 1). The survey is extensive in scope and numbers; the normative data for Fall 1983 were based on the responses of approximately 190,000 freshmen at 350 institutions (p. 109).

Table 11–1. Percentage of College Freshmen Indicating an Interest in Teaching-Related Careers, Selected Years, 1966–84

Year	College Teacher			Scientific Researcher			Elementary or Secondary Teacher		
	Men	Women	All	Men	Women	All	Men	Women	All
1966	2.1%	1.5%	1.8%	4.0%	1.9%	3.5%	11.3%	34.1%	21.7%
1968	1.3	0.9	1.1	3.8	1.7	2.9	12.7	37.5	23.5
1970	1.2	0.9	1.0	3.5	1.6	2.6	9.6	31.0	19.3
1972	0.7	0.6	0.6	3.1	1.5	2.3	5.7	19.5	12.1
1974	0.7	0.8	0.7	2.7	1.4	2.1	3.8	11.9	7.7
1976	0.4	0.3	0.4	3.0	1.7	2.4	3.8	12.5	8.0
1978	0.3	0.3	0.3	2.7	1.7	2.2	2.4	9.8	6.2
1980	0.2	0.2	0.2	2.2	1.3	1.7	2.4	9.3	6.0
1981	0.2	0.2	0.2	2.0	1.2	1.6	2.0	8.8	5.5
1982	0.2	0.2	0.2	1.8	1.2	1.5	1.8	7.4	4.7
1983	0.3	0.2	0.2	1.8	1.2	1.5	2.1	7.8	5.1
1984	0.3	0.2	0.3	1.9	1.2	1.5	2.2	8.6	5.5

SOURCE: A. W. Astin *et al.*, 1966–84; 1984 report, pp. 18, 34, 50.

dropped to a scarcely visible 0.2 percent, where it remained until 1984, when it edged up to 0.3 percent. (Interest in elementary and secondary teaching also declined precipitously.) Similarly, 3.5 percent of the 1966 freshmen—but only 1.5 percent of the 1982, 1983, and 1984 freshmen— said they planned to become scientific researchers. As Table 11–2 shows, the career choices of college teaching and scientific research have declined in popularity among freshmen of all interests (as indicated by their intended major fields), although a slight rebound since 1979 is evident among students in some fields.

Table 11–3 shows trends in the popularity of several competing professions. In stark contrast to their declining interest in teaching careers, freshmen have become more interested in business, engineering, medical, and law careers, although the growth in the popularity of these fields is attributable chiefly to changes in the aspirations of women. These survey results may well disturb proponents of affirmative action for college and university faculties: Even as freshman women were becoming steadily more attracted to careers in law (from 0.7 percent in 1966 to 3.7 percent in 1984), engineering (from 0.2 percent to 2.9 percent), business (from 3.3 percent to 14.4 percent), and medicine/dentistry (from 1.7 percent to 4.0 percent), their interest in careers in college teaching (from 1.5 percent to 0.2 percent) and scientific research (from 1.9 to 1.2 percent) receded sharply. Obviously, expanding career opportunities for women have eroded their early interest in professorial, as distinguished from professional, careers.

These shifts in the career preferences of freshmen are consonant with their diminishing interest in a liberal arts education—a portent that bodes ill for future graduate school enrollments. The past ten years

Table 11–2. Percentage of Four-Year College Freshmen Indicating an Interest in Faculty or Research Careers, by Intended Undergraduate Major, 1974, 1979, and 1983

| | College Teacher | | | Scientific Researcher | | | Percent Change, 1974–83 | |
| | | | | | | | College Teacher | Scientific Researcher |
	1974	1979	1983	1974	1979	1983		
Traditional Fields								
Arts & Humanities	2.7%	1.0%	1.4%	0.1%	0.1%	0.1%	−48	0
Biological Sciences	0.4	0.2	0.3	15.5	21.6	17.3	−25	12
Physical Science/								
Mathematics	2.2	1.2	1.6	26.3	22.5	20.0	−27	−24
Social Sciences	0.8	0.4	0.5	0.8	0.4	0.6	−38	−50
Professional and Applied Fields								
Business	0.1	0.0	0.0	0.0	0.0	0.0	−100	0
Computer Science	0.2	0.1	0.1	1.3	0.2	0.3	−50	−77
Education	1.9	1.1	0.8	0.1	0.1	0.0	−58	−100
Engineering	0.1	0.0	0.1	1.3	0.8	0.7	0	−46
All Students	0.9	0.3	0.3	2.8	1.9	1.9	−67	−32

SOURCE: Kenneth C. Green, "Entering Freshmen and the Migration of Talent Across Careers and Academic Disciplines, 1973–1984," 1985, unpublished.

have witnessed a dramatic decline, among freshmen entering four-year colleges and universities, in the proportions planning to major in an arts and sciences field (Table 11–4). The shift toward pre-professional/applied fields is particularly pronounced among abler students (that is, those who reported A averages in high school).

One finding from a Harvard University survey of entering freshmen is similarly unsettling. During the summer of 1975, entering freshmen were surveyed to ascertain their "vocational goals." At that time, 8.9 percent indicated that college teaching was their intended career; seven years later, in 1982, the proportion had fallen to 1.7 percent.[2]

The outlook is not entirely bleak; several signs appear to be hopeful. First, freshmen interest in obtaining a doctorate has remained fairly constant, as is evidenced by the responses of successive cohorts to a CIRP freshman survey item asking, "What is the highest academic degree that you intend to obtain?" Even though the proportion of men saying they aspired to a Ph.D. or Ed.D. fell from 12.7 percent in 1966 to 9.6 percent in 1984, the corresponding proportion among women rose from 5.2 percent to 8.7 percent, and the total for both sexes declined only slightly from 9.8 percent to 9.2 percent (Table 11–5). Thus, the potential doctoral pool for the early 1990s seems to have shrunk much less than

2. The precise wording of the question, and the method used to code responses, changed between 1975 and 1982, but there is no evidence to suggest that those changes had the effect of depressing the 1982 figure.

Table 11–3. Percentage of College Freshmen Citing Selected Professions as Their Anticipated Career Choices, Selected Years, 1966–84

	Businessman[1]			Engineer		
Year	*Men*	*Women*	*All*	*Men*	*Women*	*All*
1966	18.5%	3.3	11.6	16.3%	.2%	8.9%
1968	17.5	3.3	11.3	14.6	.2	8.3
1970	17.4	4.2	11.4	13.3	.4	7.5
1972	15.4	4.8	10.5	9.6	.3	5.3
1974	17.6	8.5	13.2	8.5	.8	4.7
1976	20.9	11.6	16.4	13.7	1.1	7.8
1978	16.6	9.6	13.0	16.5	2.2	9.1
1980	16.3	11.7	13.9	19.1	2.9	10.7
1981	16.4	12.1	14.1	19.5	2.9	10.9
1982	15.8	12.8	14.4	20.6	3.6	12.0
1983	16.4	12.8	14.5	18.8	3.3	10.8
1984	18.3	14.4	16.2	18.5	2.9	10.4

	Physician or Dentist			Lawyer or Judge		
Year	*Men*	*Women*	*All*	*Men*	*Women*	*All*
1966	7.4%	1.7%	4.8%	6.7%	.7%	3.9%
1968	5.6	1.3	3.7	5.5	.6	3.4
1970	5.9	1.5	3.9	6.2	1.0	3.8
1972	7.9	2.8	5.5	7.1	2.0	4.5
1974	6.9	3.5	5.3	5.3	2.3	3.9
1976	6.3	3.3	4.8	5.5	3.0	4.3
1978	5.7	3.4	4.5	5.3	3.4	4.3
1980	5.3	3.6	4.4	4.8	3.5	4.1
1981	4.9	3.4	4.1	4.5	3.4	3.9
1982	5.0	3.6	4.3	4.7	3.9	4.3
1983	5.4	4.0	4.6	4.2	3.6	3.9
1984	5.2	4.0	4.6	4.4	3.7	4.1

[1] Data may be skewed since the item changed from the single option of "Businessman" to the options "Business Management" and "Business Other," in 1973 only. Data were aggregated to include the following options: "Business Executive," "Business Owner or Proprietor," and "Business Sales Person or Buyer," from 1977 to 1984.

SOURCE: A. W. Astin *et al.*, 1966–84; 1984 report, pp. 18, 34, 50.

early interest in college teaching careers, a development consistent with the fact that an increasing proportion of doctorate-holders are finding employment outside academe.

Second, when responses are classified by type of institution, it becomes clear that, as one might expect, freshmen attending "highly selective" or "very highly selective" colleges and universities were much more likely than were those enrolled at less selective institutions to anticipate careers in college teaching.[3]

3. The proportion of freshmen anticipating college teaching careers ranged from roughly 0.2–0.3 percent at institutions low in selectivity, to about 0.5 percent at those of medium selectivity, to about 0.5 percent at highly selective institutions, and 1.0 percent at the few colleges classified as very highly selective. These data are interpolated from A. W. Astin et al., 1966–84; 1983 report, pp. 64, 80.

Table 11–4. Percentage Change Among All Four-Year College
Freshmen and Among "A" Students Planning to Major in Selected
Fields, 1983, Compared with 1974

Traditional Fields	All 4-Year Students	"A" Students Only
Arts and Humanities	− 29%	− 36%
Biological Sciences	− 43	− 44
Physical Science/		
Mathematics	− 35	− 41
Social Sciences	− 33	− 34
Professional and		
Applied Fields		
Business	+ 59	+ 87
Computer Science	+ 512	+ 420
Education	− 44	− 46
Engineering	+ 49	+ 70

SOURCE: Kenneth C. Green, "Entering Freshmen and the Migration of Talent Across
Careers and Academic Disciplines, 1973–84," 1985, unpublished.

Third, the reduced interest in academic careers among entering college students may be more closely related to their perception of the market than to their appraisal of the attractiveness of an academic career.

Fourth, we suspect that the correlation between the career aspirations of freshmen and their eventual career choices is slight. Courses
in differential equations and organic chemistry have served to divert
innumerable undergraduates from their initial career dreams. One may
safely assume that these data on freshman career preference are not
reliable predictors of the future occupations of successive freshman
classes. Nonetheless, so precipitous a decline in the popularity of college teaching as a freshman career choice cannot be dismissed as incon-
sequential.

Career Preferences of Graduating Seniors. According to an important
study of "highest achievers" among college seniors (Goldberg and
Koenigsknecht, 1985), the proportion of superior students choosing to
attend graduate school (either master's or doctoral programs) has
declined sharply in recent years: from 59 percent in 1966 to 35 percent in
1976. Focusing on the proportion of superior students choosing to
matriculate as doctoral candidates, the following figures show a drop
from 44 percent to 21 percent. Meanwhile, the proportion opting for
professional schools increased from 37 percent to 53 percent over the
same ten-year period.[4]

4. The findings of the study, conducted by the Graduate Research Project of the Consortium on Financing Higher Education, are to be released in 1985. The "highest achievers"
were essentially the top 3–5 percent of graduating seniors (classes of 1956, 1966, 1976, and
1981 attending a small number of highly regarded (mostly private) research universities.

Graduating Class of "Highest Achievers"	No Advanced Study	Graduate School		Professional School
		Master's and Doctoral	Doctoral	
1956	13%	54%	29%	33%
1966	4	59	44	37
1976	12	35	21	53

Using a different criterion to compare 1976 and 1981 "highest-achieving" graduates, the study suggests that the sharp decline in graduate school enrollments may have been arrested (up from 27 percent in 1976 to 28 percent in 1981) while the gravitation toward professional schools may have been reversed (down from 46 percent to 37 percent).[5]

According to the study, the highest-achieving baccalaureate recipients were more likely to attend graduate school than were those in a control group of randomly selected graduates from the same classes and universities. By 1976, however, the proportion of highest achievers who went on to professional school (53 percent) very nearly equaled the proportion of control-group counterparts who made that choice (54 percent). That the non-academic professions are becoming more attractive to the very best students is further evidenced by preliminary data showing that an even greater proportion of top students (37 percent) than of "average" students in the control group (29 percent) are enrolling in professional schools.

Table 11–6 shows trends in the responses of Harvard seniors to an

Table 11–5. Percentage of College Freshmen Planning to Get a Ph.D. or Ed.D., Selected Years, 1966–84

Year	Men	Women	All
1966	13.7%	5.2%	9.8%
1968	14.0	6.1	10.6
1970	12.3	6.5	9.7
1972	10.6	6.8	8.9
1974	10.0	6.9	8.5
1976	9.8	7.6	8.7
1978	9.8	8.1	8.9
1980	8.5	7.3	7.9
1981	8.7	7.2	7.9
1982	8.8	7.6	8.2
1983	9.0	8.0	8.5
1984	9.6	8.7	9.2

SOURCE: A. W. Astin, *et al.*, 1966–84; 1984 report, pp. 20, 36, 52.

5. These percentages, which take into account only those students who enrolled in post-baccalaureate study within two years after college graduation, are not comparable to the percentages yielded by the survey of the classes of 1956, '66, and '76.

Table 11–6. Percentage of Harvard University Seniors
Identifying College Teaching as Their "Eventual
Vocation," 1973–82

Year	Men	Women	All
1973	7.8%	12.3%	8.8%
1974	8.9	10.9	9.3
1975	10.0	9.8	10.0
1976	9.6	11.0	10.0
1977	8.1	8.3	8.1
1978	8.5	5.7	7.6
1979	6.1	5.3	5.9
1980	8.4	5.8	7.5
1981	6.9	6.4	6.7
1982	6.4	6.5	6.4

SOURCE: Martha P. Leape (Director, Office of Career Services and Off-Campus Learning, Faculty of Arts and Sciences, Harvard University). *Report on the Class of 1982*, Harvard College, unpublished, 1982, p. 17.

item on the Annual Senior Survey asking them to identify their "eventual vocation." Again, a gradual decline in preference for college teaching is manifest. But it is also worth noting that, as in the Goldberg-Koenigsknecht study (1985), the proportion is considerably higher among graduating seniors with the strongest undergraduate records: 17.4 percent of Harvard's Phi Beta Kappas and 22.5 percent of those who graduated summa cum laude designated college teaching as their eventual vocation (Leape, 1982, pp. 8, 11). Similarly, a study of seniors graduating from eight very selective colleges and universities in 1982 found that those who planned to attend graduate school tended to have better undergraduate grades than those who did not intend to enroll in graduate school (Consortium on Financing Higher Education, 1983, Appendix C). On the other hand, interest in attending graduate school was inversely related to concern about earning a handsome living: In the choice between graduate and professional school,

> the importance of a career's high income potential is the principal predictive value; students who consider high income to be unimportant . . . are three times as likely to enroll in graduate school (over professional school) as students who consider high income to be essential or very important. [p. 54]

Taken together, these data on the career proclivities of college freshmen and the post-baccalaureate preferences of graduating seniors suggest that college teaching has lost much of its appeal for young men and women. To be sure, many able students who, at the outset of college, have only the faintest notion of what an academic career entails, may be encouraged along the way—often by mentoring professors—to consider the genuine rewards of the life of the mind. Nonetheless, at this juncture, it is clear that the academic profession attracts only a very small number of undergraduates. This declining popularity is not, by itself, alarming, given the tight academic job market now and in the foreseeable future. Leaving aside the qualitative question, we can conclude that the supply, however weak, will suffice for a while to meet the demand. But however understandable, the near collapse of interest among college students in an academic career is distressing.

The Quality of Graduate Students Contemplating Academic Careers

The quality of graduate students currently contemplating academic careers is a subject of both concern and conjecture. Whether this quality has diminished in recent years is difficult to determine; the available evidence suggests that it has.

The Declining Popularity of Arts and Sciences. Over the past decade and a half, the proportion of highly able college graduates choosing professional school over graduate work in the arts and sciences has increased substantially.[6] Assessing the period from 1971–72 to 1980–81, one study concluded:

> In direct contrast to the period of relative stability in the high demand for higher education as a whole and for graduate professional programs in law, business, and medicine, graduate school demand displayed a marked decline. Overall, both the number of graduate applicants and first-time graduate enrollments have declined. [Consortium on Financing Higher Education, 1982, p. ix]

The experience varies, of course, from area to area and, within areas, from discipline to discipline. Clearly, the humanities and social studies have been hardest hit. Declining numbers of applicants provide one measure of decreasing popularity. Measures of quality are more elusive. Nevertheless, the difficulty of attracting top talent to graduate study in the social sciences and humanities is implied in a recent report on graduate admissions records from 1972 to 1980 at twenty very highly regarded research universities (Garet & Butler-Nalin, 1982, pp. 18–19 and Tables 11 and 17 of Appendix C). In the social sciences, the average

---

6. "The best evidence we have about 'most qualified' undergraduates tells us that they are choosing to pursue graduate education in far fewer numbers than a decade ago (and, conversely, more are choosing professional education)" (Butler-Nalin, Sanderson, & Redman, 1983, p. 1.7). The authors further assert that "the present environment in which graduate education takes place presents real barriers, real dissuasion, and a climate of general malaise for many potential graduate students, graduate students currently enrolled, faculty members and universities" (p. i).

number of applications at those institutions declined by 28 percent during that period, while the proportion of applicants accepted for admission rose from 25 percent in 1972 to 33 percent in 1980. In the humanities, applications fell by 47 percent, but acceptances increased from 38 percent in 1972 to 43 percent in 1980.[7] In other words, these outstanding universities were offering admissions to a larger proportion of their shrinking applicant pools.

Standardized Test Scores. For years, scores on the Graduate Record Examinations (GRE), administered by the Educational Testing Service, have been declining in most fields. Although these scores by no means dispose of the quality issue, they do suggest that today's graduate students in most fields are, on the whole, less well qualified, in terms of what the tests measure, than were their counterparts in previous years.[8] At the same time, many GRE test-takers do not enroll in graduate school. Furthermore, as already noted, the proportion of graduate students, and of eventual doctorate-recipients, who ultimately enter academic careers, has dropped steadily over the last decade or so, but neither the reasons for the decline in test scores nor its implications are clear. While the GRE data are not sufficiently focused for our more narrow purpose of gauging the quality of future academicians, they provide little basis for optimism.

In an ambitious effort to determine whether graduate schools have been experiencing a "brain drain"—and professional schools a "brain gain"—in the competition for able undergraduates, Hartnett (1985) compared the Scholastic Aptitude Test (SAT) scores of those who eventually earned arts and sciences doctorates with the scores of those who earned first professional degrees in three areas: business, law, and medicine.[9] The combined SAT-Verbal and SAT-Math scores, shown in Table 11–7, indicate that arts and sciences doctorate-holders had consistently scored higher on the SAT than had their professional school counterparts. The data further suggest that, on the whole, doctoral pro-

---

7. Data were analyzed for six fields in the social sciences, five in the sciences, two in engineering, and nine in the humanities. Seventeen of the universities were independent, and three were public. (See also Consortium on Financing Higher Education, 1982.)

8. As noted in a recent report, "student performance on 11 of 15 major Subject Area Tests of the Graduate Record Examination declined between 1964 and 1982. The sharpest declines occurred in subjects requiring high verbal skills" (Study Group on the Conditions of Excellence in American Higher Education, 1984, p. 9).

9. The SAT (or the American College Testing (ACT) program test) is required for admission to many colleges and universities and is typically taken by high school students in their senior year. In the Hartnett study, SAT scores for about 10,000 graduates and ACT scores (converted to SAT scores) for another 1500 were used. Eight arts and sciences disciplines were included. For both A & S and professional schools, test scores were analyzed in the main only for graduates of programs (a) that, cumulatively, generated 50 percent of the advanced degrees, or (b) that were highly ranked on one of several national reputational studies.

Table 11–7. Comparison of the Scholastic Aptitude Test Scores of Arts and Sciences Doctorate-Recipients and Professional School Graduates, Selected Years, 1966–81

	Arts and Sciences[a]			*Professional Schools*[b]			*Differences*		
	SAT–V[c]	*SAT–M*[d]	*Comb.*	*SAT–V*	*SAT–M*	*Comb.*	*SAT–V*	*SAT–M*	*Comb.*
1966[e]	577	626	1203	538	585	1123	+39	+41	+80
1971	595	638	1233	568	615	1183	+27	+23	+50
1976	610	649	1259	576	621	1197	+34	+28	+62
1981	599	645	1244	570	619	1189	+29	+26	+55

[a] Includes doctorate-holders in eight fields: chemistry, electrical engineering, English, history, mathematics, philosophy, physics, and psychology.
[b] Includes business (M.B.A.), law (LL.B./J.D.), and medicine (M.D.).
[c] SAT-Verbal
[d] SAT-Mathematical
[e] Year of receiving Ph.D. or professional degree
SOURCE: Hartnett, 1985.

grams have not lost significant ground to professional schools in the competition for talented students.[10]

Campus Site Visits. In our campus visits, we sought information on a wide range of topics, including perceptions of various segments of the pipeline: the quality of graduate students planning academic careers, the merits of recent applicants for faculty positions, and the quality of those actually appointed to tenure-track faculty positions.[11] At the fifteen campuses in our sample that had substantial doctoral training programs, we attempted to learn what faculty and academic administrators thought about the quality of those of their graduate students who were contemplating an academic career. More specifically, we asked: "How would you rate the quality of current graduate students in your (department/school/institution) who are seeking faculty positions compared with graduate students here 5 to 10 years ago?"

The responses reveal a mix of enthusiasm and apprehension. There was widespread concern that highly talented individuals were not being drawn to academic careers, especially in the humanities and social sciences. The dean of the undergraduate liberal arts school of a prominent research university put it this way:

10. A strong cautionary note is in order here. The data comparing doctorate-recipients and professional school graduates are not strictly comparable, since the time it takes to earn a graduate or professional degree varies considerably, both by field and by individual student, and members of the same cohort of graduates may have taken the SAT in different years. Further, the most recent cohort for which data are available (those who received the Ph.D. in 1981) graduated from college around 1974, if not earlier (that is, before the alleged "brain drain" from the graduate to the professional schools had gained momentum).

11. The process by which we conducted 532 interviews at 38 campuses is described in Chapter 8 and in Appendix A.

> Ten years ago, a good second-rank department could get excellent gradu-
> ate students, but not today. Of those who graduate in the top quarter of
> their undergraduate classes, only a few seek A & S graduate work.

And the history chairperson at a good research university (the department is well regarded, but not in the top echelon) gave voice to the
anxiety of many humanities scholars when he said:

> In the last couple of searches, there were some top people, but the pools
> are getting thinner. We were snatching the last of the very good ones.

At the same time, many respondents expressed admiration for
those brave souls who persisted in the face of a daunting labor market.
This sentiment was captured by the graduate dean of arts and sciences
at a research university:

> We still have that small core of *good* students who still aspire to faculty
> careers despite the problems. They're the hard-core maniacs.

A similar point was made by a senior professor of English at another
research university:

> When we had loads of National Defense Education Act students, they
> weren't all that committed. Now the only Ph.D. students who are here are
> anxious to make an academic career.

Responses at the six doctorate-granting universities reveal some disappointment, occasionally approaching dismay, in the quality of doctoral
students.[12] Of the academic administrators, department chairs, and
senior faculty to whom we put our question, most (45 percent) saw
signs of improvement, but a sizable minority (22 percent) thought that
the quality of the graduate students contemplating academic careers
had declined over the past five or ten years.

Survey of Leading Graduate Departments. In another approach to the
quality issue, we sought to learn what "the very best" academic departments have been experiencing with respect both to their own graduate
students and to the tenure-track junior faculty they have hired in recent
years. Our reasoning was as follows: If a serious problem exists at those
institutions accustomed to attracting as graduate students the ablest
undergraduates and to hiring as neophyte professors the best young
talent available in the market, then there is indeed cause for considerable alarm. Data on the perceptions of those at the leading academic
departments help to throw light on the question.

---

12. These six institutions did not meet the stricter Carnegie criteria for research university designation.

Using unpublished standardized scores from the National Research Council's *Assessment of Quality-Related Characteristics of Research-Doctorate Programs in the U.S.* (Jones, Lindzey, and Coggeshall, 1982), we first identified the highest-ranking 15 percent of departments in each of the thirty-two academic fields encompassed by the NRC study. A questionnaire was addressed individually to each of the chairpersons of the 404 departments so identified.[13] We received 316 usable responses, for a rate of 78.2 percent. See Appendix B for details.[14]

The questionnaire asked chairpersons to compare "current advanced graduate students" with their counterparts "during the 1968–72 period" in terms of quality and number. As Table 11–8 shows, the chairpersons in all fields reported that current graduate students were of higher quality than the students doing graduate work in the 1968–72 period—a very encouraging sign.[15] One humanities chairperson observed:

> This year's applicants are the best I have seen for many years. My impression is that the tight job market has helped; those who go on to graduate studies are deeply committed.

According to another humanist:

> Although applications have declined substantially, the pool of highly qualified applicants has not, and those who eventually come are at least as good overall as students over the past twenty years.

Still another respondent said:

> The best students now enrolled are as good as the best ten years ago; there are just fewer of them in proportion to the total enrolled group.

In all, almost three times as many chairs of leading graduate departments believed that the quality of their advanced graduate students had improved (49 percent) as held that a decline in quality had occurred (17

13. The number of departments included in the NRC assessment ranged from 150 in psychology and 145 in chemistry to 35 in linguistics and in classics; accordingly, the number of departments we surveyed—the top 15 percent—ranged from 22 each in psychology and chemistry to 5 each in linguistics and classics. The authors acknowledge the valuable assistance of Eileen Heveron, Claremont Graduate School, in connection with this survey.

14. On the questionnaire, we defined "advanced graduate students" as "full- or part-time doctoral students having been, or about to become, advanced to candidacy." We noted: "By 'quality' is meant excellence in those characteristics which your department traditionally has regarded as the best predictors of a productive academic career."

15. These findings are consistent with the results obtained by the Higher Education Panel, American Council on Education, in two surveys that asked academic administrators for their opinions on changes (1) in the quality of graduate school applicants over a recent five-year period (1976–77 to 1981–82) and (2) in the quality of 1981–82 doctorate-recipients compared with 1976–77 recipients. The surveys did not address comparative numbers of applicants or degree recipients (Andersen, 1984; Atelsek, 1984).

Table 11–8. Change in Quality of Advanced Graduate Students, 1983–84, Compared with 1968–72, as Reported by Leading Graduate Departments

	Much Better			About the Same		Much Worse		No Response[a]	Mean
	7	6	5	4	3	2	1		
Social/Behavioral Sciences	4%	18%	21%	27%	20%	6%	0%	5%	4.4
Mathematics/ Physical Sciences	4	16	31	35	13	1	0	0	4.6
Humanities	10	12	25	32	13	0	3	5	4.6
Engineering	14	23	26	26	11	0	0	0	5.0
Biological Sciences	10	18	22	37	9	3	0	2	4.7
All:	8	17	25	32	14	3	1	3	4.6

[a] Includes "no opinion" or no usable answer.

SOURCE: Survey of Chairpersons of Leading Graduate Departments, 1984, by the authors.

percent). Thirty-two percent reported no change. Within areas, the disciplines reporting the most and least satisfaction with the quality of their advanced graduate students are shown in Table 11–9. Wide variation in the quality of advanced graduate students is evident among the disciplines comprising each of the five broad areas. Most striking is the spread of the mean scores for the humanities disciplines, ranging from classics (which ranks near the top) to five fields clustered near the bottom. These findings suggest that generalizations concerning the broad areas are risky.

The survey also revealed that the number of doctoral students in these leading departments was, in the aggregate, approximately the same in 1984 as during the 1968–72 period. By disciplinary area, the number in the humanities and the social and behavioral sciences had decreased noticeably, the number in engineering had increased, and the number in mathematics, the physical sciences, and the biological sciences had changed very little. Table 11–10 displays the data for these five areas.

To get a glimpse of the near future, we asked the chairpersons to anticipate "the quality of those who will become doctoral students in this department during the period 1987–90" relative to the quality of their currently enrolled doctoral students. Their predictions were generally optimistic, as Table 11–11 shows.

The four tables (11–8—11–11) presented in this section give an encouraging picture. The quality of doctoral students, as reported by respondents in these elite academic departments, has held up well. Signs of slippage are few. In those areas where a drop in quality might have been expected—the humanities and the social/behavioral

Table 11–9. Change in Quality of Advanced Graduate Students Enrolled, by Discipline, 1983–84, Compared with 1968–72, as Reported by Leading Graduate Departments

Rank	Discipline	Area	Mean Score
1	Chemical Engineering	Engineering	5.6
2	Civil Engineering	Engineering	5.3
3	Classics	Humanities	5.2
4T	Computer Science	Physical Science	5.2
4T	Geography	Social/Behavioral Science	5.2
6	Zoology	Biological Science	5.2
7	Anthropology	Social/Behavioral Science	5.1
8	Microbiology	Biological Science	5.1
9	Biochemistry	Biological Science	4.9
10T	English	Humanities	4.9
10T	Geosciences	Physical Science	4.9
12	Philosophy	Humanities	4.9
13	French	Humanities	4.8
14	Electrical Engineering	Engineering	4.8
15	Physiology	Biological Science	4.7
16	Chemistry	Physical Science	4.7
17T	Economics	Social/Behavioral Science	4.6
17T	Physics	Physical Science	4.6

MEAN for 32 disciplines = 4.62

Rank	Discipline	Area	Mean Score
19	Art History	Humanities	4.6
20	Mechanical Engineering	Engineering	4.4
21T	Botany	Biological Science	4.4
21T	Sociology	Social/Behavioral Science	4.4
23	Cellular/Molecular Biology	Biological Science	4.4
24	Statistics	Physical Science	4.3
25	Psychology	Social/Behavioral Science	4.2
26T	German	Humanities	4.2
26T	Mathematics	Physical Science	4.2
28	History	Social/Behavioral Science	3.9
29	Political Science	Social/Behavioral Science	3.9
30	Spanish	Humanities	3.8
31	Music	Humanities	3.7
32	Linguistics	Humanities	3.0

NOTES:

(a) On the seven-point scale, "7" was "Much Better," "4" was "About the Same," and "1" was "Much Worse."

(b) Standard deviation for disciplines is 1.25. Only Linguistics fell more than one standard deviation below the mean. No discipline is more than one s.d. above the mean.

(c) For additional details, including response rates by discipline, see Appendix B.

SOURCE: Survey of Chairpersons of Leading Graduate Departments, 1984, by the authors.

sciences—the *number* of doctoral students has been sharply curtailed. Apparently, this decrease has enabled the departments not merely to maintain but, on average, to improve the quality of their graduate students.

Table 11–10. Change in Number of Advanced Graduate Students, 1983–84, Compared with 1968–72, as Reported by Leading Graduate Departments

	Much Larger			About the Same		Much Smaller		No Response[a]	Mean
	7	6	5	4	3	2	1		
Social/Behavioral Sciences	1%	5%	7%	27%	22%	20%	15%	4%	3.1
Mathematics/ Physical Sciences	10	14	14	30	17	10	6	0	4.2
Humanities	2	8	5	17	20	27	20	2	2.9
Engineering	6	23	34	20	6	11	0	0	4.7
Biological Sciences	10	6	19	29	24	7	3	2	4.2
All:	6	10	14	25	19	15	10	2	3.7

[a] Includes "no opinion" or no usable answer.

SOURCE: Survey of Chairpersons of Leading Graduate Departments, 1984, by the authors.

Post-doctorates: Way Station to the Professoriate. A post-doctoral appointment constitutes one section of the pipeline linking graduate study and faculty employment. While many "postdocs" go on to non-academic careers or, in some cases, to non-teaching research careers within academe, the post-doctoral experience frequently precedes a faculty appointment, especially in the natural sciences. Thus, the conclusions reached by a recent comprehensive study of post-doctoral recipients should cause concern about the quality of those entering upon academic careers in some fields. According to Zumeta (1984), the study shows

> for the first time clear evidence of substantial declines over the 1970s in the quality of the postdoctoral cadre (as measured by the stature of their Ph.D. departments) in the biomedical disciplines, psychology, the social sciences and the humanities. . . . The negative trend in the quality of the "input" to postdoctoral training, where it continues, must be cause for concern. [p. xviii]

In sum, the campus site visits, spanning a diverse group of doctorate-producing institutions, suggested an unevenness in the caliber of those seeking academic careers. But our survey of leading graduate departments indicated that quality is being maintained at the very top. Moreover, even though higher education will make relatively few new hires in the proximate future (see Chapter 10), these top departments exude confidence that they can avoid any erosion in quality. They seem to be saying that they will be able to remain highly selective, while lesser departments will have to scramble to recruit worthwhile graduate students.

Table 11–11. Predicted Change in Quality of Advanced Graduate Students To Be Enrolled, 1987–90, Compared with 1983–84, as Reported by Leading Graduate Departments

	Much Larger		*About the Same*		*Much Smaller*			
	7	6	5	4	3	2	1	No Response[a]	Mean
Social/Behavioral Sciences	4%	15%	34%	31%	9%	2%	0%	6%	4.7
Mathematics/ Physical Sciences	4	13	31	44	1	0	0	7	4.7
Humanities	0	13	15	53	8	2	0	8	4.3
Engineering	3	17	34	34	9	0	0	3	4.7
Biological Sciences	0	9	44	34	7	0	0	6	4.6
All:	2	13	32	39	7	1	0	6	4.6

[a] Includes "no opinion" or no usable answer.

SOURCE: Survey of Chairpersons of Leading Graduate Departments, 1984, by the authors.

As they become more familiar with the harsh realities—as well as the satisfactions—of academic life, how many reasonably capable graduate students will look elsewhere for a fulfilling career? We cannot begin to answer this question, but we understand the tensions pervading academic life that impelled a university arts and sciences dean to shrug and say:

> I'm glad I chose the academic profession, but I seriously counsel my own Ph.D. students about doing something else besides teaching because it just isn't as much fun to be a faculty member any more.[16]

The Quality of Junior Faculty

Whatever may be said about the quality of doctoral students and of the horde of applicants for the relatively few assistant professorship vacancies that exist, the most important consideration—maybe the only surpassingly important one from an institutional point of view—is the quality of persons actually hired into junior tenure-track positions. Both the site visits and the survey of leading graduate departments produced evidence on this crucial issue.

Campus Site Visits. Each of the thirty-eight campuses we visited had been engaged in some hiring. While financial constraints had obliged some of them to hire more part-time than full-time faculty, all the campuses were making at least some tenure-track appointments.

16. See also Butler-Nalin, Sanderson, & Redman, 1983, pp. 4.23–4.24.

Perhaps no other inquiry evoked such unequivocal enthusiasm from respondents. Almost uniformly, campuses reported that their recent junior faculty appointments were outstanding: "dynamite classroom teachers" (a community college math chair), "superb" (president of a regional university), "spectacular" (a biologist at an eminent university). One dean at a doctorate-granting university beamed, "We are able to hire top Yale Ph.D.s in the humanities!" The chairperson of a life sciences department at another university put it this way:

> The present application pool is the best I have ever reviewed. If one has one faculty vacancy, one scarcely makes a mistake in appointments. My concern is directed to those applicants who are merely superior—they may well not obtain a quality appointment.

Sometimes the enthusiasm approached euphoria, as when the chair of history at a well-regarded eastern university described his experience after the department had waited for years for the authorization to appoint an assistant professor:

> When we recently filled a position we had 240 applicants. Ten applicants had books either already published by reputable presses or about to come out. Our new hires are better than anyone already on the faculty.

This elation applied across the board; it was evident at the research universities, long accustomed to hiring fine junior faculty, but also at the community colleges, the less selective liberal arts colleges, and the comprehensive colleges and universities, where good but not spectacular new hires have long been the norm. Today, everyone is feasting— except, of course, the new appointees, who commonly find themselves at institutions about which they had known little or nothing and which they would have studiously avoided in a more favorable academic market. But this topic is discussed elsewhere.

The only variation in the responses was whether the new tenure-track faculty members were merely better or were much better than those who had been hired five or ten years ago. On many campuses, so few new hires had been made (except in a few high-turnover fields) that the responses must be understood for what they are—enthusiasm for a mere handful of new entrants into the profession. The political science chairperson at a top-ranked university expressed the matter succinctly: "It may be that they're better because they're fewer." Nevertheless, the consistency of enthusiastic responses was very impressive. Of 285 interviewees, only twenty (8 percent) replied that recently appointed junior faculty were "worse" or "much worse" than those hired five or ten years before, and quite a few of these comments applied to high-demand fields. By contrast, 195 (68 percent) opined that the recent junior appointments were "better" or "much better." About a quarter reported that there had been no change.

The vice president for academic affairs at a sprawling urban university, which in the past has had difficulty landing top talent, sounded as though a Brinks truck had just backed up to his garage. He smiled and said, "If an institution doesn't get the cream of the crop, it's doing something wrong." "The job market," added the liberal arts dean at a first-rank university long accustomed to having little competition, "is a golden opportunity."

Although the great majority of respondents acknowledged that recent new hires had stronger formal academic credentials, their assessments were occasionally tempered by reservations and doubts. For instance, administrators sometimes discerned a lack of institutional loyalty, of commitment. The president of a small but growing master's-granting campus brooded:

> There's no question but that, in terms of degrees and publications, our newly hired junior faculty are better. But if you ask whether they are better teachers, more devoted teachers, no, they're not necessarily better.

And an associate chief academic affairs officer at a struggling public institution added:

> They're better in terms of paper qualifications, but personally, I'm not impressed. You have to go beyond degrees.

Perhaps new faculty members share the characteristics frequently ascribed to today's undergraduates: they are less vital and idealistic, more preoccupied with the material and the mundane. And surely the numbing threat of not attaining tenure forces many of them to confine themselves to their immediate academic tasks. According to a humanist at a major university:

> They have a different orientation. Now, they're older and more sober; then, they were more avant-garde.

Survey of Leading Graduate Departments. In our departmental survey, we asked chairpersons to compare "the quality of junior faculty here during the 1968–72 period" with "the quality of current junior faculty" in their department. The results were impressive, tending to dispel any notion that recent initiates into the guild, at the most respected academic centers, are weaker than their predecessors. As Table 11–12 shows, the proportion of academic departments reporting a drop in quality is small, only about 7 percent, and almost all of those said that the drop was modest. To be sure, the pattern varies among the five disciplinary areas: Predictably, the social/behavioral sciences and the

Table 11–12. Change in Quality of Junior Tenure-Track Faculty, 1983–84, Compared with 1968–72, as Reported by Leading Graduate Departments

	Much Better		About the Same			Much Worse		No Response[a]	Mean
	7	6	5	4	3	2	1		
Social/Behavioral Sciences	6%	24%	23%	29%	7%	5%	0%	5%	4.8
Mathematics/ Physical Sciences	3	31	27	31	3	0	0	6	5.0
Humanities	8	17	18	37	8	3	0	8	4.7
Engineering	6	31	34	26	3	0	0	0	5.1
Biological Sciences	9	15	25	40	0	2	1	10	4.9
All:	6	23	25	34	4	2	*	6	4.9

[a] Includes "no opinion" or no usable answer.

* less than 0.5 percent.

SOURCE: Survey of Chairpersons of Leading Graduate Departments, 1984, by the authors.

humanities were more likely than the other areas to report some diminution in quality. Nonetheless, these numbers contrast dramatically with the 54 percent of departments reporting an improvement and the 34 percent reporting no change. In all, about seven-eighths of the nation's most prestigious academic departments said they had made improvements—or, in any event, had not lost ground—in the caliber of new hires.

Some surprises cropped up among the disciplines at both ends of the distribution (Table 11–13). These apparent anomalies include the strength of geography and classics, with respect both to graduate students (Table 11–9) and new faculty; the high quality of new English professors; the curious phenomenon of civil engineering, whose graduate students rank very near the top (Table 11–9) but whose faculty appointments rank near the bottom; and the relatively poor showing of economics in terms of improved faculty quality. In the humanities, the relatively good showing of English is probably in large part a function of very few new hires; the same is true, though to a lesser extent, of classics. Ranking at the bottom were new hires in philosophy. In short, in fully 30 of the 32 disciplines, department chairs told us that the quality of their assistant professors had improved over the past decade and a half.

We also asked the chairs of leading graduate departments whether the current number of tenure-track junior faculty was larger or smaller than during the 1968–72 period. As Table 11–14 shows, the number of such faculty had declined in all disciplinary areas except engineering, with the decrease in numbers being particularly marked in the humani-

Table 11–13. Change in Quality of Tenure-Track Junior Faculty, by Discipline, 1983–84, Compared with 1968–72, as Reported by Leading Graduate Departments

Rank	Discipline	Area	Mean Score
1	Geography	Social/Behavioral Science	6.0
2	Microbiology	Biological Science	5.7
3	Geosciences	Physical Science	5.4
4	Classics	Humanities	5.4
5	English	Humanities	5.3
6	Electrical Engineering	Engineering	5.2
7	German	Humanities	5.2
8	Anthropology	Social/Behavioral Science	5.1
9	Mechanical Engineering	Engineering	5.1
10	Chemistry	Physical Science	5.1
11T	Botany	Biological Science	5.0
11T	Mathematics	Physical Science	5.0
11T	Zoology	Biological Science	5.0
14	Chemical Engineering	Engineering	4.9
MEAN for 32 disciplines = 4.85			
15T	Music	Humanities	4.8
15T	Statistics	Physical Science	4.8
17	History	Social/Behavioral Science	4.8
18	Art History	Humanities	4.8
19	Physics	Physical Science	4.8
20	Sociology	Social/Behavioral Science	4.7
21	Physiology	Biological Science	4.7
22	Computer Science	Physical Science	4.7
23	Psychology	Social/Behavioral Science	4.6
24	Cellular/Molecular Biology	Biological Science	4.6
25	Biochemistry	Biological Science	4.4
26	Spanish	Humanities	4.3
27	Political Science	Social/Behavioral Science	4.2
28	French	Humanities	4.2
29	Economics	Social/Behavioral Science	4.2
30	Civil Engineering	Engineering	4.1
31	Linguistics	Humanities	4.0
32	Philosophy	Humanities	3.3

NOTES:
(a) On the seven-point scale, "7" was "Much Better", "4" was "About the Same," and "1" was "Much Worse."
(b) Standard deviation for disciplines is 1.14. Only Geography is more than one standard deviation above the mean and only Philosophy is more than one standard deviation below the mean.
(c) For additional details, including response rates by discipline, see Appendix B.

SOURCE: Survey of Chairpersons of Leading Graduate Departments, 1984, by the authors.

ties and the social/behavioral sciences. In all, 44 percent of the leading departments said that they now have fewer junior tenure-track faculty members, 20 percent reported increases, and 30 percent reported no change. These figures parallel the information concerning the number of advanced graduate students. Clearly, institutions, with fewer billets

Table 11–14. Change in Number of Junior Tenure-Track Faculty, 1983–84, Compared with 1968–72, as Reported by Leading Graduate Departments

	Much Larger			About the Same		Much Smaller		No	
	7	6	5	4	3	2	1	Response[a]	Mean
Social/Behavioral Sciences	1%	4%	6%	29%	20%	26%	9%	6%	3.2
Mathematics/ Physical Sciences	4	6	14	30	20	10	11	6	3.6
Humanities	0	2	13	27	13	13	23	8	3.0
Engineering	9	14	14	40	9	14	0	0	4.3
Biological Sciences	3	10	12	31	12	16	9	9	3.7
All:	3	6	11	30	16	17	11	6	3.5

[a] Includes "no opinion" or no usable answer.

SOURCE: Survey of Chairpersons of Leading Graduate Departments, 1984, by the authors.

to fill, have a built-in opportunity to be more selective. The big numerical gainers and losers are shown in Appendix B.

Given that the proportion of all full-time faculty who are tenured has increased considerably in recent decades (see Chapter 3), we wanted to know whether the leading graduate departments had been similarly affected; that is, whether the numbers of non-tenured assistant professors at these leading departments were too few, arguably, to replenish their respective disciplines. To find out whether there was still "room at the top," we asked about tenure rates. As Table 11–15 indicates, 42 percent of the leading graduate departments reported that at least 80 percent of their faculty members were tenured, and 65 percent reported that seven out of ten of their full-time faculty members had tenure. Table 11–15 also demonstrates that this tenuring-in level has risen significantly over the past decade.

As for the proximate future, these department chairpersons took a somewhat sanguine view of their ability to expand the number of full-time tenure-track junior faculty. Table 11–16 shows that, in the aggregate, they foresaw a slight rebound in the number of such faculty between now and 1987–90. In only one area, engineering, was marked expansion anticipated. Departments in the other four areas tended to feel they could do no more than preserve the status quo (that is, maintain the current number of junior faculty). When one considers the ground lost between 1968–72 and 1984 (the year of the survey), however, then the idea of merely hanging on must be rather reassuring. The prevailing mood was captured by the chair of a romance language department at Berkeley when he wrote: "The trough has bottomed out—mild and guarded optimism is in order."

Table 11–15. Proportion of Leading Graduate Departments
Reporting Tenure Rates at Various Levels, Cumulative Percentages,
1968–72 and 1984

Percentage of Faculty on Tenure	1968–72	1983–84
At least 50%	82.3%	88.3%
At least 60%	67.1	82.6
At least 70%	42.1	64.9
At least 80%	17.4	42.4
At least 90%	4.1	15.2
100%	1.6	2.5

SOURCE: Survey of Chairpersons of Leading Graduate Departments, 1984, by the authors.

Taken together, these assessments and projections point to the likelihood of relatively few vacancies in the top-ranked departments in most disciplines, at least until 1990 and possibly into the mid-1990s. Many respondents in these academic centers of excellence indicated that they felt themselves to occupy "privileged" positions within American higher education, given their departments' elite ranking within their respective disciplines. Although some said they had problems attracting and retaining both high-quality graduate students and high-quality faculty, quite a few surmised that their institutions were among the fortunate few.

In sum, in this very strong buyer's market, many institutions are

Table 11–16. Predicted Change in Quality of Anticipated Junior Tenure-Track Faculty To Be Employed, 1987–90, Compared with 1983–84, as Reported by Leading Graduate Departments

	Much Better		About the Same			Much Worse		No Response[a]	Mean
	7	6	5	4	3	2	1		
Social/Behavioral Sciences	2%	6%	23%	43%	13%	6%	2%	5%	4.1
Mathematics/ Physical Sciences	3	4	23	47	16	1	1	6	4.2
Humanities	2	0	15	58	10	10	0	5	3.9
Engineering	3	29	14	43	11	0	0	0	4.7
Biological Sciences	2	6	18	46	10	9	2	10	4.0
All:	2	7	19	47	12	6	1	6	4.1

[a] Includes "no opinion" or no usable answer.

SOURCE: Survey of Chairpersons of Leading Graduate Departments, 1984, by the authors.

convinced that they have never done better in recruiting new faculty. Their enthusiasm is understandable. For the immediate future, these provosts, deans, and department chairs need not fret about the thinness of the applicant pools from which their prize young faculty are drawn. But the currently submerged threat will surface in the not-too-distant future: As faculty vacancies increase, most of academe will be confronted with some harsh realities. For now, however, the faculty hiring process simply piles one exciting success upon another.

Career Choices of Highly Talented Populations

Further evidence bearing on the ability of the professoriate to attract highly able persons to academic careers comes from two separate investigations which were conducted for this study and which examined the career choices, over many years, of American Rhodes Scholars and Phi Beta Kappa members.[17]

American Rhodes Scholars. The career choices of Americans designated as Rhodes Scholars, from 1904 (when Americans were first included in the program) to 1977, were coded and plotted. Ordinarily, 32 Americans are awarded Rhodes scholarships each year; over the seven-decade period examined, some 2,120 were named, and sufficient career-related data existed for 1,984 of them to permit classifying their careers.[18]

As Table 11–17 shows, interest among Rhodes Scholars in academic careers began to erode around 1960. In the five-year periods between 1904 and 1959, the proportion of these Scholars entering professorial careers ranged from roughly 33 percent to 49 percent. (These percentages include those identified as higher education administrators, almost all of whom had had faculty careers.) In each half-decade, these percentages were greater, often substantially greater, than those for the next most popular career undertaken by the Scholars. In the post-World-War-II period from 1946 to 1959, the proportions choosing academic careers held steady at around 48–49 percent. Then the appeal of

17. Very few longitudinal data exist on the career choices of highly able segments of the population. For assistance in these investigations, the authors acknowledge the valuable help of Barbara Koolmees Light, Whittier College.

18. Our access to these data was facilitated by the American Secretary of the Rhodes Scholarship Trust, David Alexander, president of Pomona College. Careers were coded on the basis of biographical entries in *A Register of Rhodes Scholars, 1903–1981.* In some instances, these entries were updated, using information received by the office of the American Secretary subsequent to the publication, in 1981, of the most recent comprehensive *Register.* We chose 1977 as the cut-off date; scholars designated in that year would not normally complete their two-year scholarships until 1979, and many entered graduate or professional school after that, thereby deferring entry into an occupation. Thus, career information on scholars named after 1977 was not sufficient to permit appropriate trend analysis.

Table 11-17. Percentage of American Rhodes Scholars in Selected Careers, by Year of Selection, Aggregated by Five-Year Periods, 1904–1977*

Selected Careers	1904–1909	1910–1914	1915–1919	1920–1924	1925–1929	1930–1934	1935–1939	1946–1949	1950–1954	1955–1959	1960–1964	1965–1969	1970–1974	1975–1977
Higher Education														
Faculty	32.5	29.3	33.0	42.2	28.8	35.5	39.1	44.3	43.6	42.8	35.5	21.8	23.1	18.2
Administration	1.8	3.7	1.7	3.3	4.5	2.0	2.6	3.3	5.8	5.2	3.3	0.7	0.7	0.0
Total	34.3	33.0	34.7	45.5	33.3	37.5	41.7	47.6	49.4	48.0	38.8	22.5	23.8	18.2
Law	22.9	25.6	20.8	22.3	19.9	15.2	19.2	12.3	14.7	14.9	23.7	32.0	34.0	25.8
Business	15.8	16.9	12.1	10.4	19.2	17.8	11.3	13.9	7.1	7.1	11.8	13.6	8.8	13.6
Medicine	2.6	3.0	5.2	4.5	1.9	2.0	4.6	0.8	1.3	3.9	3.9	4.8	7.5	10.6

*NOTE: The first period, 1904–09, spans six years; the eighth period, 1946–49, four years; and the most recent period, 1975–77, three.

SOURCE: Study of American Rhodes Scholars' Careers, 1984, by the authors.

an academic career began to wane. During the decade of the 1960s the proportion dropped to around 31 percent, and during the most recent period (1970–77) to 20 percent. Indeed, for 1975–77 the proportion had slipped to 18.2 percent.

Over the period 1945–59, almost twice as many Rhodes Scholars chose academic careers as opted for a career in the competing fields of law, medicine, and business. But for the period 1965–79 the situation has been reversed: 2.3 times as many Scholars have taken up careers in one of the three competing fields as have chosen an academic career.

Phi Beta Kappa Members. Phi Beta Kappa members constitute a much larger pool of talent than Rhodes Scholars. Since 1967 some 13,000–14,000 undergraduates (about 12 percent of them juniors and the remainder, except for a few graduate students, seniors) have been elected to membership each year from (as of 1984) 234 campus-based chapters.[19] These campuses account for roughly a tenth of our baccalaureate-granting institutions, generally the most respected. The thousands of students elected to Phi Beta Kappa each year—from among the million or so Americans who annually earn baccalaureates—constitute a sizable proportion of the top 1 or 2 percent in academic talent. Thus, Phi Beta Kappa members are an intellectual elite, and their distribution across occupations, besides being intrinsically interesting, reveals something about the shifting values of American society.

A two-page questionnaire was distributed to 2,513 randomly selected Phi Beta Kappa members who had been elected to membership between 1945 and 1983. About 1400 usable responses were received, constituting a 56 percent return rate.[20]

For a full quarter-century from 1945 to 1969, these highly able undergraduates gravitated toward academic careers at a fairly consistent rate, slightly more than one in five (Table 11–18). The choice of college educator peaked at 24.2 percent in the 1960–64 period. During the 1970s, however, the proportion dropped sharply to around 8 percent. Looking again at the relative drawing power of the competition, one finds that, in the years from 1945 to 1969, for each Phi Beta Kappa member choosing an academic career, 1.2 members opted for a career in business, law, or medicine. But during the 1970s, 4.7 times as many Phi Beta Kappa members chose one of those three occupations as entered upon an academic career.

Trends and Implications. The trends, depicted in Figure 11–1, are obvious: The interest on the part of these two elite populations in aca-

19. Founded at the College of William and Mary as a secret social and literary fraternity in 1776, Phi Beta Kappa had spread to three other campuses in 1800 and to a total of eleven campuses by the middle of the nineteenth century. The element of secrecy eventually dropped away, and it became an honorary society based on scholarship. Expansion accelerated after World War II, as higher education institutions and enrollments mushroomed.

20. We are grateful for the assistance of the national Phi Beta Kappa office, particularly Kenneth M. Greene, Secretary of the United Chapters of Phi Beta Kappa.

Table 11–18. Percentage of Phi Beta Kappa Members in Selected Careers, by Year of Election, Aggregated by Five-Year Periods, 1945–83.

Selected Careers	1945–49	50–54	55–59	60–64	65–69	70–74	75–79	80–83
Education								
Educator (college)	20.6	22.6	19.4	24.2	21.7	8.6	7.2	2.4
Educator (elementary)	1.3	4.3	2.0	1.7	1.2	3.9	2.9	4.1
Educator (secondary)	5.2	8.6	6.1	8.3	4.2	3.5	1.0	4.9
Educator (school principal, supervisor)	1.3	1.1	1.0	.8	—	.4	.7	—
Subtotal, K–12[a]	7.8	14.0	9.1	10.8	6.4	7.4	4.3	9.7
Subtotal, all educators[b]	28.4	36.6	28.5	35.0	28.1	16.0	11.5	12.1
Other Selected Careers								
Accountant/actuary	1.3	1.1	—	3.3	2.4	2.4	1.0	4.1
Business[c]	7.8	14.0	9.1	9.2	9.0	12.3	9.5	13.1
Clergy[d]	3.9	2.2	3.1	2.5	—	2.4	0.7	0.8
Computer Programmer/Analyst	—	1.1	3.1	3.3	3.0	3.5	4.2	5.7
Engineer	2.6	2.2	2.0	.8	.6	.8	2.0	3.3
Homemaker	14.3	11.8	5.1	6.7	8.4	4.7	4.2	3.3
Journalist/writer	5.2	1.1	3.1	1.7	1.2	2.7	3.6	4.1
Lawyer/judge	5.2	6.5	5.1	5.8	9.6	15.3	20.2	6.5
Physician/dentist	3.9	10.0	15.3	10.0	7.8	16.1	18.9	5.7
Social/welfare worker	1.3	1.1	2.0	.8	3.6	.8	1.6	1.6
Other occupation	26.1	12.3	23.6	22.4	27.1	22.6	22.3	40.0
	100.0	100.0	100.0	100.0	100.0	100.0	100.0	100.0

NOTE: Percentages for 1980–83 and to a lesser extent, for 1975–79, show sharp declines in some areas, e.g., "educator (college)," "lawyer/judge," and "medical," undoubtedly because some aspirants to such careers are still in graduate/professional school.
[a] Includes only the three K–12 entries (elementary, secondary, and school principal/supervisor).
[b] Includes all four "educator" entries.
[c] Includes business clerical, business executive, business proprietor, and business sales.
[d] Includes minister/priest and clergy miscellaneous.
SOURCE: Survey of Phi Beta Kappa Recipients, 1984, by the authors.

demic careers has declined rapidly. One might argue that, however interesting, these trends are of little consequence. Rhodes Scholars are so few in number that whatever they choose to do has little impact on any sector of society. As for the much larger number of Phi Beta Kappa members, they still constitute only a tiny group compared with the huge college and university teaching force—some 700,000 strong, counting full- and part-time faculty members. Moreover, brilliant scholars are not needed to staff many of our college classrooms, where skills other than scholarship may be more useful in teaching the many underprepared students. Clearly, the necessity of staffing college classrooms adequately goes far beyond the hypothetical prospect of attracting *every* Rhodes and Phi Beta Kappa recipient to an academic career. So where does that leave us?

For one thing, the loss of the very best talent is a real loss to higher

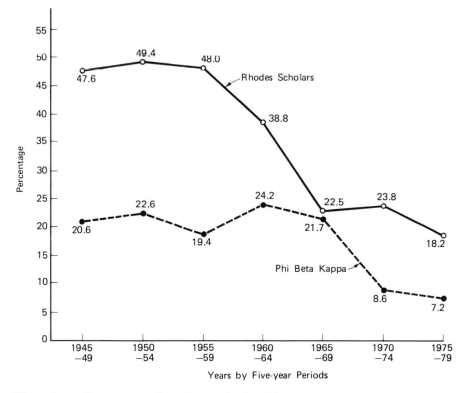

Figure 11–1. Percentage of American Rhodes Scholars and Phi Beta Kappa Recipients Choosing Academic Careers, by Five-Year Periods, 1945–1979

education. We do not disparage the need for highly able men and women in other careers. But higher education suffers to the extent that it cannot compete successfully for a substantial share of the most academically gifted. Beyond that philosophical observation lies a more practical one. The populations we studied—the Rhodes Scholars and the Phi Beta Kappa recipients, along with the "highest achievers" referred to earlier—have many options, far more than most college graduates. No doubt the current glut of academics in most fields weighs heavily in the career choice of any reasonably acute college graduate. Nonetheless, it is sad to see how far the academic profession has slipped in its ability to recruit the very ablest graduates. The symbolism may be even more disturbing than the actual loss of talent. The academy, it seems, grows less and less attractive as a house of intellect, a nurturing and stimulating environment for the gifted and creative. We have not yet felt the pinch; there are too few jobs. As noted earlier, almost all campuses are pleased, even delighted, with their recent recruits. But the future, when demand begins to turn sharply upward, may present a very different picture.

Concluding Thoughts

As noted at the outset, the flow of exceptional talent into the professor-iate is a many-faceted phenomenon. The last decade-and-a-half has not been kind to a professoriate which had grown in stature, in influence, and in real earnings during the preceding boom years. The recent past has dealt out many frustrations familiar to all associated with academe.

It is heartening to find that the quality of persons currently entering the profession is excellent. Young men and women with fine academic training are taking jobs at institutions unaccustomed to such good for-tune. Almost everywhere, the new recruits are busily raising academic standards and expectations (and anxieties, too) at their new academic abodes. No wonder that the hiring institutions are rejoicing, even if the attitudes of the newly minted professors themselves are often much more guarded.

Nor is it just the junior faculty who seem so promising. When we look at the quantity and quality of advanced graduate students who seem to be contemplating academic careers, we find many encouraging signs. There are, to be sure, pockets of qualitative and quantitative vulnerability. And, as we have noted, the persistence of quality may be largely a function of fewer excellent graduate students seeking still fewer professorial vacancies. Nonetheless, the available evidence is compelling: No pervasive crisis in the supply of exceptional talent is immediately upon us. But the emphasis here is on "immediately."

This leads to the tougher and more elusive question: What lies ahead? It is not difficult to find evidence that general interest in an aca-demic career, including interest among the exceptionally talented, is waning. No single source of data, considered by itself, is conclusive: the career preferences of freshmen, the choices of highly able graduating seniors, the evidence afforded by standardized test scores, the loss of interest in college teaching among several elite populations. But taken together, these data are disturbing, if not alarming. Furthermore, we suppose that the frustrations and stresses currently attendant to aca-demic careers will become all too apparent to many bright young men and women contemplating such a career, causing them to lose sight of the quieter, deeper satisfactions. We can predict that the exceptional talent now in the pipeline will thin out, certainly by the time that the forthcoming bulge in retirements and the predicted upsurge in enroll-ments lead to a sharp expansion in faculty vacancies (see Chapters 9 and 10). We are particularly concerned that the supply of talented minority-group members drawn to academic careers may be in acute, even cata-strophic, shortage in the foreseeable future (Chapter 8). Moreover, as opportunities for women in a variety of attractive professions continue to escalate, their interest in academic careers is likely to dwindle.

For all these reasons, we believe that a serious problem exists. Though not immediately tangible, it is a problem with the most far-reaching consequences. It must be dealt with soon, not a decade or so down the line. If we postpone remedial action, our nation, and the higher educational community in particular, will fall heir to an inadequate instructional staff and, accordingly, will pay a heavy social price.

Part Four

AGENDA FOR A VITAL PROFESSION

Our argument in the preceding sections of the book has been that the American professoriate is an endangered profession. It is endangered, not in the sense that there will be many unfilled positions, but rather that many positions may be filled with persons of inadequate qualifications. In this section, we consider ways and means to assure that competent faculty will be available and affordable in sufficient numbers. Our recommendations are directed toward institutions of higher education, government, and faculty members themselves.

CHAPTER TWELVE

The Basic Faculty Contract

The prevailing contractual relationships between individual faculty members and employing colleges or universities are unusual, perhaps unique. They resemble in some respects, however, the contractual arrangements applying to civil service workers. For both groups, the contract conveys a considerable amount of job security, and for both groups the objective is protection—in the case of civil service workers against the vicissitudes of the spoils system, and in the case of faculty against assaults upon freedom of thought, speech, and publication. And in both cases, a secondary purpose is to provide careers secure enough to attract capable appointees at reasonable rates of compensation. For faculty members there are also two other reasons for the special contractual relationship: the important role of faculty in the governance of college and universities and the critical need for collegiality in academic institutions. In this chapter we propose to consider the contractual relationships for faculty and to offer comments and recommendations for the future. In so doing we do not confine ourselves to written, formal contractual relationships. We include also traditions and other informal understandings which are not necessarily spelled out in writing.[1]

The usual contract between faculty members and their institutions involves several critical stages as individuals progress from their original appointments to resignation or retirement. Ordinarily these are: (1) initial appointment as instructor or assistant professor for a limited term; (2) a probationary period (including the initial appointment)

---

1. A formal, that is, written contract often—especially at private colleges and universities—consists of no more than a brief letter of appointment with pertinent salary information. At the other end of the spectrum, at campuses where faculty are represented by an exclusive bargaining agent, the collective bargaining contract between the union and the employer may be very lengthy, specifying, for example, detailed procedures for evaluating faculty teaching, elaborate point systems for determining promotion and merit pay, complex salary formulas, and highly specific means for determining work load.

ranging in various institutions from one to seven years or more; (3) promotion to associate professor and advancement to tenure; (4) eventually, in most cases, advancement to full professor; and (5) provision for retirement either at a predetermined age or on some other basis. The appointment is terminated if tenure is denied at the end of the probationary period. It may be terminated at any other time for grave malfeasance on the part of the faculty member, for financial exigency on the part of the employing institution, or in case of discontinuation of a program or department for educational reasons. At each decision point, due process according to established rules is observed and these procedures include an important role for the judgments of peers. In most instances, the institution tacitly agrees to provide a traditional pay scale and various monetary fringe benefits, to provide specified non-monetary fringe benefits and working conditions, and to maintain academic freedom.

The above contractual arrangements are the norm in the mainstream of American higher education. Most, but not all, institutions adhere to them. Under these arrangements, about two-thirds of all full-time faculty are tenured[2] and many others are in the probationary stage. However, a substantial minority of faculty work under different contractual arrangements. Most institutions which confer tenure on the main body of their faculties also employ so-called "non-tenure-track" persons who work under no contracts or contracts of short duration. These contracts may or may not be renewable, but they do not ordinarily lead to tenure. Some of the non-tenure-track people may be employed full time, some part time. A few institutions, mainly community colleges, do not grant tenure to any of their faculty. Most of these offer short-term contracts of at least one year, some of which are renewable and some non-renewable. The "non-tenure-track" arrangement in its various forms is convenient for the employing colleges and universities. It lends flexibility in meeting changing educational needs and demands, and it avoids the long-term financial commitments implicit in tenured appointments.[3]

In the rest of this chapter, we propose to consider several policy questions of contemporary concern regarding the basic faculty contract: tenure, involuntary faculty separation, faculty leaves, compensation, collective bargaining, and retirement. These are controversial matters. We have tried to sort out the issues and in doing so have consulted widely.

2. The percentage varies from about 50 to 80 percent in most institutions.

3. However, the difference between tenure-track and non-tenure-track positions is sometimes slight. If few tenure-track people are ever granted tenure at the end of their probationary periods, as happens in some institutions, the difference becomes negligible. Or if non-tenure-track people are kept on indefinitely, as happens in some institutions, the difference also becomes negligible.

Tenure

Tenure literally refers only to job security. A faculty member on tenure has a commitment from the employing institution, barring exceptional circumstances, that he may hold his position throughout his career until retirement. However, in practice, tenure is part of a wider contractual system that relates to academic freedom and to the participation of peers in personnel decisions involving faculty.

The practice of granting lifetime tenure to professors has long had its critics. It is often asserted that tenure is a one-sided arrangement which binds a college or university but does not obligate the faculty member; that it harbors incompetence, weakens incentives for excellence, and tolerates sloth (W. S. Light, 1984); that it protects radical activists who use their professorial office to advocate partisan political views and to foment unrest; that it impairs the ability of colleges and universities to adjust their programs to advancing knowledge and to fluctuations in the demand for different fields of study; that it clogs the entry of vigorous young people into the profession, including especially women and minorities (Shapiro, 1983, p. 3a); that it impedes desirable or essential retrenchment at a time of financial decline; that it is not essential to its stated purpose, which is to guarantee academic freedom. Moreover, it has been observed that the relatively few institutions that do not grant tenure seem to have suffered few ill consequences. Finally, in a thoughtful and eloquent address on tenure, Harold Shapiro, president of the University of Michigan, rhetorically asks (1983, p. 6a):

> How effective has tenure been in helping to ensure that all colleges and universities are introduced to alternative visions of our reality in a way that induces them to consider alternative visions of our future? How effective has tenure been in helping ensure an open campus environment characterized both by a toleration for alternative approaches and an intellectual discipline in evaluating contending hypotheses?

All of the above criticisms and questions contain at least an element of truth. They are not wholly idle or irresponsible. The countervailing arguments must be compelling if the case for tenure is to be sustained.

We would offer four arguments in favor of tenure; in our opinion, the practice of tenure must stand or fall on the basis of them. Unfortunately, they are subtle arguments that are neither easily explained nor readily grasped, and perhaps have about them a tinge of unreality and special pleading.

The principal argument is that the primary function of a college or university is to discover and disseminate the *truth*, and that this purpose requires a professoriate that is free to seek, discover, teach, and publish without interference from any source either from within or outside the institution. This principle of academic freedom was eloquently stated by the Board of Regents of the University of Wisconsin in 1894 in words that were later engraved on a campus plaque:

> Whatever may be the limitations which trammel inquiry elsewhere, we believe that the great State University of Winconsin should ever encourage that continual and fearless sifting and winnowing by which alone the truth can be found.

The truth as known by human beings is of course never complete and never absolute. It is always subject to error and dispute and is always in need of revision and amplification. New truths are often upsetting of old beliefs and of old vested interests. But "the truth" can be approached only through untrammeled inquiry and "continual and fearless sifting and winnowing." The professors, upon whom we depend in our society for the unending search for truth and the teaching and publication of truth, are often engaged in a perilous undertaking and they need protection. Tenure is intended to provide that protection. It may be claimed that tenure is not the only or even the best way to protect them, but it cannot reasonably be asserted that the protection is not needed. Indeed, a generation that has lived through McCarthyism, student revolts, racial and religious intolerance, intolerance toward alien ideologies, the rise of "creationism," the environmental movement, the controversies over nuclear power and nuclear weapons, and the increasing public control of colleges and universities, can hardly deny that the need is great. And the need has become even more urgent than usual at a time when the mobility of faculty members has diminished.

A second argument for academic tenure is that the job security it provides is conducive to dispassionate and sustained thought. Faculty members as scholars, researchers, teachers, advisers, and publicists need long periods of unbroken time and freedom from distractions to perform their duties well. Thinking and communicating are exacting tasks that require concentration and peace of mind. As one British professor commented, "I don't subscribe to the theory that uncertainty and instability keep people on their toes." One of the most costly aspects of the current anxiety among faculty about future job security is the adverse effect on their productivity. If the future professoriate were continuously insecure, it is not likely that the amount of important achievement would increase.[4]

A third argument favoring tenure is that it is conducive to collegiality. A college or university to be maximally effective must be a community to which people belong and which they care about, not merely

---

4. One critic has suggested that anxiety has made the young untenured faculty more productive of research, but has made the older faculty less productive.

an entity for which they work and from which they receive paychecks. In this kind of community, faculty members are the central figures. They are largely responsible for establishing the degree requirements, approving the courses of study, setting the admission standards, advising the administrators, counseling of students, recruiting new faculty, conducting teaching and research, and participating in the extra-curricular and social life of the campus. They are the core of the academic community. They should be not employees in the usual sense, but managers. They exercise their influence partly as individual professors, partly as members of departments and other subdivisions, and partly as members of committees, academic senates, or assembled legislative bodies. As we have already indicated, it is only a slight exaggeration to say that they *are* the university. A college or university is likely to prosper in proportion to the commitment of the faculty to the campus community. Institutions of higher education require the sharing of common ideals, people of ability to work together toward distant goals, their capacity to weave ideals and goals into the fabric of the institution and thus to convey to students what the institution is and stands for, not merely what it teaches. Such collegiality is not created instantly or spontaneously. It is developed over time through the presence of committed faculties. Tenure is a powerful tool for enlisting that commitment. McPherson and Winston (1983) extend this third argument by drawing upon the literature of analytical labor economics, of which one of the central insights is that the productivity of an organization depends heavily on the character of the work environment it is able to provide (p. 166). McPherson and Winston also point out (p. 164) that "stable long-term employment relationships" have "desirable efficiency properties" in an employment that places "expensively trained and narrowly specialized people" in "narrowly specialized tasks."

Finally, in the appraisal of tenure, we would point out that colleges and universities have seldom had the money to pay their faculties at rates competitive with other industries. One way of overcoming the effects of inadequate finance has been to provide various fringe benefits, one of which is job security. If this security were impaired, the relative attractiveness of academic work would be reduced, the caliber of talent drawn to the field would in the long run be lowered, and the cost of operating our colleges and universities would be increased. Even President Shapiro, who raised penetrating questions about the efficacy of tenure, concluded (1983, p. 6a):

> On balance, however, I would say that the institution of academic tenure has played an important and positive role. . . . Our first task as members of the higher education community, therefore, is not periodic evaluation of tenured faculty members, but an evaluation of the general teaching and research environment of the university community. We must ensure the

academic freedom of the college or university as a whole, as well as defending our prerogatives as individuals under the tenure system.

Alternatives. In our opinion, these arguments favoring academic tenure are persuasive. But before coming to conclusions on tenure, we decided to consider alternatives. Would it be possible to discover a different arrangement that would minimize the drawbacks of tenure as we now know it and enhance its advantages?

The alternative most often proposed is the granting of tenure in short, renewable increments of five or ten years. One example is the proposal of Shirley Hufstedler, former federal judge and former U.S. Secretary of Education, "to limit future tenure to ten years, with options by the university or college to renew it in five year increments." She added that, "As an incentive for particularly gifted and productive faculty members, longer renewal periods could be prescribed for those professors named to specially-created chairs" (Address to the Association of American Colleges, January 13, 1983). We are dubious about this and other similar plans (some of them actually in operation). We think they would provide inadequate protection to faculty.[5]

A variant of the incremental approach toward tenure is a periodic review of tenured faculty, perhaps every five years, as part of a program of faculty development, not as a device for abrogating tenure in some cases. We see considerable merit in this approach. The dean of one college that is considering such a plan wrote:

> The Board of Trustees has requested the Administration to develop a systematic method of reviewing tenured faculty members. If this is approached constructively as a means of faculty development, I think it would be a useful step. Obviously any review can readily become counter-productive if it is seen as threatening rather than supportive. My interest in it would primarily be to help tenured faculty plan their professional priorities so that the variety of resources such as sabbatical leave, grant support, special teaching or research assisgnments, etc. can be made available in the most effective way. Often this can better be done by planning periods of several years rather than a year-to-year response. Similarly, considerations of promotion, merit salary increases, and eligibility for endowed professorships can be made in a more orderly way. Conversely, in the rare instance in which a faculty member has become less effective, alternatives can be explored in a more systematic way to the benefit of all concerned.

In considering alternative plans, one seemed to us to have some interesting features. This alternative plan contains several parts:

---

5. In commenting on this passage, Michael McPherson observed that renewable contracts do not work well for practical reasons: they turn out to be close to instant tenure because it is never worth the grief not to give a person one more chance.

1. Faculty appointments would no longer carry lifetime tenure but would be made for one year at a time or for longer periods of perhaps three to five years but with no guarantee of renewal.
2. The employing college or university would guarantee to provide academic freedom. The guarantee would be in the form of a clause in a written contract for each faculty member including those employed part time and full time and those of all ranks. The guarantee would carry a definition of academic freedom and would specify penalties for infringement.
3. General surveillance of academic freedom would be provided by an impartial, non-governmental, statewide or regional commission. The members would include distinguished faculty, academic administrators, and laymen.
4. The Commission would publish periodic reports with specific statements about the state of academic freedom at each institution in its region.
5. Complaints of specific violations would be referred to the Commission for investigation and recommendation as to the remedy.
6. Decisions of the Commission would ordinarily be final, but appeals could be presented to a court of law in the form of an action based on an alleged breach of contract.
7. Due process including peer involvement would be provided in connection with all faculty personnel decisions.
8. The plan would contain a grandfather clause so that the vested rights of current faculty would not be impaired. However, faculty members might be encouraged to give up tenure in exchange for a corresponding benefit, for example, a pay raise, increased leave, bigger retirement benefits.[6]

We have not tried to elaborate all of the details of such a plan. In particular, we have not dealt in depth with the questions of how the members of the proposed commission would be selected and of how the commission would be financed. One possibility (which we neither recommend nor oppose) would be to place the policing of academic freedom in the regional accrediting associations.

Our alternative plan, which seemed the best we could devise, might be in some sense workable, but in our opinion it would almost surely dilute academic freedom as it has been traditionally known. Faculty members could not count on the decisions of an external commission as confidently as they can now count on tenure in a single college or university. The proposal would almost surely weaken the attachment of faculty members to their institutions and thus impair the benefits of tenure related to collegiality and to the encouragement of sustained thought. Also the elimination of tenure would seriously

6. As part of this plan, the common "up or out" provision might be abolished. This is a provision that faculty members who are not promoted to associate professor after a probationary period are automatically terminated. However, there would be nothing in the plan to prevent dropping people who were not promoted.

reduce the attractiveness of the academic profession and might therefore either lower the caliber of people drawn to it or increase the cost of
attracting people of traditional caliber. To sum up, we were simply not
able to discover alternatives to the present tenure system that we could
recommend.

Conclusions. Perhaps the strongest argument for the continuation of
the tenure system is that it has proven to be a pretty durable institution.
It is widely prevalent, it is buttressed by an ancient and honorable tradition, it has proved to be resilient against attack, it has generally been
upheld by the courts, it has been embraced within collective bargaining, and it commands the support of most faculty (Chait and Ford, 1982,
pp. 10–11).[7] Moreover, it is not opposed by an appreciable number of
academic administrators and trustees—though many of them would
favor modifications that would reduce its rigidity. As we shall indicate
in the next section, we also would favor cautiously introducing some
flexibility.

Our argument has one glaring omission. It avers that tenure is fine
for those who have it, but it is silent regarding those who do not have it.
These include the thousands of college teachers and research workers
who are not on a tenure track and other thousands who are tenurable
but still on probation. These people have as much right to freedom of
thought and expression as fully-tenured senior faculty, but they have
much less protection in asserting this right. Their chief protection is the
general philosophy and attitudes of their superiors and the general
atmosphere of the employing institutions. In institutions where academic freedom is secure these people probably are not very much at
risk, but in institutions that are less sophisticated and less financially
secure, their position is likely to be vulnerable. We believe that colleges
and universities should incorporate into the contracts of the non-
tenured and non-tenurable a binding and enforceable commitment to
academic freedom.

Involuntary Separation

Tenure, as it exists in American higher education, has never been an
absolute right of faculty members. In the widely accepted tenure statements on academic freedom and tenure promulgated by the American
Association of University Professors and the Association of American
Colleges (AAUP, 1977), termination of faculty appointments is clearly
admitted as appropriate for any of four reasons: (1) medical disability,

7. One study has shown, however, that a substantial proportion of faculty members
indicate, given the choice, that they would trade off tenure for increased pay (Kuhlmann,
1982, p. 2). In this connection, the experience of Webster College is of interest. Faculty
members there are presented with the option of higher salaries or tenure (Chait & Ford,
1982, p. 72).

(2) "cause" meaning incompetence, a serious offense, or dereliction of duty, (3) discontinuance of a program or department based on educational considerations, and (4) retrenchment due to "financial exigency" (AAUP, 1983a, pp. 16a–18a). In each of these cases, the recommended procedures are intended to ensure that the dismissal is not decided upon lightly, that it is based on legitimate grounds, and in particular that it does not abridge academic freedom. The procedures are spelled out in detail and include consultation with representatives of the whole faculty.

The question presented by these four escape clauses is this: How strictly should the escape clauses be administered? For example, how disabled must a faculty member become before he or she is asked to resign or take extended leave with or without pay? What offenses are serious enough to lead to dismissal? What "educational considerations" are sufficient to justify termination of a program or department? What constitutes a program or department for purposes of termination? How serious and how prolonged must be a state of financial exigency before it justifies termination of faculty members (Garland, 1983, pp. 24–26). These are all difficult questions. The answers may vary from time to time with changes in social mores, financial practice, and educational missions and methods. These questions are all subsidiary to the larger issue of the strictness with which the loopholes relating to tenure should be handled to prevent infringement of academic freedom and to provide appropriate collegiality and job security in the academic profession.

In the past, at least in the mainstream of higher education, tenure has usually been administered quite strictly in the sense that termination of faculty positions has been a course of last resort. For example, faculty members with medical disabilities have usually been treated considerately; and only grave and well-documented offenses have led to dismissal for cause. Moreover, few faculty members have been discharged in connection with the discontinuance of programs or departments; and financial exigency has seldom been invoked as a reason for discharging tenured faculty members. With the onset of financial stringency in higher education, however, exigency has become more common though still not widespread. If depressed conditions worsen, many institutions may feel obliged to discharge faculty whether or not they can claim financial exigency as defined by the AAUP. The question arises whether, under current tenure practices, institutions must allow themselves to sink into a state of financial exigency before they may drop tenured faculty, or whether they may act before the fact to prevent exigency.

Most presidents and boards of trustees place a high priority on maintaining the soundness and continuity of their institutions. At the same time, most acknowledge the justification of faculty tenure and are reluctant to discharge tenured faculty. In coping with hard times most institutions try to avoid dropping tenured faculty by first cutting non-personnel costs (e.g., energy use), trimming non-academic and administrative personnel, maintaining a low tenure ratio and allowing the appointments of non-tenured faculty to lapse, and all the while improving their admissions and fund-raising efforts (Mortimer and Tierney, 1979). When these expedients are not enough, or when they raise serious questions of fairness as between non-tenured and tenured personnel and questions of operating efficiency, institutions are likely to begin making cuts in tenured faculty.

We have considered modification of the tenure contract to provide a slightly more lenient interpretation of financial exigency, an interpretation that would permit institutions in good faith to act in advance of exigency rather than after the fact. Such an interpretation might be in the interests of both faculty and institutions, and also in the public interest. To demand that boards and presidents wait until disaster is upon them before taking remedial action may be both unrealistic and unwise. More academic jobs might ultimately be lost and more damage to institutions imposed by this demand than if remedial action could begin sooner. In that event, the boards and the presidents should of course be expected to protect academic freedom, to consult faculty, to encourage collegiality, and to provide feasible job security, as the present AAUP statement requires, but not to have their hands tied in such a way as to rule out prudent and timely institutional decisions. To achieve such flexibility would require only the changing of a few words in the AAUP statement on financial exigency (AAUP, 1983a, p. 16a). The revised version might read as follows (deleted language in square brackets and new language in italic type):

> Termination of an appointment with continuous tenure, or of a probationary or special appointment before the end of the specified term, may occur under extraordinary circumstances because of [a demonstrably bona fide financial exigency, i.e., an imminent financial crisis] *adverse financial trends* which *if uncorrected* would threaten the survival of the institution as a whole and which could not be alleviated by less drastic means.

With this small change in meaning and tone, the rest of the AAUP statement on the subject of financial exigency could be preserved. This statement provides for faculty participation in the decision to declare financial exigency and in its implementation. It also calls for appeals and hearings for terminated faculty, assistance in relocating them, and a ban on immediately employing new faculty to replace those discharged.[8]

8. Our proposed revision of the AAUP statement may be compared with a statement on tenure of the American Association of State Colleges and Universities (*Chronicle of Higher Education*, Feb. 27, 1985, p. 29). The AASCU statement, as we understand it, would allow considerably more flexibility to the institutions than ours.

We have also considered relaxation of the AAUP tenure regulation pertaining to dismissal of tenured faculty members for "cause." It is an embarrassment to the academic profession that it harbors idlers, incompetents, frauds, and persons of odious personal character. There are few of these in the profession but there are some and there should be a way of getting rid of them. Their protection under the umbrella of rules and procedures intended to guarantee academic freedom weakens the credibility of these rules and procedures and of the whole concept of academic freedom (Cahn, 1983, p. 64).

The AAUP statement on dismissal for cause is clear and to us acceptable. It reads as follows:

> Adequate cause for a dismissal will be related, directly and substantially, to the fitness of faculty members in their professional capacities as teachers or researchers. Dismissal will not be used to restrain faculty members in the exercise of academic freedom or other rights of American citizens.

The problem is that the procedures for ridding the profession of misfits are so arduous and so embarrassing that few administrators are willing to take the time of themselves and of the faculty to prosecute the cases. The procedures take on the flavor of a trial for murder. Moreover, faculty persons as members of review committees have been notoriously reluctant to assume their full responsibility for self-policing and enforcement of high levels of professionalism and ethics. Thus, from the point of view of the president or board, it is usually more efficient, in terms of time and energy, to put up with the offending person than to remove him or her. We believe that the procedure could be greatly simplified without infringing on the rights of individual faculty members or compromising academic freedom. Without going into detail, we believe that a procedure which gives a prominent place to the judgment and advice of informed and impartial peers but which takes on less of the trappings of a trial would be adequate. For example, binding arbitration might be applicable to cases of dismissal. This is a form of dispute settlement which is increasingly recommended as a substitute for litigation (Rand Corporation, 1984, pp. 3, 6; Birnbaum, 1984). It is widely used in Canadian universities in cases involving dismissal for cause. Arbitration might also have a place in settling issues relating to financial exigency. Indeed, one might envision the formation of a corps of experienced arbitrators who would help in the resolution of issues arising out of the special requirements of academic freedom.

We propose no change in the AAUP regulation regarding termination of appointments "as a result of *bona fide* formal discontinuance of a program or department of instruction" (AAUP, 1983a, p. 17a). However, this provision has been subject to increasing use as American higher education has been going through a period of retrenchment. For example, the regents of the University of Colorado have formally adopted a policy "allowing all four of the university's campuses to dismiss tenured faculty members when academic programs are abolished for educational reasons" (*Chronicle of Higher Education*, Feb. 2, 1983).

Taken as a whole, the present rules and practices relating to tenure, because of their rigidity, threaten to undermine the whole tenure system and are therefore endangering academic freedom, collegiality, and legitimate job security. The tenure system should be strong enough to give protection but flexible enough to be workable. When it is too rigid, as we think it may be today, institutions are tempted or forced to abandon it outright or to bypass it. They may bypass it in any or all of several ways. For example, they may resort to provisions relating to discontinuance of programs, they may give up lifetime tenure and adopt a system of short-term appointments, they may appoint increasing numbers of non-tenure-track persons, or they may deny tenure regardless of merit to persons who have served their probationary period in tenure-track positions. The bypass method of circumventing tenure has become epidemic in American higher education. It is inhibiting the entry into the tenured professoriate of many capable younger faculty persons, it is creating a class of dissatisfied and despairing intellectuals, and it is undermining collegiality and legitimate job security.

We have noted that in some institutions, mainly community colleges, tenure is granted with skimpy probationary periods. For example, in the more than one hundred community colleges of California, the probationary period is by state law one year, and in practice the tenure decision must be made even before a full year of service is completed. We believe that such a probationary period is inadequate for judging the character and qualifications of candidates and should be lengthened—preferably to the seven years customary among four-year institutions.

In our opinion, many administrators, trustees, and members of the public are overreacting to the issue of tenure in the belief that faculty redundancies will be more numerous than they are likely to be. As shown in Chapter 9, normal annual turnover of faculty from their current institutions is of the order of 7 to 8 percent per year. This figure includes persons who leave the professoriate and persons who merely shift to other institutions. About 4 percent leave faculty positions altogether. Barring such events as war, in no future year is the total student enrollment in higher education likely to decline by as much as 4 percent. The decline, if any, is more likely to be on the order of 1 to 2 percent a year. Therefore, much of any necessary cut in the overall number of faculty can be managed without firings. There may be exceptions of course in particular institutions where the need for retrenchment is exceptionally acute or where the turnover rate is exceptionally low, and there will also be exceptions in particular academic fields. But the future does not hold out the prospect of vast numbers of tenured faculty who will have to be purged. Given the likely rate of decline in enrollments of 1 to 2 percent a year over the period 1985 to 1994, the overall need to reduce faculty numbers could with patience and good planning be met through normal turnover—supplemented perhaps by early retirement schemes. In that case, the academic community's paranoid preoccupation with tenure might be allayed, and the institutions could grant tenure more freely to outstanding younger faculty.

Two final comments on involuntary separation are in order. First, many people in thinking about financial retrenchment focus their attention on aggregate faculty compensation as though it made up all or most of the institutional budget. In fact, faculty compensation is on the average only about 25 percent of the overall budgets of colleges and universities, and is probably not much over 20 percent for tenured faculty. Therefore, equitable and efficient programs of retrenchment should surely include other items in the budget such as the salaries and wages of non-academic personnel, purchases of goods and services, student financial aid, and capital expenditures (H. R. Bowen, 1980b, p. 7). The faculty should not disproportionately bear the brunt of retrenchment.

The second comment refers to the mode of retrenchment if and when cutbacks in tenured faculty become inevitable (H. R. Bowen, 1983, pp. 21–24).[9] In recent years, a widely accepted view has been that financial adversity should be an occasion for sloughing off programs and people deemed to be of low priority, and that limited resources should be reserved for programs and people of higher priority. In some situations, this mode of adjustment may be desirable, practicable, or inevitable (Kerchner & Schuster, 1982). However, the selective approach has drawbacks because of its impact on morale. Selective retrenchment, no matter how meticulously fair or economic the selection process may be, will seem to many as arbitrary and unjust (H. L. Smith, 1984, p. 19). Not everyone agrees on what is dispensable and what indispensable. To some, the selective approach will be viewed as the breaking of a trust if not of a contract. Given the slack academic labor market, it may utterly destroy the careers of some people. It tends to place the chief executive officer in a vulnerable position. The controversy surrounding selective retrenchment depletes his or her capacity for leadership and uses up precious administrative capital. It breeds growing insecurity not only among those affected but also among others who will be wondering when their time will come. In short, selective retrenchment if pushed very far is likely to be highly damaging to organizational morale, and morale is something that is very much needed in hard times.

A possible alternative to selective retrenchment is across-the-board salary reductions. These are not viewed by those affected as a cause for jubilation, but they have advantages. Across-the-board cuts tend to spread the burden of retrenchment widely. They are seen as fair, they

9. See also: Brown, 1983, pp. 4–8; Furniss, 1974; Glenny, 1982; Kauffman, 1982; Melchiori, 1982; Morgan, 1982; Volkwein, 1984.

are supportive of collegiality ("we are all in this together"), and they are compatible with those faculty attitudes that are as much a basis of institutional effectiveness as the "economic" allocation of resources.

They leave institutions with their organizations intact and with people to carry on work that would otherwise be neglected. Their prospective use cuts the risks involved in granting tenure to rising younger faculty. They free decision-making on educational policy from financial pressure and allow it to proceed with normal deliberation (Kaplan, 1982). Even when the amount to be cut overwhelms any one mode of adjustment, across-the-board cuts have a place among the many available options.

Finally, it should be observed that across-the-board cuts have been the most common method of effecting financial retrenchment in colleges and universities. For example, during the Great Depression of the 1930s, they were commonly used. David Henry (1975) reports that 84 percent of all colleges and universities reduced faculty salaries at least once between 1931 and 1936. Moreover, during the 1970s and early 1980s adjustments having the effect of across-the-board cuts were almost universal as the rate of inflation systematically exceeded the rate of dollar increases in expenditures. This was retrenchment by subterfuge, but it *was* retrenchment and it resulted in a substantial decline in the real earnings of faculty. Indeed, should rapid inflation reoccur in the 1980s and beyond, it will continue to be the primary method of retrenchment whether anyone wills it or not. Obviously, faculties were not joyful because of the pay cuts in the Great Depression or in the 1970s and 1980s, but they accepted the inevitable when the burden was spread equitably. Adverse effects on the organization, operation, and morale of the institutions were largely avoided until selective cuts were resorted to.[10]

Leaves of Absence

When a person joins the faculty of a college or university, the prevailing pattern of leaves of absence in that institution becomes part of the implicit contract. Leaves provide opportunities for refreshment, new experience, uninterrupted study and research, or travel. Leaves are particularly important because the work of faculty members, though interesting, tends to be draining of energy and to have about it a lifelong sameness. As a result, faculty members are vulnerable to both burnout and staleness.

Many institutions provide leaves with pay, the most common form being the sabbatical (seventh year) leaves with full pay for a half-year or

10. It is worth noting also that some highly visible corporations resorted to across-the-board cuts during the deep recession of 1982–83.

half-pay for a full year. The expectation is that each person going on sabbatical leave will use the opportunity in a way that meets the twin objectives of personal refreshment and professional development, and that his or her value to the institution will be enhanced thereby. Almost all institutions (including those that provide sabbatical leaves) also, within reason, grant leaves without pay. These enable faculty members to take advantage of special opportunities for new experience, such as visiting professorships, government service, other temporary positions outside academe, or pursuit of personal interests. There are many variants on the two types of leaves and both can serve the purpose of refreshment and professional development. Most faculty members, however, cannot afford leaves without pay and temporary income-earning opportunities are not available to all. The sabbatical leave is therefore an essential ingredient of an institution-wide leave program.

The academic community seems agreed that occasional leaves are of great potential benefit to the institutions as well as to the faculties. At least, there are few complaints about the practice. The availability of leaves is one of the features of academic life that attracts and holds capable people, and helps them to remain energetic and alive throughout their careers. However, the potential benefits to faculty members and to the institutions are not always realized. Some leaves become little more than long vacations without any pretense of professional development. In our opinion, an opportunity as rare and as valuable as substantial time off, a privilege that is seldom enjoyed by non-academic members of our society, should be based on carefully thought through (but not inflexible) policies of the institutions and at the same time on responsible planning by the faculty members. Given adequate provision for mutual planning, leaves of absence are an excellent investment in the improvement of higher education and should become even more widely available throughout the academic world.

The Pay Scale[11]

An important part of the faculty contract is the salary and monetary fringe benefits offered to each individual who joins the faculty of a college or university. This part of the contract consists not only of the specific salary offer presented to the new faculty member, but also of the whole pay scale under which the institution operates. The pay scale includes the relative compensation for the several academic ranks, the proportion of compensation in the form of cash and the proportion in the form of fringes, the rate at which seniority and merit are reflected in compensation, and the general competitiveness of the pay scale with other colleges and universities and other industries and occupations.

11. For our general discussion of compensation, see Chapter 6.

The general form of the pay scale is likely to be rather similar for most institutions. The main difference is likely to be in comparative levels of compensation. Some types of institutions simply pay more on the average than others. Whatever the pay scale may be at the time an individual joins the faculty, that scale usually gives some indication of what the faculty member can expect in the future. The pay scale as a whole may rise or fall, but its form and its amount relative to other institutions are likely to be fairly persistent. Changes in the pay scale that may be contrary to the expectations of faculty or that may affect various elements of the faculty differently are sometimes sources of dissatisfaction and controversy. There are two such matters before the American professoriate today. One is change in "pay differentials" and the other is "outside earnings." These will be considered in the remainder of this section.

Differential Pay. Traditionally, the pay scales of individual institutions have been quite compressed. Differences in compensation between the lower and higher ranks have not been pronounced and have reflected long-standing custom more than the ebb and flow of market forces (Tuckman and Chang, 1983; H. R. Bowen, 1968, pp. 10–12). The norm in most institutions has been such that professors receive roughly twice the pay of instructors and lecturers (see Table 12–1). This implies that a young person entering the academic profession may expect to earn by time of retirement around twice the earnings of a beginner. Moreover, pay differentials among faculty members in different disciplines have usually been quite small. Market influence has tended to appear in the form of accelerated promotion or greater research support rather than in salary. The principal exception to these generalizations has been the field of medicine in which clinical faculties have been paid (in one way or another) considerably more than persons of comparable rank in other disciplines. Faculty pay has also varied somewhat by types of colleges or universities reflecting differences in the kind of work required and also in financial means of the institutions. Generally, major research universities and wealthy private colleges have paid higher salaries than community colleges or less affluent four-year institutions. But with few exceptions, the salaries of even famous professors in major universities do not exceed three or four times the salaries of anonymous assistant professors in institutions of lesser resources and prestige. In short, pay differentials have been noticeably less in academe than in other professions such as law, medicine, or business.

The rather compressed pay scale has reflected a traditional concept of the academic profession. The profession has been regarded both by aspirants and incumbents as a vocation, akin to the ministry, to which one is "called." It has been thought of as leading to a life of service, of personal satisfaction flowing from the "life of the mind," of the rewards from association with young people and participation in a "company of scholars." The profession might in some cases also lead to recognition

Table 12–1. Index Numbers[1] of Weighted Average Compensation of Faculty Members by Rank, 1982–83

		Professor	*Associate Professor*	*Assistant Professor*	*Instructor*	*Lecturer*
Doctoral Level Institutions	I	100	73	60	46	54
Comprehensive Institutions	IIA	100	80	65	52	56
General Baccalaureate Institutions	IIB	100	80	66	55	63
Specialized Institutions	IIC	100	81	67	53	59
Two-year Institutions[2]	III	100	85	71	60	52
All Combined[2]		100	76	62	49	57

[1] Compensation of each rank as a percentage of the compensation of professors. Compensation includes both salary and countable benefits.
[2] Excludes institutions without ranks.
SOURCE: W. Lee Hansen, "A Blip on the Screen." *Academe*, July–Aug. 1983, p. 13.

from one's peers for outstanding contributions to knowledge. The financial expectations have been modest. The main requirements have been enough income to support a family in decency and to provide reasonably for old age. The rewards of the profession, then, have consisted partly of a valued way of life and partly of modest—not necessarily parsimonious—remuneration.

The pay scales of most colleges and universities have been consistent with this concept of the academic career. Ideally, they have provided enough pay to recruit capable and committed young men and women and to retain them as career members of the profession.[12] Moreover, colleges and universities have provided pay-scales in mid-career sufficient to recruit some people from government, business, and other industries. But almost always new faculty, whether young or mid-career, have been attracted in part by non-monetary rewards in which the campus way of life has been an important element.

The traditional emphasis on non-monetary rewards has been reflected in the pay scales by relatively small differences in pay based on merit. The academic community has not looked kindly on wide differentials among persons of approximately the same seniority. It has resisted the "star system" which pays people of exceptional brilliance and productivity substantially more than the majority of faculty. It also has resisted differential pay among academic disciplines among which there are differing market conditions. The prevailing attitude has been that all faculty are members of a collegial organization, that each contributes in his or her own way, and that the financial rewards should be more or less similar for persons of equal seniority. Quality of personnel

12. The ability to attract young recruits has been greatly augmented by the heavy subsidy of graduate education.

and incentives for individual excellence have depended not so much on differential pay as on rather strict procedures for appointment, probation, and advancement in rank. Special merit and special bargaining power have been rewarded mainly by acceleration of advancement through the ranks. Otherwise, merit pay has been simply a little whipped cream on the pie.

The traditional concept of the academic community, which we have been describing, and its financial counterpart have prevailed widely until recently. Indeed, they have been reinforced by the rigid and bureaucratic pay scales common to many of the larger campuses and also imposed by most central offices of multi-campus systems or by statewide coordinating bodies. However, from our field studies, we have found many signs that the traditional pay scales are under pressure and are being compromised. Some of this pressure is due to the general weakening of academic traditions and loyalties—especially in the larger institutions and in some of the newer types of institutions such as large community colleges and regional universities. Most of this pressure, however, derives from recent trends in technology and industrial organization which have rather suddenly increased labor market demands for workers trained in fields such as accounting, finance, business management, engineering, and computer science. This demand has in turn increased the flow of students seeking entry into these fields and has resulted in a shortage of faculty members to teach them. Many reports from schools of engineering and business indicate that academic salaries are too low to attract and retain the requisite number of teachers. Indeed some reports suggest that recent graduates with bachelor degrees in these fields can earn more than teachers who have had many years of experience. Because these same fields are widely believed to be of critical importance in the economic progress and military defense of the nation, the case for overriding traditional pay scales is further strengthened.

A recent study conducted by a committee of the American Assembly of Collegiate Schools of Business (1982) revealed the following findings pertaining to the nation's accredited business schools:

> Doctoral production has declined 20 percent since 1975. During the same period, the number of MBA graduates has risen 52 percent, and business baccalaureates 40 percent.

> Twenty percent of authorized positions for doctorally-qualified faculty were unfilled in 1981–82.

> At the current annual rate of net gain in business doctorates, it would take over six years to fill vacancies existing in 1981–82 alone.

Similarly, in an editorial in *Science* (Oct. 28, 1983) Daniel C. Drucker, Dean of the College of Engineering at the University of Illinois at Urbana-Champaign, declared:

Across the country a terribly overloaded faculty strives valiantly to educate a growing number of the brightest and best students ever to enter engineering schools. Almost every school of engineering is unable to provide what it views as an adequate quality of education because of obsolete instructional laboratories and facilities and an increasing shortage of qualified faculty.

Given no other choice, many schools have already granted tenure to the best of the inadequate people available and will continue on this downward path. Yet engineering graduates will be turned out in appreciable numbers, although many schools are restricting their enrollment. It is the quality of education, not the quantity, that progressively degrades. That is the insidious nature of the crisis in engineering education, devastating to the economic welfare and defense capability of the country on a time scale of less than a decade ahead.

As these quotations testify, economic forces have been overriding academic traditions and in fact have produced increasing pay differentials among various types of institutions, among various disciplines, and also among individual professors in particular departments. For example, as shown in Table 12–2, academic salaries in 1983–84 were quite similar for social sciences, humanities, and other low demand areas, but substantially higher for engineering and business. Another example is found in a 1983 directive of the Chancellor's Office of the California State University relating to faculty appointments in engineering, computer science, and business administration:

RESOLVED, By the Board of Trustees of The California State University that from April 1, 1982 until June 30, 1983 faculty newly hired in the rank of Assistant Professor in the Disciplines of Engineering, Computer Science, and Business Administration in those cases where it is necessary to offer competitive salaries, following the normal consultative process as required by Title 5, California Administrative Code, may be placed in Range 4, steps 1 to 5, for salary purposes only, and be it further

RESOLVED, That under the same restrictions and during the same time period, Assistant Professors in Range 3, step 5 may be advanced to Range 4, step 1 while remaining in the rank of Assistant Professor . . . .

On the whole, however, the customary flat pay scale still survives—but with increasing pressure to abandon it and with increasing numbers of exceptions.[13]

The persistence of the traditional pattern of faculty compensation

13. Medicine and law should be added to the list of fields for which student demand has been strong. In the case of medicine the problem has been at least partially solved by the long tradition that medical school pay scales should exceed those in other disciplines, and that enrollments should be strictly controlled. In the case of law, the problem has been dealt with partly by relatively high pay scales and partly by the ease with which facilities and faculty for legal education could be expanded.

Table 12–2. Average Salaries by Disciplinary Groups, 1983–84

	Professor	Associate Professor	Assistant Professor	All Ranks
Engineering and Computer Science	$41,108	$32,964	$29,041	$36,787
Business and Economics	38,884	31,412	27,046	32,652
Science and Mathematics	37,513	28,282	22,532	31,339
Social Sciences	35,091	27,005	22,657	29,974
Humanities	35,453	27,169	20,545	29,226
Arts, Fine and Applied	34,076	26,344	21,734	28,102
Physical Education	33,966	27,590	23,663	27,820
Vocational Education	33,443	30,016	22,426	27,611

SOURCE: John Minter Associates, *Chronicle of Higher Education,* March 7, 1984, p. 28.

indicates that non-monetary factors are still very important to faculty members. The sense of vocation and collegiality, and the love of academic life have continued to be powerful influences in their job decisions. These explain in part why there has been no major flight from the profession during recent years of deteriorating working conditions, declining real compensation, and pessimistic projections for the future. However, one glaring consequence of the modest and flat pay scale has been to encourage faculty to supplement their earnings through outside employment. For most, these outside earnings have been small, but for a sizable minority they have been substantial. Thus, while the differences among basic salaries may be small, the differences in total earnings may be considerably greater. The eminent physician, the renowned scientist, the gifted artist, the successful author, the business consultant, the brilliant public lecturer may be content with modest academic salaries simply because their supplemental earnings are substantial. The classical scholar, the medieval historian, or the anthropologist may be fully as eminent as their colleagues in the more worldly fields, but their total earnings are likely to be less.

Many observers of academe believe that the faculties would be strengthened if the differentials in the base salaries could be widened. It is seldom made clear whether or not this proposal is based on the assumption that increasing funds for the purpose would be forthcoming. If not, then the higher salaries for some would be at the expense of the lower salaries for others and would presumably weaken the power of the institutions to attract and retain people in the rank and file. Be that as it may, arguments are frequently advanced in favor of an expanded pay scale. The most important of these is that it would enable colleges and universities to attract and retain increasing numbers of highly gifted people. It would reduce the pressure on faculty members to seek outside employment. It would lend flexibility in that it would enable institutions to compete for faculty in fields that are in high demand. By the same token, it would permit institutions to save money by paying lower salaries to persons in fields of low demand. An expanded pay scale would also permit institutions to offer a super-grade rank, in effect above the rank of professor, and thus provide stronger incentives to high performance for persons at the present top rank of professor. In addition, it is argued that collegiality, the service motive, and the love of teaching and learning, are passé and can no longer be relied upon for attraction and retention of faculty or for incentives to hard and effective work among senior faculty. Money, it is said, is becoming the mainspring that powers the academic enterprise.

These arguments in favor of the widened pay differentials cannot be dismissed lightly. They are the arguments of the practical world, of people who believe in the free market as the best allocator of resources. However, there are strong arguments also in favor of retaining the traditional pay scale with all its implications. Most of these arguments revolve around the concept of the college or university as a collegial community. The major premise is that the values of the academic community are not identical with the values of the free market. A community to which people belong because they wish to serve, because they love learning, and because they want to teach, it is argued, will provide a more favorable environment for the personal growth and development of students and for the advancement of learning and culture than one in which the members are there primarily for money. The amount of money paid faculty should, it is said, be related primarily to personal or family need rather than to the amount that can be commanded in the labor market. Salaries should be set according to the valuations of the academic community, not necessarily according to the valuations of the outside world. For example, professors of classical languages, political science, or anthropology should receive as much as professors of computer science, business administration, or engineering. Any other arrangement, it is said, will impair collegiality, will convey mistaken academic values, will create a sense of injustice among faculty, and will damage morale.[14] Moreover, it is said, the academic community cannot be expected to respond to every fluctuation in demand. Most of these prove to be temporary. Shortages of faculty in engineering, or accounting, or computer science as well as surpluses in physics, or political science, or history come and go. To try to respond to every change in technology, or every variation in student interests, or every change in intellectual fashion would lead to a chaotic pay scale especially because pay increases are easier to effectuate than decreases. Finally, it is noted that most of the colleges and universities in America—including many of the strongest and most admired ones—continue to operate with a traditionally compressed pay scale.

14. As noted in Chapter 8, we found in our campus visits considerable indignation voiced by arts and sciences faculty members who resented accelerating pay differentials.

The arguments on both sides are impressive. Neither can reasonably be ignored. They represent two competing views of the academic enterprise. Our opinion leans toward the traditionally flat (though generous) pay scale which implies reliance on motives and values associated with the concepts of "vocation" and "collegiality." We believe that the institutions will attract and keep better professors and conduct better education and scholarship if there is a major element of vocation and collegiality than if money and bureaucratic status become the main rewards of academic life. However, we recognize that the pressures of the market are powerful and that they sometimes represent important values that the academic world might otherwise neglect. But if money is essentially all that colleges and universities have to offer, they will probably be outbid by other industries for the best talent and will not perform as well as they do now. One evidence of the power of vocation and collegiality is the way faculties have stood by their institutions in periods of depression and discouragement. Examples include the time of the Great Depression of the 1930s, the depressed period in the 1950s after the departure of the GIs, and the recent period of declining real pay and deteriorating working conditions. Throughout these troubled times, there were no wholesale departures from the profession, and no major falling off in the number of applicants for academic positions. We are not saying that hard times are not discouraging to faculty or that their morale is always strong regardless of events. Rather we are saying that the non-monetary incentives are powerful and conducive to good performance, and that monetary rewards and incentives are not sufficient. One of the most important tasks of the academic leader is to strike the right balance between monetary and non-monetary benefits in the motivation and reward of faculties.

Finally, we believe that the wide fluctuations in demand for faculty members of different disciplines could through mild adjustments be dampened down. For example, the undergraduate curricula could be modified, with good educational effect, to give greater emphasis to liberal education; enrollment in some of the burgeoning disciplines could be mildly restricted, thus reducing faculty loads; faculty members could be encouraged and helped to retool and shift from low demand to high demand disciplines; the rate of faculty promotion might be accelerated moderately for faculty members in the high-demand disciplines; and persons in the high-demand disciplines might be favored somewhat in the apportionment of supplemental earnings for research and summer teaching. These are mild palliatives that would not upset too drastically the traditional arrangements. They would be consistent with good education and would be reversible over time with new shifts in demand which will inevitably occur.

Supplemental Earnings. A discussion of differential pay leads directly to the controversial question of supplemental earnings. The possibility of supplemental earnings tends not only to increase the average amount earned but also to widen the differences among faculty in total earnings. Before launching into this subject, however, it would be well to pause briefly for some definitions.

There are primarily four sources of faculty earnings: (1) the base contract pay for the academic year of nine months; (2) the extra contract pay for those faculty members who are appointed for 11 and 12 months; (3) extra pay *ad hoc* for special services rendered to the institution outside the basic contract—services such as summer work or overloads during during the academic year; and (4) earnings from sources outside the employing institution. Frequently when statistics on faculty pay are gathered, they refer only to compensation (contract salary plus fringe benefits) for nine-month appointments.[15] These nine-month statistics are useful in tracing the trend of faculty pay from year to year but understate total earnings. Moreover, since these earnings are not distributed equally over the whole faculty, the available figures understate the range of earnings differentials among faculty members.

Supplemental earnings for faculty are not a new or recent phenomenon. A careful survey conducted in 1961–62 (Dunham et al., 1966, pp. 145–49) reported that about three-quarters of faculty on 9 to 10-month contracts and half of those on 11 to 12-month contracts, received supplemental pay, and that the supplements amounted to 19 percent of contract salaries for those on 9 to 10-month contracts and 11 percent for those on 11 to 12-month contracts. However, it was clear that much of the supplemental earnings were from the home institutions for summer teaching and other services rather than from outside sources. The percentages of faculty members reporting supplemental earnings from various sources were as follows (Dunham et al., pp. 145–49):

	Faculty on 9 to 10-month contracts	Faculty on 11 to 12-month contracts
Mainly from Home Institution		
Summer teaching	44%	10%
Other teaching	13	7
Research	7	4
Mainly from Other Sources		
Royalties	8	9
Lectures	9	9
Consulting fees	13	16
Other professional earnings	10	10
Non-professional earnings	8	6
Other summer employment	11	4

15. For instance, the statistics published by the American Association of University Professors are stated in terms of nine-month appointments and exclude pay for "summer teaching, stipends, extra load, or other form [sic] of remuneration. Where faculty members are given duties for eleven or twelve months, salaries are converted to a standard academic year (or nine-month) basis . . . " (American Association of University Professors, 1983a, p. 20).

Table 12–3. Full-Time Faculty Members on 9 to 10-Month Appointments with Supplemental Earnings from Sources Other Than Base Salary, 1980–81

	From Within	From Outside	Total
Four-Year Private Institutions			
Percent reporting			
supplemental earnings	52.6%	54.8%	78.8%
Average amount	$3,321	$4,053	$5,301
Average amount as percent			
of base salary	15.8%	18.5%	24.8%
Four-Year Public Institutions			
Percent reporting			
supplemental earnings	67.0%	50.9%	82.1%
Average amount	$3,792	$2,969	$4,935
Average amount as percent			
of base salary	16.6%	13.0%	21.9%
Two-Year Public Institutions			
Percent reporting			
supplemental earnings	67.1%	35.1%	76.0%
Average amount	$2,715	$2,771	$3,677
Average amount as percent			
of base salary	13.5%	13.5%	18.0%

SOURCE: Minter, 1981, pp. 91–92. Supplemental Earnings refers to income received for personal services and excludes receipts from dividends, interest, rents, gifts, and other sources not derived from work. The average amount and percent of supplemental earnings includes only those faculty members who received such earnings and excludes faculty members whose supplemental earnings were zero.

A report prepared by W. John Minter (ca. 1981) provides recent and detailed data on supplemental earnings. Though this report is not strictly comparable to the Dunham study, it suggests a moderate increase in the percentage of faculty receiving supplemental income, and perhaps also a moderate increase in the constant dollar amounts. The Minter data are summarized in Table 12–3. They indicate that about 80 percent of all full-time faculty members receive supplemental earnings, and that on the average these earnings augment the base salaries by about 20 percent. In considering the financial welfare of faculty members, one must take into account these supplemental earnings, and the fringe benefits as well, rather than concentrating only on the base salary. The two together add around 40 percent to the base salary, a not inconsiderable percentage. In interpreting this percentage, however, we note that generous salary supplements are also available in large corporations and in the professions.

There are wide differences among the several types of institutions and the various academic disciplines in the amount of supplemental earnings. These differences are shown in Table 12–4, which presents the Minter data on average supplemental earnings as a percentage of base salary. Business and economics, and engineering and computer science,

Table 12–4. Supplemental Earnings as Percent of Base Salary, Full-Time Faculty Members on 9 to 10-Month Appointments with Supplemental Earnings from Sources Other Than Salary, by Type of Institution and Academic Discipline, 1980–81

	Supplemental Earnings from Outside Sources			Total Supplemental Earnings		
	Four-Year Private Institutions	Four-Year Public Institutions	Two-Year Public Institutions	Four-Year Private Institutions	Four-Year Private Institutions	Two-Year Public Institutions
Arts, Fine and Applied	12.7%	14.1%	9.2%	15.8%	18.1%	13.8%
Humanities[1]	12.5	6.2	4.5	16.5	13.3	11.7
Social Sciences	12.4	10.6	12.2	22.7	23.0	17.0
Business and Economics	41.3	23.3	21.0	45.5	36.7	29.3
Science and Mathematics	14.5	10.0	7.6	19.8	20.1	16.6
Engineering and Computer Science	30.0	22.7	32.7	42.4	30.6	27.6
Vocational Education[2]	7.3	12.5	20.0	15.5	16.8	22.1
Physical · Education	14.5	11.3	9.7	20.7	18.7	16.0

[1] Includes language, literature, communications.
[2] Includes home economics, nursing, health.

SOURCE: Minter, ca. 1981, pp. 94–116. For description of the data, see Table 12–3.

as would be expected, stand out in the amount of supplemental earnings, but the differences among the other disciplines are fairly small.

There has been considerable criticism and much uneasiness about the outside work of faculty members. The grounds for this concern are that it takes time, that it shifts the focus of faculty members from the campus and from teaching to outside interests, that it may divert research activity from basic to applied subjects, and that it may lead to entanglements that will impair scholarly objectivity (Linnell, 1982). Others argue that faculty should be encouraged to take part in outside remunerative activities on the grounds that these activities enhance the skills and knowledge which faculty members bring to their teaching and research, that these activities serve society, and that they enhance the prestige of the colleges and universities (Boyer & Lewis, 1984).

The main types of earnings from outside sources have been: royalties from writing; royalties from inventions; sales of works of art; artistic performances; lecture fees; and temporary teaching in other institutions. These have long been accepted as part of the established academic way of life. Indeed, they have been regarded as an almost mandatory part of a distinguished academic career. Fundamentally, they involve teaching, that is, the dissemination of learning or art. They involve teaching just as much as classroom instruction in the home institution, except that they serve different student bodies.

It has been generally understood that these outside activities should not be carried on excessively to the detriment of the local students and the local academic community. But on the whole they have been encouraged and applauded. Through these activities, a few professors have been able to become wealthy, but even that has not been opposed or resented on the campuses when the success was based on merit and did not bring about neglect of local responsibilities. Studies of the matter have shown that faculty who conduct off-campus activities teach as much and publish more than their colleagues (cf., Bok, 1982, p. 154).

Two other quite different types of outside earnings are sources of current uneasiness. One of these is consulting—usually in the form of private professional practice involving scientists, lawyers, physicians, engineers, economists, management experts, and other specialists. Private consulting often differs from the other income-producing activities in that it tends to be a private activity. It does not result in the wide dissemination of learning and thus tends to be incompatible with the traditional objective of colleges and universities. Private consulting often involves the sharing of knowledge or skill with particular clients who are in a confidential relationship and are able to pay. Private consulting sometimes develops into formal tie-in relationships, for example, the organization of new private business enterprises by professors and the development of joint ventures between corporations and universities for executing research projects.[16] All of these arrangements would appear to encourage applied research at the expense of basic research, and to influence the selection of lines of research according to clients' interest rather than the advancement of knowledge on a broad front untrammeled by commercial considerations. Because of the radical implications of consulting by professors, especially as it has developed in the past few years, it is no wonder that it is viewed with reservations by many people both in and out of academe.

Another quite different source of outside earnings is the acceptance by full-time faculty of teaching assignments at institutions other than their home college or university. This practice has long been acceptable in the case of appointments during the summer vacations, but raises obvious practical and ethical issues when full-time faculty members at one institution accept part-time (or even full-time) appointments at other institutions.

A third type of outside earnings that is widely criticized and is at

---

16. For a succinct discussion of these issues, see "Selling Research May Be Dangerous," *Times* [London] *Higher Education Supplement,* June 25, 1982.

the opposite pole from high-level consulting, is the increasing practice among full-time faculty members of accepting miscellaneous outside work unrelated to their professional expertise, for example, driving cabs, door-to-door selling, tending bar, and selling insurance or real estate. It would be unfair to criticize anyone for trying to make a decent living, and there may be faculty persons who are ill paid to the point of desperation. But such employment is an unfortunate way of solving the problem. As any academic person knows, to teach a full load, to keep up-to-date in one's field, to carry one's weight in institutional affairs do not allow time and energy for unrelated employment that affords no professional enrichment. This kind of outside work has no redeeming features beyond the compensation it affords, and is almost certain to detract from the quantity and quality of faculty performance.

The conclusions we draw concerning outside earnings are by no means rigid or dogmatic. The opportunity to earn outside income encourages faculty to disseminate their learning to the world beyond the campus, to take part in practical affairs, to gain practical experience, and to become proficient in their academic disciplines. The opportunity also offers a special incentive to ambitious persons because it provides the chance within the academic profession to earn substantial amounts and even in a few cases to get rich. It provides some relief from the ceilings on earnings attainable through regular salaries and thus strengthens the incentives for a sprinkling of adventurous and imaginative people to enter the profession. Moreover, outside work may be an antidote to the boredom or burnout that afflicts some faculty members in mid-career. As Boyer and Lewis have concluded, "faculty consulting has been overestimated and underappreciated" (p. 195).

There are of course disadvantages and dangers in outside work: some faculty overdo it to the neglect of their academic duties, some lose their loyalty to their institutions, some misuse their professional status in the public arena, and some get involved in questionable ties with private interests or in relationships that are not consistent with the dissemination of learning. Certainly an institution has a right to expect that faculty members will meet their classes punctually, will be prepared for teaching assignments, will be available to students, will take part in the affairs and activities of the institution, and will be loyal to the academic ethos. Faculty members who allow outside activities to dominate their efforts or subvert their values should either leave the university voluntarily or should be terminated "with cause." The position of professor is not a part-time job. In view of the disadvantages and dangers of outside consulting, colleges and universities should work out broad general guidelines relating to outside work and should be watchful that external activities do not get out of hand to the detriment of the institutions (H. R. Bowen, 1980, pp. 72–73).

The faculty and administrative officers of many leading universities have been debating these issues recently, especially as related to university research sponsored by corporations. Several institutions have adopted guidelines forbidding agreements that limit publication or discussion of research findings, and that require faculty members to disclose all corporate ties.

Derek Bok, the president of Harvard University, has offered some wise counsel on these matters (1982, pp. 167–68):

> My conclusion is surely not that rules are useless. . . . My point is simply that in an academic community everything will turn on how such rules are regarded by the faculty to whom they apply. Even the most sensible guidelines will succeed only as they gain understanding and support through a process of wise application to individual cases. This is a task that must be carried out with the active help of faculty members who appreciate the institutional problems involved while commanding the respect that professors will accord only to their peers.
>
> No problem better illustrates the true meaning of the university as a community of scholars. In such a community, leadership has an important but distinctly limited role to play. Presidents and deans can avoid industrial entanglements that produce intolerable strains; they can point to dangers and encourage active debate. They can even propose appropriate guidelines for the faculty to consider. But only if professors care enough about their institutions to participate fully in the task of maintaining proper academic standards can universities contribute to the useful application of knowledge without eventually compromising their essential academic values.

The Work Environment[17]

Still another part of the faculty contract is the work environment in which individual faculty members are placed. This includes the teaching load; the kinds of colleagues; the qualifications of students; the facilities, equipment and supplies; secretarial and research assistance; the physical, cultural, and social ambience; support of professional travel; sabbatical and other leaves; and non-monetary fringe benefits such as use of recreational facilities and scholarship funds for faculty children. These emoluments are very important to faculty and often weigh more heavily in decisions to join or to leave the academic profession, or to join or leave a particular institution, than monetary compensation. And it is not uncommon for faculty members to accept less pay in order to attain a better work environment. Indeed, it may be the generally superior work environment of academe that accounts for its relatively low compensation.

Just as monetary compensation as reflected in the prevailing pay scale is part of the faculty contract, so also is the work environment.

17. See Chapter 7 for a full discussion of the work environment.

When a person joins a faculty, he or she is entering into a tacit contract that the institution will maintain or improve the then prevailing work environment. One of the current sources of faculty unrest is that in many institutions the work environment has been deteriorating. Often this is more disturbing than the cuts in real monetary earnings.

The maintenance of a favorable work environment is especially important because the kind of environment that is attractive to faculty is also attractive to students and conducive to good education.

Collective Bargaining

An increasingly important feature of "the faculty contract" is collective bargaining. Where this occurs, the contract, which traditionally has been largely unwritten and even unspoken, becomes more formal and detailed. It is then committed to writing and many of the relationships between faculty and administration are governed precisely by the written contract. Collective bargaining, once it is established and the parties become accustomed to it, appears to work smoothly in many institutions, and in a surprising number of cases both parties express satisfaction with the arrangement. Our view, nevertheless, is that collective bargaining is not the optimal arrangement for people in a profession in which collegiality and community are essential. Faculty unions are symptoms of, and responses to, the declining circumstances in which many faculties find themselves. Indeed, collective bargaining is seldom sought after and adopted in institutions where faculty members are treated with respect, occupy a significant role in decisions affecting them, receive reasonable compensation within the financial ability of the employing institution, and are part of a genuine academic community. For institutions where these increasingly uncommon conditions exist, we agree with the proposition advanced by the U. S. Supreme Court in the Yeshiva case that faculty should be regarded as "management" and not as "labor."

Retirement

An important part of the faculty contract is the provision for termination through retirement. Over many decades, it had been common among American colleges and universities to fix a particular age, usually between 65 and 70, for retirement. In a few cases additional part-time or full-time service beyond the retirement age, on a year-to-year basis, has been permitted, but in virtually all cases tenure has ended at the official retirement age.

Among the provisions for retirement, most institutions have established retirement annuities accumulated throughout the working life of each individual faculty member and payable after retirement in install-ments throughout the remaining life of the retiree and of his or her spouse. The annuity plans have in most cases been compulsory and have required regular monthly contributions of both the individual faculty member and the employing college or university. For most pri-vate institutions and many public institutions, the annuity funds have been administered by the Teachers Insurance and Annuity Association and its companion organization, the College Retirement Equities Fund, known jointly as TIAA-CREF (Mulanaphy, 1981)[18] For other public institutions the annuity funds have usually been incorporated in state employee retirement systems (Cook, 1983). Whatever the plan, the basic presumption has been that the general security and stability of the faculty throughout their years of employment required that they be able to look forward to a reasonably secure old age. Also, it has been assumed that compulsory retirement would be practicable only with adequate financial provision for old age.

A weakness of many retirement systems of the distant past was that they were tied to particular institutions or states and did not allow for full portability and continuity of pension plans when individual pro-fessors moved from one institution (or state) to another or moved out of academe to other industries. This situation was corrected by TIAA-CREF, which from the beginning provided full vesting and full porta-bility of accumulated pension rights. Later the problem of forfeited ben-efits was ameliorated in many public institutions when the state retirement system which covered them adopted shorter vesting periods or TIAA-CREF became an optional plan.

On the whole, the retirement arrangements existing in higher edu-cation have been viewed favorably by institutional leaders and faculty alike. This does not mean there have been no problems. For example, a vesting period longer than five years persists in some state institutions, and some retirement plans still provide quite modest benefits. But dis-satisfaction relating to the age of retirement or to the financial provision for it has been relatively low key and issues have been of modest dimension. In the 1970s, however, several changes occurred which gave rise to controversy and upset the previous equilibrium. Problems aris-ing from these changes have not yet been resolved.

Perhaps the most fundamental of these changes was demo-graphic—a notable increase in longevity, a growing percentage of older people, and a sharp decline in the number of births. From 1950 to 1979, average life expectancy at birth rose from 68.2 to 73.8 years and at age 65 from 13.9 to 16.7 years—the latter, a spectacular increase of one-fifth.

---

18. TIAA was founded in 1918 by the Carnegie Foundation for the Advancement of Teaching for the specific purpose of administering annuities and insurance services for teachers in schools and colleges.

During this period, the percentage of the population age 65 and over rose from 8 to 11 percent, and by the year 2000, it is expected to reach 13 percent. Moreover, with the drop in births in the 1960s and 1970s, the time was not far off when a dwindling number of young people would be called on to help support the increasing number of old people. The demographic developments, along with some other factors, had, and will continue to have, important effects on legislation. Growing awareness of the long-term national issues of financing Social Security and Medicare coincided with equity concerns and mounting political pressure to protect the employment rights of older people. In 1978 Congress amended the Age Discrimination in Employment Act, raising the minimum age for mandated retirement from 65 to 70 and outlawing age discrimination in employment for persons of ages 40 to 70.

Enactment of this legislation came just at the time when the higher educational community believed it was facing a need to retrench to meet an impending decline in enrollments, and also when it was worrying about the rising average age of faculty. In response to representations from members of the academic community, mainly administrators, colleges and universities were granted a temporary exemption; they would not have to comply immediately with the new minimum fixed retirement age for their tenured employees. The reprieve expired in 1982. Now a new proposal lingers in the halls of Congress. It would outlaw age-mandated retirement and, from age forty, prohibit discrimination in compensation, terms, or conditions of employment for reasons of age. If the proposed law is passed and if it includes higher education, American colleges and universities would have to adjust policies to conform with the new legal environment.

Demographic factors combined with inflation brought about another change that is now in an early stage but is almost certain to become an important development, namely, the revision of Social Security. Already part of the retirement benefit has become subject to income taxation. In addition, the government plans to raise the starting age for unreduced Social Security payments, and other adjustments in benefit structures are likely. The modifications will affect after-tax income and probably will alter attitudes as to the appropriate age for retirement. They may also affect the design of retirement plans in higher education, though it is too early to say in what specific ways.

A third area of change relates to the increasing restiveness of a minority of faculty and other higher education personnel concerning the investment, administration, and even the use of retirement funds. Their interests and concerns range over a wide area. Although many of the expressed desires and concerns are focused on TIAA-CREF, they often relate to the basic philosophy of retirement plans—to the purposes they are to serve and how they are to be achieved—and to the conflict and reconciliation of diverse (and often multiple) interests of institutions and of employees during their working life and in retirement. Some individuals claim that elements of the retirement system constitute unwarranted paternalism—that they should have wide latitude in choosing their investments and complete freedom to switch funds among them and to withdraw their capital accumulation in a single sum at termination. Some accept the annuity system as a necessary outgrowth of the institutions' interest in the plan, but desire various modifications, most often in investment policy. Others seem to desire little or no change. In response to the growing interest in reform, the Carnegie Corporation has taken the lead in establishing a Commission on College Retirement to review some of these matters. Discussions about features and administration of retirement plans will likely continue for some time.

Although it is still too early to know what the Commission will have to say, we should review here some of the indispensable features of a satisfactory retirement scheme. First, it should provide financial adequacy from the likely or "normal" retirement age throughout the rest of life. A guide often used in higher education is "an after-tax income equivalent in purchasing power to approximately two-thirds of yearly disposable income (after taxes and other mandatory deductions) during the last few years of full time employment" (Association of American Colleges and American Association of University Professors, "Joint AAC-AAUP Statement of Principles on Academic Retirement and Insurance Plans," 1980). Second, the amount made available for retirement should be geared to the likely retirement age so that financial considerations related to the basic plan do not unduly outweigh other factors in determining the time of retirement. The financial situation as retirement nears should be neutral; it should not militate too strongly against early retirement and should not build up pressure for delayed retirement. Third, expert retirement counseling should be available to faculty members. As they approach retirement, the counseling should inform them about the features of the plan in which they are enrolled and apprise them of possibilities and options connected with their retirement annuities, and at a still earlier age, advise them about tax-deferred securities for additional retirement savings (King, in Fuller, 1983, pp. 81–97; Jenny, Heim, and Hughes, 1979). Fourth, if the retirement age is to be flexible at the option of the individual there is great need of careful evaluation of faculty members to determine who should be encouraged to remain and who should be encouraged to leave. Since the Age Discrimination in Employment Act prohibits age-triggered evaluations, more general programs that include all faculty, or those of a given class, such as all full-time faculty, may have to be used.

The future of academic retirement is quite uncertain as it depends on the outcome of proposed legislation that would end mandatory retirement related to age. We believe, as we have shown in Chapter 3, that the problem of aging of the faculties has been exaggerated in two respects. First, it is based on a static conception of the faculties that minimizes the effect of steady turnover which year after year opens up places for younger persons. Second, it is based on an exaggerated notion of the decline in productivity of faculty members as they grow older. It should be noted that higher education is not the only industry that faces the prospect of aging staff. In fact, the whole society is aging. Organizations of all types will have to learn to conduct their affairs with older work forces, and many are already dealing with the problem (Lazear, 1983). There is nothing new about this. Human longevity has been increasing and health improving over centuries, and the labor force has steadily and routinely assimilated workers from ever older age groups.

The problem, as we see it, is *not* that the average age will be too high, or that there will be too many old people. The problem is that the age distribution may be uneven as we move from one decade to the next. There may be concentrations of new recruits in the youthful ages from 26 to 35 and in the older ages of 56 and above, and perhaps too few in the prime ages of 36 to 55. In our opinion the unevenness of the age distribution is something higher education can accept without great hardship. Moreover, there are partial remedies other than special treatment in pending retirement legislation. One of these is to provide retirement annuities on terms that will allow individual faculty members to retire gracefully at a time that is consistent with the mutual interests of the institution and of the faculty members. Another is to plan the recruitment of new faculty to fill in the age groups in which numbers are deficient. If there should be a "lost generation of scholars" as we move through the next several decades, perhaps some of them could be attracted to academe in mid-career. Perhaps also, if there should be a flight from the profession, the institutions could make special efforts to retain those most talented. Finally, the institutions could institute more effective faculty development which would improve the abilities, morale, and performance of the given faculty. As we see the situation, the problem is to recruit, retain, motivate, and train excellent people. The achievement of some particular age distribution is a minor—almost trivial—objective in comparison. Though admittedly there are particular institutions where the concentration of faculty in the older age groups may become serious.

There has been considerable speculation about the effect of mandatory retirement at age 70, or of no specific retirement age, upon the actual ages at which people would retire. In fact, the actual retirement ages will probably be more largely determined by people's attitudes about leisure and work and by their expectations about inflation than by changes in the law respecting retirement. A study based on a questionnaire directed to older participants of Teachers Insurance and Annuity Association revealed that 70 percent had a specific retirement age in mind and that the expected ages were as follows (Mulanaphy, 1981, p. 18);

60–64	11%
62–65*	5
65	37
66–69	11
65–70*	8
70	15
Other age	13
	100%

* The overlapping age groups simply reflect the ranges as specified by the respondents.

From these figures, one might infer that no more than 20 to 25 percent would elect to retire at age 70 or later.

To conclude, the arrangements for retirement are a major element of the faculty contract. They are of interest to young faculty recruits as well as to older faculty members. If they are adequate, predictable, and trustworthy they will go a long way toward providing the sense of security and collegiality so essential to academic freedom and scholarly excellence. These arrangements are the capstone of the major personnel provisions of academe. American higher education has over the years worked out enlightened retirement arrangements and has been an innovative leader in this field, especially through the efforts of TIAA-CREF. Though these retirement arrangements have not been without their detractors, they have worked well. What is now at stake is the ability of the higher educational community to attract and hold highly qualified people in the faculty corps. The retirement system is a critical factor determining the attractiveness of the holding power of the academic community.

Concluding Comments

We have considered the several elements of the traditional faculty contract which controls the relationships between most American colleges and universities and their faculties. This contract includes the following elements: appointment, tenure, involuntary separation, leaves of absence, compensation, the work environment, collective bargaining, and retirement. Whenever a person joins the faculty of a college or university, he or she becomes subject to the several elements of this contract. The contract is usually partly written and partly embodied in unspoken understandings and traditions.

In recent years, at least since 1970, all the elements of the contract have been under serious scrutiny. The element that has received the most intense criticism has been *tenure* in all its aspects. It has been inevitable that in a time of financial stringency hard-pressed administrators and boards would be searching for ways to cut costs, enhance flexibility, and unload contingent liabilities. The difficulty is that in their laudable effort to streamline and economize in the short run, they are likely to raise costs and cut quality in the long run. The weakening of tenure would be certain to reduce the power of academe to attract and retain staff. Furthermore, because it would undermine security and collegiality, it would impair morale even in the short run. The tactless references to "dead wood" and "new blood," the unconcealed glee sometimes shown over arranging an early retirement, and the operation of a revolving door policy in the recruitment and probation of young faculty are all unnecessary and ultimately counter-productive departures from well-understood tradition and are viewed by some, morally, if not legally, as breaches of contract. The accumulation of these events across the country simply makes the academic profession less attractive and bids up the price of recruiting new talent in the future. We think this is short-sighted policy.

Our position is that the traditional features of the basic faculty contract are well suited to the academic profession and to the kind of people whom it ought to employ, and we regard with deep regret major deviations from it. We recognize of course that involuntary separations are sometimes unavoidable. We ourselves have recommended loosening up the provisions of the contract relating to separations for reasons of financial exigency, cause, and poor health. But in each of these cases we would emphasize the importance of good faith and attention to correct procedures. Our judgment after a great deal of thought and consultation is that the higher educational community would be well advised to hold fast to tenure and to the related personnel policies.

CHAPTER THIRTEEN

Recommendations for the Recruitment and Retention of Faculty

The purpose of this study has been to describe the current condition of the American faculty and the principal changes that have occurred over the years 1970 to 1985. We have concluded that the professoriate is imperiled. In arriving at that conclusion, we have not claimed that the faculty or American higher education is now in grave danger. To the contrary, in some respects the faculty condition has improved. For instance, in recent years the quality of new entrants into the profession has remained steady or even improved. Moreover, American colleges and universities have held up remarkably well throughout a prolonged period of financial stringency. But despite the absence of a dramatic crisis, we have argued that an insidious deterioration of the quality of faculty life has taken place.

In our view three congeries of problems today press hard upon the professoriate and affect also those who contemplate academic careers: inadequate compensation, a deteriorating work environment, and an inhospitable academic labor market. The most visible of these three is the steep decline in real faculty compensation over the past dozen years. As serious as the erosion of faculty earnings is, less tangible developments pose an even greater threat to the faculty. After all, those who are attracted to "the life of the mind" rarely value income potential above other considerations.

Although less measurable than the fall-off in earnings, the decline in working conditions, as discussed in Chapter 7, is both multifaceted and widespread. This steady decay of various aspects of the work environment undermines both faculty performance and esprit.

Finally, the constraints in the current academic labor market impede mobility within academe and serve to discourage would-be professors from seeking academic careers. Even so, the existing imbalance between the copious supply of faculty aspirants, and the limited demand for them, will continue for another decade or so. By the mid-

1990s we foresee a changing academic labor market. We have contended that demand for new faculty will expand principally as a function of two developments: vacancies created by numerous retirements from the professoriate and an anticipated resurgence in college enrollments due to an increase in the number of 18-year-olds by the mid-1990s. However, we fear that by that time the corresponding supply of high quality aspirants to faculty careers may be inadequate.

The confluence of factors outlined above has made the academic profession less attractive relative to other prestigious but more lucrative professions. We believe that there will be no shortage of individuals willing to assume academic careers; to our knowledge colleges and universities have never been unable to recruit adequate *numbers* of faculty. The crux of the challenge is, and will be, one of quality, not quantity.

As we have repeatedly asserted, the nation's colleges and universities must maintain the ability to recruit excellent new faculty as openings occur. Especially, they should enter the period of heavy recruiting from 1995 on, when several hundred thousand persons will have to be appointed, prepared to attract faculties of outstanding quality. There is no academic objective more imperative than this for the institutions themselves, for their students, and for the nation. But it would do no good to recruit outstanding new talent if the most gifted faculty members—both old and new—were slowly drifting away as a result of inadequate compensation, substandard working conditions, and depressing prospects. With the smaller intake of new faculty and the higher risk of losing the best older talent, the need becomes urgent to make every position count—those filled and those unfilled. The care and encouragement of capable members of the present faculties should go right along with recruitment of gifted new faculty persons, and both should be among the highest priorities of academic leadership.

We believe the prevailing situation is dangerous but not yet desperate. Interventions that can allay harmful consequences are possible and, indeed, are essential. The changes in policy that we deem most important are discussed below. We have organized our recommendations according to the sector most directly responsible and able to make the desired change. For these purposes we have directed our recommendations to colleges and universities and to government (the federal government in particular). We conclude our discussion of an agenda for revitalization with an examination of the current educational reform movement and its implications for the professoriate.

The Role of Institutions

For more than a decade, colleges and universities have experienced financial stringency, and have also been bombarded with predictions of declining enrollments and commensurately declining revenues. In this environment, they have become intensely risk conscious. They are hesitant to make commitments for long periods lest enrollments fall, tuition revenues decline, and appropriations diminish. This risk consciousness lies beneath the many efforts to avoid or sidestep or delay the granting of tenure. Among the expedients are: increasing the employment of non-tenure-track faculty (some part-time and some full-time), lengthening the period of probation for tenure-track appointees, increasing the severity of the standards for advancement to tenure, converting tenure-track positions to non-tenurable appointments bearing designations such as "clinical" and "research" faculty, and slowing up the rate of promotion through the ranks to full professor.

All of these practices when applied excessively tend to impair the attractiveness of the academic profession to young aspirants and to discourage talent from entering the profession or staying in it. As the risk for institutions is lessened, the risk for new appointees and younger faculty is increased. Perhaps even more serious, the overall caliber of the faculty is diminished as compared with what it might have been if a larger proportion of the faculty could have been firmly attached to the institutions as career members. This is not to say that there is no place for part-time or temporary full-time people, but rather that in a time when vacancies are scarce they should be filled so far as possible with capable career persons who are worthy of tenure and who have a large stake in the institution and in the profession. The revolving door is not an appropriate metaphor for faculties whether in good or bad times.

There are at least four possible ways of reducing the institutional risk connected with employment of career faculty. The first is implicit in our recommendations regarding faculty tenure. We have proposed (Chapter 12) that two provisions of the current rules on tenure should be eased to give the institutions somewhat greater latitude. These were the rules on dismissal of faculty in case of financial exigency and on dismissal for cause. These changes would tend to reduce slightly the institutional risk of adding to tenured faculty and thus open up a few more career opportunities. Such latitude is subject to administrative abuse, and, accordingly, carefully specified criteria would be necessary to serve as guidelines.

A second possibility is that institutions might evaluate the risks more realistically. We suspect that most academic people, given the continual warnings about declining enrollments, have underestimated the number of positions likely to open up through the turnover of faculty. Careful estimation of turnover might contribute toward more realistic estimates of the number of positions that will fall vacant. Third, we urge colleges and universities, where possible, to use strategies to maintain vitality in academic departments that are heavily tenured-in with little prospect for turnover in the near future. At some campuses we found that the campus administration had authorized an academic department to recruit a junior faculty member even though current enrollments were insufficient to justify adding an additional faculty member. As part of the bargain the department understood that it would not be authorized to replace a senior faculty scheduled to retire in several more years. Though initially costly to the campus, the strategy appears to have much potential for reinvigorating departments whose members have been growing older together without the benefit of freshly trained younger colleagues.

A fourth possibility is to raise funds specifically designated for the appointment of younger faculty to tenure-track positions.[1] It is fortunate that leading corporations have shown a strong inclination to provide funds for supporting young faculty in fields of special interest to business such as science, engineering, computer science, and management. Corporations' and universities' interests converge with programs that provide corporate support to help universities recruit and retain highly able faculty in critical shortage areas (so long as university autonomy is not compromised). For example IBM has provided substantial support to enhance compensation and to provide equipment for faculty in computer science, electrical engineering, and materials science. Another leading example is the Faculty Development Programs designed by Hewlett-Packard Company, in conjunction with the American Electronics Association, to help develop new electrical engineering and computer science faculty.[2] Corporate gifts of company equipment and other products, especially in the technological realm, are critically important to replace outmoded instrumentation and to enable faculty (and students) to stay more nearly abreast of rapidly changing technologies.[3]

Funds for new faculty, or to provide chaired professorships in other disciplines, might be raised from corporations and individuals. We recognize that it will be more difficult to demonstrate to corporations the mutuality of interest so evident in the cultivation of faculty in technological and management-oriented fields.

Other Campus Initiatives. The recruitment of a career faculty member represents a potential investment of more than a million dollars. It deserves the same amount of attention and care that would be given to the design and construction of a small academic building or to the pur-

1. For example, in announcing a $400 million fund-raising effort to increase endowment at Columbia University, President Michael Sovern declared that the proposed additions would include $28.7 million "to support positions for younger faculty members as well as $79.5 million for professorships" (*Chronicle of Higher Education*, Nov. 17, 1982).

2. A "Catalog of Industrial Programs To Aid Graduate Engineering Education," produced by the Project on Engineering College Faculty Shortage, Eleven Dupont Circle, Suite 200, Washington, DC 20036, concisely describes about three dozen corporation and foundation programs, many of which are intended to support promising young faculty members.

3. Corporate giving to education reached record heights of nearly $1.3 billion in 1983; approximately $1 billion of that amount went to colleges and universities, a very significant sum (*Chronicle of Higher Education*, Nov. 28, 1984, p. 22).

chase of a major piece of equipment such as an advanced computer. The recruitment process demands of institutions care, imagination, and aggressiveness. This is true year in and year out, regardless of the market. But it is especially true in the years ahead when the academic profession may not spontaneously attract the cream of American talent.[4]

Under present and prospective conditions, the academic profession as a whole is less than attractive even to persons who are exceptionally talented for a life of teaching and scholarship and strongly inclined toward it. The outlook, as they perceive it, is that after many years of advanced study, getting a suitable job will be highly uncertain, and even if a job is found, the chance of advancement to tenure and up through the ranks will be problematical. Moreover, working conditions will likely be less than ideal. This situation might be improved if individuals could establish relationships with future employing institutions before making the commitment to graduate study (or perhaps during the early years of graduate study) and could maintain these relationships until they are ready for appointment to full-time positions. We are suggesting that through *early* "recruitment" of prospective faculty and through maintaining continuing relationships with them during their graduate years, institutions might attract outstanding talent that would otherwise escape to other professions.

The phrase "continuing relationship" is intentionally ambiguous. Certainly it does *not* mean a binding contract. It might be likened in some ways to "farm systems" of professional baseball. It does mean that talented young academics would be identified while they are still in college or in graduate school, that their progress would be encouraged and observed, and that entrée would be established for discussion of an appointment even though the certainty of a contract would be missing.

Another fruitful source of talent that might be cultivated would be the thousands of highly educated and capable people in industry, government, and the independent professions. Efforts might be made to identify some of those who might be attracted to the academic profession. Indeed, many of these might already have been in academe if the market had been stronger.

Our suggestions concerning faculty recruitment are intended to apply to a period when the number of new appointments will be relatively few and first-class talent will be scarce. Our message is simply that aggressive and imaginative recruiting will help to bring about the strengthening of the faculties even in a period of declining faculty numbers. Faculties could become stronger even though smaller. But this would not happen without sustained effort and attention to faculty-building.

4. In this connection, an interesting source is the *1982 Idea Handbook* of Sidney G. Tickton (1982), in which he reports what colleges and universities are doing toward attracting and retaining highly qualified young faculty members.

Leaving aside the recruitment imperative, we believe that campus administrations must take more initiative in addressing the problems of their existing faculties. We have repeatedly spoken to the urgency of improving both faculty compensation and working conditions. While these objectives are paramount, we recognize that administrations and faculty together cannot effectuate major improvements by themselves. But efforts must not be spared to impress upon public agencies and private benefactors, from individual alumni to corporate giants, the urgency of the situation.

Beyond these obvious needs, we recommend that other measures be taken. We believe that campus administrations, too often overworked and underappreciated, sometimes lose sight of the human element. Accordingly, campus administrators, from the chief executive officer through department chairs, should be sensitive and humane in relations with faculty. As institutions of higher learning, colleges and universities are great repositories of knowledge about enlightened human relations and employee-sensitive management techniques. But as employers, colleges and universities too often behave toward their professional staff in an inconsiderate manner. We perceive widespread distress among faculty members, junior and senior alike, on many campuses. Some of the anxiety can be alleviated by intensified efforts to be supportive of faculty wherever possible.

We believe that faculty career counseling programs, such as that pioneered at Loyola University of Chicago, hold promise for enriching faculty careers. With emphasis on one-to-one career consulting, the use of support groups, and career-related workshops, such programs extend beyond the more limited objectives of conventional faculty development programs. Skilled faculty career counselors can enable faculty to address key career issues knowledgeably and creatively.

We urge that universities wherever possible mount special efforts to increase financial aid to potentially outstanding Ph.D. students. The newly established Regents Fellowship Program of the University of Michigan provides a promising example, one that is all the more noteworthy because of the financial constraints with which the University has had to cope in recent years.

As noted in Chapter 12, we believe that campuses should be cautious about resorting to differential pay policies for faculty. We believe that a strong presumption should obtain in matters of academic compensation: controlling for rank, experience, and quality, one faculty member is equal to another. We fully recognize that the marketplace has always influenced academic compensation. But we fear that both campus administrators and faculty members in favored fields underestimate the adverse consequences of opening wide this Pandora's box. We have seen too many examples of ill will, even bitterness, among faculty, as well as between faculty and administrators, that derives from unrestrained differential pay policies.

We urge also that administration and faculty together strive to invigorate faculty senates and other instrumentalities for involving faculty in the difficult choices inherent in the distribution of scarce resources. Too often the faculty role in governance has been allowed to deteriorate. Meaningful participation in the governance process is important for faculty morale and vigor, and steps ought to be taken to revitalize significant faculty involvement in the governance process.

The Role of Government

We believe that conditions in the academic labor market are serious enough to justify governmental action toward faculty recruitment. We recognize, as we write in 1985, that the prospects for new governmental initiatives seem less than bright. The federal deficit, the military build-up, and the inhibitions against raising taxes tend to close off federal funds for social programs. The states also are threatened by the withdrawal of federal funds and by the shifting of fiscal responsibility from Washington to the state capitals. Nevertheless, the nation cannot go on indefinitely ignoring its educational and social obligations any more than it can indefinitely neglect to maintain its physical infrastructure. We believe, therefore, that it is worthwhile to suggest some modest public programs to help maintain and improve the nation's professoriate.

Others have commented in detail about changes in federal policy that would serve to reinvigorate the higher education enterprise and to strengthen the American faculty in particular.[5] We do not intend to restate here the numerous meritorious proposals that range from protecting the charitable deduction to resuscitating nearly extinct federal fellowships and traineeships, and from expanding federal research support in the social sciences and humanities to providing support for colleges and universities to renovate decaying facilities and to upgrade outmoded instrumentation. Rather, we set forth now four specific suggestions.

Bridging the Trough. As most observers envision the coming situation of colleges and universities, many or most will experience declining enrollment beginning in the mid-1980s, reaching a low point or trough around 1995–96, and recovering thereafter until about 2005 or 2010 (see Figure 13–1). Not all institutions will have the same experience. Some will have deeper troughs than others, some will have none at all, and the timing of the downward and upward swings will vary.

5. For other proposals see National Research Council, 1979, pp. 84–85. See also, for example, Anderson and Sanderson (1982); Butler-Nalin, Sanderson & Redman (1983); National Commission on Student Financial Assistance (1983); Council of Graduate Schools in the U.S. (1984).

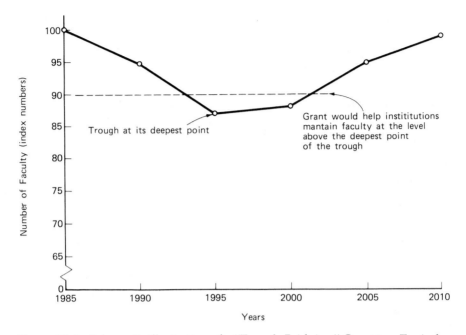

Y-axis: Number of Faculty (index numbers)

X-axis: Years

Labels: Trough at its deepest point

Grant would help instititutions mantain faculty at the level above the deepest point of the trough

Figure 13–1. Schematic Illustration of a "Trough-Bridging" Grant to a Typical College or University

But most will probably go through a cycle of decline in number of students followed by recovery. They will be in a position of having to reduce faculty numbers during the downward phase of the cycle and then to turn around immediately and increase the numbers during the upward phase. The proposal is simply to provide special grants to enable them to reduce the amplitude of the swing, that is, to "bridge the trough" during the years when it is at its deepest (Radner & Miller, 1975, pp. 340–46). The faculty members supported by the proposed grants would be expected to contribute toward qualitative improvement of the institutions. Institutions would qualify for the grants on the basis of formal proposals.

One can imagine many possible administrative variations connected with such grants. We would favor a simple plan in which an appropriate government agency, state or federal (or a donor, individual or corporate), would agree to make grants to recognized colleges or universities on conditions such as the following:

1. The grants would be designed to help institutions maintain or increase the number of their faculties during a uniquely difficult period.
2. The grants would be competitive; they would be awarded on application of the institutions provided that the applicants would:
   (a) Indicate how the extra faculty afforded by the grant would be used for improvement of educational quality, and

    (b) Provide evidence that the timing of the grant would be appropriate to the demographic situation of the particular institution.

3. The grants would be made for periods of up to five years.

4. The grants would support only full-time faculty on tenure or tenure track or having other long-term attachment to the institution.

5. The amount of the grant for each institution would be calculated as a percentage (perhaps 2 to 5 percent) of the salary and fringe benefits for all tenured and tenure-track faculty.

We estimate the cost of this plan on a nationwide basis at roughly $140 million annually or $700 million over five years.[6]

The proposed grants could be administered in various ways. They might be provided under a nationwide federal program, or by the states individually, or by some combination. They might not become part of a formal and visible program, but rather might simply be incorporated in the regular higher educational appropriations of state governments—as incremental, not replacement funds. There are already indications that some state legislatures are not cutting appropriations in proportion to enrollment declines.

Recruiting and Retaining Exceptional Talent. As a more comprehensive—yet economical—way of strengthening America's faculties over the next several decades, we propose a national program designed to facilitate the recruitment, education, placement, and professional development of persons of exceptional talent who are interested in and suited to academic careers. There is considerable precedent for such an initiative in programs of the Woodrow Wilson Foundation, the Danforth Foundation, the Mellon Foundation, corporations such as Exxon and Bell Labs, and the many training grants of the federal government (Scully, 1982). These programs have provided awards to help finance graduate study for persons of exceptional promise who profess an intent to pursue an academic career. What we are suggesting is an integrated plan with several related purposes each designed to lift the quality of the faculties.

The proposal is in four parts:

A. Fellowships for graduate students
B. Faculty appointments for young scholars
C. Research support for young scholars
D. Support for personal development of faculty in mid-career

---

6. These figures assume: 425,000 full-time tenured or tenure-track faculty; average compensation per person at $33,000; grants at 3 percent of the total compensation of all tenured and tenure-track persons in institutions qualified for grants; and one-third of all institutions qualifying. Thus, the proposal envisions that a small college with, say, 150 full-time faculty members might, if its application were approved, receive $148,500 (calculated at 3 percent of faculty compensation) for each year the bridging plan were in effect; a university with 800 full-time faculty might receive $792,000.

The proposed plan might conceivably cover all academic disciplines. We suggest, however, that it be limited to persons whose academic disciplines lie in the arts and sciences. This limitation does not imply that other fields do not merit assistance but only that the greatest danger of faculty deterioration lies in the arts and sciences. The plan would be devised so that all eligible academic fields, all types of institutions, and all geographic areas of the country would be represented. The plan would be administered by a public or private foundation, and financed mainly with federal funds supplemented by private gifts.

Part A would provide fellowships for outstanding men and women who wish to prepare for faculty careers in the arts and sciences. The objective would be to increase the flow of exceptional talent to college and university faculties beginning on a small scale about 1986–87 and expanding to the period of heavy faculty-rebuilding between about 1995 and 2010. The program would taper off as the regular financing of higher education improved, and would end when regular financing had been restored to proper levels. The program might start with 300 fellowships awarded in 1986 and build up to perhaps 2000 a year by 1995. The objective would be to make available to the academic community each year about 2000 bright, well-educated, and fully-qualified people. The number 2000 would be roughly equal to one percent of the full-time arts and science faculties in all institutions of higher education in the United States.

The fellowships would be available to carefully selected graduates of accredited four-year institutions[7] who are admitted to an approved graduate school as full-time students in the humanities, social studies, or natural sciences. Approximately 75 to 100 graduate schools would be selected as eligible for participation. They would be chosen for excellence of programs but with concern for representation of all major regions of the country. The fellowships would be distributed equitably among the participating graduate schools and among the several disciplines. Under this arrangement, each participating graduate school would be assigned a pre-arranged number of fellowships in designated fields. For example, each graduate school on the average might enroll twenty to thirty fellows—enough for a critical mass.

The administering foundation would select the participating institutions, assign fellowships by fields to each institution, recruit and select the applicants, and arrange for the placement of those selected. Placement would of course be final only when an applicant had been admitted to a participating institution.

To qualify for the program, participating graduate schools would offer high quality graduate instruction appropriate to the Ph.D. in a major field or fields designated, would provide supervised apprentice-

---

7. The fellows would be college graduates, but the awards would not necessarily be confined to recent graduates.

ships in teaching, and would offer a continuing seminar relating to the philosophy and practice of higher education. This seminar would be supplemented by regional or national conferences and workshops sponsored or conducted by the foundation. Institutions would receive, in addition to regular tuition from students enrolled, grants to cover the costs of these special services. The end product should be capable young men and women who are well educated, and well prepared for teaching, scholarship, and academic citizenship. Though not all of them would choose to enter academe, most of them would. A steady flow of these people into our colleges and universities as faculty members would almost surely make a substantial difference in the quality and morale of the academic community. The flow might begin on a small scale in the early 1990s and expand after 1995 when increasing faculty recruitment would become necessary.

Part B of the plan would be a counterpart of Part A. It would help provide demand to match the supply created by Part A. Part B would provide grants to colleges and universities (including all types of institutions from research universities to community colleges) for the recruitment of outstanding young faculty. The aggregate number of grants would be equal to the number of Part A awards. However, the job openings generated by Part B of the plan would not be restricted to persons whose education had been supported by Part A funds, and the part A recipients would be free to seek employment elsewhere. All accredited colleges and universities would be eligible to apply for the grants. The new appointees would be either alumni of the fellowship program or persons of comparable basic ability, general education, mastery of a major field, and knowledge of higher education. The openings would be solely full-time faculty positions on a *bona fide* tenure track.

The basic condition of the Part B grant would be a showing to the effect that the appointment would represent an urgently needed strengthening of the faculty. The amount of the grant might be on a sliding scale for three years, for example, 75 percent of salary and fringes in the first year, 50 percent in the second year and 25 percent in the third year. To this might be added a one-time payment for recruitment and moving expenses.

To achieve the purpose of the institutional grants it would be necessary to diffuse the new talent among the major universities, four-year colleges, and community colleges. Without preventive measures, the appointees would tend to congregate in the most prestigious institutions, and only a few of them would find their way to lesser institutions. To overcome this problem, two conditions would broaden the program to serve all types of institutions. One would be to offer larger grants to the second tier institutions to enable them to offer relatively higher salaries than the first tier institutions. The other would be to divide the candidates into three "non-competing" groups, those with strong research interests who aspire to places in major universities;

those with interests more focused on teaching and who are headed for other four-year institutions; and those interested in community colleges. The various constraints imposed on the plan would be designed to tilt it in favor of geographic distribution, distribution among types of institutions, and distribution among disciplines. The program is conceived as applicable to all parts of the higher educational system.

Part C of the plan would offer grants to assist exceptionally promising young scholars under the age of thirty-five to carry on important research or other scholarly and creative work. The grants would allow release from teaching amounting to half-time over as many as three years. It would also provide modest funds not exceeding $3000 for travel, equipment, books, and other expenses. The average annual cost of such grants (in 1985 wages and prices) would probably be of the order of $20,000 for salary replacement and $3000 for expenses or a total of $23,000. Perhaps five hundred such grants might be in effect each year.

Part D would provide grants for the personal development of faculty in mid-career. The grants would allow released time ranging from a semester to two years for personal development of a kind that would make faculty members more valuable to their institutions. It might involve learning new skills, shifting from a discipline in low-demand to one in high-demand, preparing for an assignment in academic administration, preparing for leadership in faculty affairs, devising new curricula or courses, etc. The cost of such grants might range from $15,000 to as much as $80,000 and would perhaps average $35,000. As many as five hundred such grants might be in effect each year.

The four parts of the proposal when combined would constitute an integrated plan designed to help recruit able people into the profession, to place them in congenial positions where they could make maximum contributions, and to encourage their development. Another purpose would be to make a public declaration to the effect that the maintenance and strengthening of higher educational faculties is an urgent goal of the nation.

The cost of the proposal would be substantial, but not prohibitive. Cost estimates for the four parts of the program are presented in Table 13–1. As the table shows, when the program had hit its stride, the annual direct costs would be nearly $195 million in dollars of 1985 purchasing power. To this should be added perhaps $15 million annually for administrative expenses, making the grand total around $210 million. One important feature of this cost is that much of it would be relief of institutional budgets. This would be true of the tuitions paid by the graduate fellows (Part A), the salary payments to new faculty members (Part B), a portion of the research grants (Part C), and a portion of the personal development grants (Part D). We estimate that at least half of the total cost would flow as income or budgetary relief to the institutions.

Recruiting Minorities. We believe that the need to recruit able per-

sons of minority background into academic careers is particularly compelling and, for reasons discussed in Chapter 8, the outlook for the
recruitment of minorities is bleak. We suggest the creation of a special
not-for-profit corporation, to be supported by public and philanthropic
funds. The organization's basic responsibilities would be the identification of very talented minority undergraduates and the coordination of
a national fellowship program. The national identification program
might be fashioned after the National Merit Scholarship Corporation's
National Achievement Scholarship Program for Outstanding Negro
Students (NASP). That program, established in the mid-1960s, was
designed to identify academically talented black high-school students
and to provide their names to colleges and university admissions
offices.[8] We propose a process that would first identify highly able
minorities, especially blacks and Hispanics, toward the end of their college sophomore years. Many campuses now house administrative units
which have special responsibilities for minority affairs. Such offices
could be effective in helping to identify prospects. Students so identified would be encouraged both by mentoring faculty and through suitable materials generated by the national corporation to contemplate
graduate study and academic careers.

A sizable number of seniors—perhaps 500 a year—would be
awarded portable fellowships for graduate study, primarily in the arts
and sciences. We envision a competition modeled after the Woodrow
Wilson Fellowship Program. These fellowships would need to be generous—possibly in the range of $12,000 to $16,000 annual stipends on top
of full tuition for several years—in order to be competitive with other
career and professional school options. The scholars designated by this
means would morally obligate themselves, as was the case of the Woodrow Wilson Fellows, to undertake academic careers.

The federal government currently supports several important programs for minorities in graduate and professional study.[9] But these programs, however meritorious, are not adequate: the amount of financial
aid a student can receive is far too limited, and the number of students
too few.

A major national effort to attract minorities to academic careers
should be launched in the near future and generously supported.
Unless this is done, we believe that affirmative action policies, even if
carried out in good faith by campuses, will be inadequate to prevent
serious erosion in the numbers and quality of minorities in the academic profession.

8. The NASP initially provided fellowships for those ultimately designated as Scholars,
similar to the program for National Merit Scholars.

9. Several existing programs are relevant and helpful, in particular programs sponsored
by the National Science Foundation and the National Institutes of Health. The Graduate
and Professional Opportunities Program (GPOP) for minorities and women is also relevant.

Table 13-1. Estimated Cost of Proposed Program of Faculty Support

Academic Year	Part A[1] Number of graduate students supported	Part A[1] Total cost at $10,000 per student (millions)	Part B[2] Number of faculty supported	Part B[2] Total cost at $14,000 per position (millions)	Part C[3] Number of faculty receiving grants	Part C[3] Total cost at $23,000 per grant per year (millions)	Part D[4] Number of persons receiving grants	Part D[4] Total cost at $35,000 per grant per year (millions)	Combined cost of four programs (millions)
1986–87	300	$3	300	$4	167	$4	500	$18	$29
1987–88	800	8	800	11	333	8	500	18	45
1988–89	1500	15	1500	21	500	12	500	18	66
1989–90	2400	24	2100	29	500	12	500	18	83
1990–91	3200	32	2700	38	500	12	500	18	100
1991–92	4000	40	3300	46	500	12	500	18	116
1992–93	4800	48	3900	55	500	12	500	18	133
1993–94	5600	56	4500	63	500	12	500	18	149
1994–95	6400	64	5100	71	500	12	500	18	165
1995–96	7100	71	5600	78	500	12	500	18	179
1996–97	7600	76	5900	83	500	12	500	18	189
1997–98	7900	77	6000	84	500	12	500	18	191
1998–99	8000	80	6000	84	500	12	500	18	194
1999–2000	8000	80	6000	84	500	12	500	18	194

[1] Assumes support of 300 new graduate students a year at the beginning of the program in 1986–87, gradual growth to 2000 a year by 1995–96, and leveling off at 2000 thereafter. Also assumes four-year average duration of the grants. The average grant would be about $10,000 per year.

[2] Assumes support of 300 faculty members at the beginning of the program in 1986–87, gradual growth to 2000 a year by 1995–96, and leveling off at 2000 thereafter. Also assumes three-year average duration of the grants with payments equal to 75% in the first year, 50% in the second year, and 25% in the third year. The *average* grant would be 50% of the recipient's compensation or about $14,000.

[3] Assumes support of 167 faculty members at the beginning of the program in 1986–87, growth to 500 in 1988–89, and leveling off at 500 thereafter. Also assumes three-year average duration of the grants. The average grant would be $23,000, $20,000 for salary replacement and $3,000 for expenses.

[4] Assumes support of 500 faculty members each year. Assumes also one-year average duration of the grants and average grants of around $35,000.

281

The Faculty and Educational Reform

Our advocacy of a series of proposals intended to strengthen a profession under stress would be incomplete without consideration of the educational reform movement that appears to be gathering momentum in the mid-1980s. We have no quarrel with the need for substantial reforms designed to improve the quality of higher education, particularly undergraduate education—quite the contrary. We are enthusiastic supporters of the reform movement. Our concern is that too often the faculty are depicted as obstacles to significant reform; too seldom is the difficult position of the professoriate adequately comprehended. In the following pages we discuss the background and origins of the reform movement and then consider several policy options.

The American people, long concerned about the need to extend educational opportunity to all, have become increasingly worried about the *quality* of the education being dispensed. Adverse criticism has become widespread. The public concern has perhaps been deepest as directed toward the elementary and secondary schools, and this concern has been eloquently expressed in the famous report, *A Nation at Risk* (National Commission on Excellence in Education, 1983). But the colleges and universities, especially their undergraduate programs, have not gone unscathed. The widespread opinion that the quality of undergraduate education needs improvement has been expressed in a series of widely publicized reports.[10]

The urge for reform has derived from a variety of allegedly unsatisfactory conditions in the higher educational system. These may be divided into roughly three categories pertaining to students, faculty, and curricula.

Students. Too many qualified high school graduates do not attend college at all, and too many of those who do attend drop out before graduation. Of those who attend, too many come with inadequate preparation—especially in the verbal, quantitative, and analytical skills—and during their college years do not find sufficient guidance or adequate remedial programs. Furthermore, an increasing proportion of students attend part time or commute from their homes to the campus and thus do not receive the total immersion in college life available to full-time residential students.

Faculty. Many faculty members are poorly prepared for the role of

10. Commission on the Humanities, 1980; Boyer & Levine, 1981; National Commission on Higher Education Issues, 1983; National Commission on Student Financial Assistance, 1983; Business-Higher Education Forum, 1983; Commission on Strengthening Presidential Leadership, 1984; Study Group on the Conditions of Excellence in American Higher Education, 1984; William J. Bennett and Study Group on the State of Learning in the Humanities in Higher Education, 1984; Project on Redefining the Meaning and Purpose of Baccalaureate Degrees, 1985; National Commission for Excellence in Teacher Education, 1985.

undergraduate teacher. The faculty in place during the 1980s is generally considered to be highly capable in terms of intelligence, energy, advanced degrees, and mastery of their special fields. Yet, in the opinion of many thoughtful observers, they have serious shortcomings: their education has been highly specialized and lacks the breadth that would be desirable for the mentors of undergraduate students; they identify more fully with their disciplines than with their institutions, and many of them are more highly motivated for research than for pedagogy; their teaching style fails to take advantage of innovative teaching methods of proven efficacy. Moreover, too much of undergraduate teaching is turned over to part-time faculty, to graduate assistants, and to temporary appointees rather than assigned to well-qualified career people. Finally, the morale of faculty members in many institutions is shaky as a result of the slow but steady deterioration of pay, working conditions, and job security.

Curricula. The undergraduate curricula in many institutions are out of kilter. This condition is due in part to unrelenting financial pressure which leads to an intense market orientation of institutions and also to keen internal competition among departments within institutions. It is due in part to the excessive demands of professional accrediting agencies. It is due also to the tendency of institutions to appoint, compensate, and promote faculty members on the basis of their achievements in research and scholarship rather than on the basis of their excellence in teaching. It is exacerbated by the failure of presidents, academic vice-presidents, and deans to provide leadership in the development of educational policies and plans.

The three-part diagnosis of undergraduate education is in our opinion applicable in varying degrees to many—perhaps most—institutions. The several national reports we have cited all contain recommendations intended to correct some or all of the ailments specified in the diagnosis. Their general thrust is toward greater emphasis on undergraduate education and toward resuscitation of liberal education. The report of the Study Group sponsored by the National Institute of Education was most specific in its recommendations. It presented twenty-seven proposals for the improvement of quality in undergraduate education. Of these, twenty-one would directly affect the work—and the instructional load—of faculty. In addition to calling for greater attention to undergraduate instruction and to liberal education, the NIE Report recommends greatly increased personal contacts between students and faculty, instructional styles that involve students as active learners, expanded remedial programs, major emphasis on assessment of learning, and greater weight given to *teaching* in the appointment and promotion of faculty.

We believe that a major effort to improve undergraduate education along these lines is urgently needed, and we support the various commissions and study groups that have so persuasively identified the need and pointed toward the solutions. We would suggest, however, that adoption of these recommendations would clearly require a substantial increase in faculty time, energy, and skill devoted to undergraduate instruction. This is, indeed, the very essence of the recommendations. How would the increased faculty time and energy be mobilized? The answer to this question is not readily apparent. Our institutions of higher education are not blessed with great quantities of surplus faculty and other resources. After fifteen years of tight budgets, most have little organization slack that could be put to new uses. As we have indicated in Chapter 5, faculty members on the average already work long hours and extraordinarily hard. To ask them to increase their effort in a good cause might bring forth a favorable immediate response, but would be unlikely to produce sustained results year after year. In the past, many calls for increased workloads in the name of educational reform have elicited extra effort in the short run, but have seldom made an appreciable difference in the long run (Grant & Riesman, 1978). The modes of instruction and the general conduct of the higher educational enterprise do not change much over the years. And the financing of higher education is generally geared to the assumption that it will continue more or less in the traditional style. What are the options?

First, in all colleges and universities, as in all other organizations, there is a small minority of people who do not carry their weight and who might be called upon to give more effort to undergraduate instruction. These, however, are the very persons who would probably not rise to the imagination and hard work required for successful educational reform.

Second, a possibility, often suggested, is to cut back on research and shift the faculty time and effort thus released to the task of planning and implementing the reform of undergraduate instruction. This would be a possibility but should be approached with caution. Much of the research that goes on in higher education is of extraordinarily high quality, and is a critical element in the progress and welfare of the nation. The persons engaged in this type of research should not be hampered in the interests of improving undergraduate instruction. However, some great scientists and scholars do engage in undergraduate instruction with profit and pleasure to themselves and to their students, and this is a practice which on a moderate scale could be encouraged. At the other extreme, some of what passes for research among academics is trivial. This kind of research might be discouraged but the people thus released would mostly not be the ones chosen for educational reform. Between the extremes is a large volume of research that is not of earth-shaking importance but is nevertheless socially valuable and could perhaps be moderately curtailed without great loss. However, the "in-between" research is valuable not only because much of it yields important contributions to knowledge, but also because it is a vehicle by which faculty members can keep abreast of their fields, can continue throughout their lives to be learned men and women, and can share in the joy of discovery.

Third, another candidate for the release of faculty time and effort is to reduce committee work and other forms of institutional service. Unfortunately, educational reform is likely to require a substantial increase in time and effort devoted to such activities, though at a time of substantial educational reform, other initiatives involving committee work might be sidetracked.

Fourth, still another possible way to shift resources to educational reform would be to "buy" the time and energy needed by reducing the amount of resources applied to conventional education. There are several possibilities. One would be to streamline the curriculum by offering fewer undergraduate majors and by reducing the number and size of graduate and professional programs. Another would be to adopt the long-neglected proposals of Beardsley Ruml and Donald Morrison (1959) by conducting some instruction in very large lecture courses at low cost per student thus releasing faculty time for the intensive and personal instruction called for in the reform proposals.

Still another idea would be to conduct more instruction through programmed independent study.[11] Experience has shown, however, that efforts to cut costs by streamlining curricula and modifying instructional methods are not attractive to institutions partly because they may adversely affect enrollments and public relations, and partly because they run counter to the personal touch in teaching—which is the very thing being sought in the current round of reform proposals. In our opinion, some economies are possible that would help finance a program of educational reform, but they would be difficult to achieve and small in amount. They should not be totally dismissed, however, as every little bit counts when mobilizing resources for a significant purpose.

The report sponsored by the National Institute of Education (Study Group on the Conditions of Excellence in Higher Education, 1984) dealt with the problem of mobilizing resources as part of their recommendation of faculty salary increases. They commented as follows (pp. 71–72):

> . . . in the context of this report, it is insufficient to suggest that simply raising salaries will automatically focus the faculty's attention on the improvement of undergraduate instruction. If our recommendations— particularly those pertaining to assessment—are adopted, faculty will have to do more than they do at present: to learn more, to stretch further, to devote more of their time and energies to involvement with students.

---

11. These and other suggestions for reducing instructional time and effort are outlined in some detail in H. R. Bowen & Douglass, 1971. As shown in this study, various modes of instruction call for considerably different amounts of faculty time and effort.

> While we have suggested incentives for some of these added activities, we regard others as so central to the purpose of undergraduate education that they should be regarded as normal duties of employment. A majority of faculty members are dedicated and hard workers, but if we are asking them for more work, we simply will have to pay for it.

This statement suggests that the proposed reforms are so urgent that they should be adopted by the institutions as faculty overloads which would be compensated by increased salaries. This solution might work in the short run, especially if, under persuasive leadership, faculty members would develop enthusiasm for the reforms. In the long run, however, faculty are not likely to continue indefinitely with an overload that is on top of an already heavy basic load. They are likely to curtail gradually their non-teaching activities such as research, "keeping up" with their fields, committee work, participation in co-curricular programs, etc., and also to cut back on their efforts in undergraduate instruction. In the end, the total workload would likely revert to its original level. This result might not be too bad, however, if in the process significant additional time and effort were permanently allotted to undergraduate teaching.

The underlying condition, which leads to difficulty in rearranging resource allocations, is that most institutions, after fifteen years of austerity, are simply underfunded in relation to their present programs and in relation to their conception of what kind of institutions they might be. The richer institutions could, if their constituencies would cooperate, make substantial shifts of resources into undergraduate education without serious detriment to their overall programs. But the poorer institutions do not have the leeway to make such shifts without crippling their basic programs and even jeopardizing their existence (H. R. Bowen, 1980b, pp. 246–53).

The effect of the proposed educational reforms on the recruitment and retention of new faculty is problematical. Would the reforms make the academic profession more or less attractive? If prospective faculty recruits would see the proposed educational reform as an exciting and rewarding way to serve the youth of America, their response might be positive. If they saw the proposals as unreasonably onerous and underpaid, the proposed reform would repel them. We do not know which way they would turn. We suspect, however, that effective institutional leadership would be a key determinant.

The various commissions and study groups which have called for new patterns of undergraduate education are really asking for sacrificial and idealistic behavior. There is nothing new about that. Great teaching has always contained an ingredient of idealism and sacrifice. But such proposals have a greater chance of success in the long run if they are accompanied by the resources to carry them out. The reforms are critically needed and should be launched with reasonable financial support.

Concluding Comments

The basic theme of this chapter, indeed, the thrust of the entire book, is the urgency of strengthening the faculties despite declining enrollments and precarious finances. We have emphasized the need for institutional planning; we have offered suggestions for dealing with the risks facing colleges and universities as they meet their personnel problems; we have made proposals relating to institutional recruitment of faculty; we have offered suggestions for governmental support to finance the infusion of new blood into academe; and we have commented on the special problems of affirmative action and educational reform. We have said more about recruitment of faculty than about retention. There was no need to dwell on the obvious truth that without retention of the best faculty, recruitment schemes will be of little avail toward building faculties of sustained power and vitality. Both recruitment and retention require adequate finances. Money is needed not only to pay competitive salaries and fringe benefits, but also to provide the working conditions and the environment essential to good academic performance.

Money comes from people who are convinced of a need. These people individually pay tuitions and make gifts, and collectively they determine or influence public appropriations. The higher educational community desperately needs to make the case for funds from tuitions, gifts, and appropriations that will sustain above all else adequate faculties.

The situation today is in some respects like that of the Sputnik era of thirty years ago. Higher education is underfinanced at a time when the educational needs of the nation, not only for economic and military purposes but also for cultural advancement, are enormous. Just as in the time of Sputnik, there is need today to raise faculty compensation to levels competitive with those for comparable personnel in other industries, to improve the working conditions and the working environment of faculty, and to give them the psychological assurance that their work is important, that it is appreciated, and that it can offer satisfying careers.

In closing, we invoke the advice given by the Sibyl to Aeneas: "Yield thou not to adversity, but press on the more bravely." Though adversity abounds, extraordinary commitment and persistence by the higher educational community and its numerous allies can accomplish much to preserve and enhance an irreplaceable national resource, the American professors.

APPENDIX A

Methodology for Campus Visits

Sample of Institutions

The task of identifying thirty-eight campuses that might be regarded as reasonably representative of the American higher education system posed challenges. Given the justly famous diversity of that system—which comprises some 3100 accredited institutions—it was clear that *no* combination of thirty-eight could be defended as truly "representative." Yet representativeness, insofar as practicable, was our objective, for we sought to explore the condition of the professoriate at large, not merely narrow segments of it. The institutions we visited, organized by classification of institution, are listed at the end of this appendix.

Nine factors were taken into account in choosing the sample of institutions: (1) diversity by type of institution, (2) type of control, (3) geographical dispersion, (4) special-mission institutions, (5) religious affiliation, (6) collective bargaining status, (7) selectivity in admissions, (8) enrollment trends, and (9) access (in the sense of being reasonably accessible to the visitor and, in some instances, having a supportive contact person on the campus).

Of the thirty-eight campuses originally invited to participate in the study, thirty-seven agreed to do so. The one campus that declined to participate was a small, church-related, single-sex, liberal arts college which had just changed presidents and was apparently experiencing considerable turmoil. The administration declined to participate, but we were easily able to find a substitute campus that matched the originally invited campus on all of the basic criteria (type, control, location, size).

The distribution of our sample of campuses by type of institution, compared with the distribution of all U.S. two- and four-year institutions, and the rationale for deliberately overrepresenting some classifications of institutions (and, thereby, underrepresenting others), as well as explanations for the indices, are set forth in our report on site visit methodology (Schuster, 1985b).[1]

The Interviews

We interviewed 532 people—as individuals rather than in groups. The interviewees were of three categories, subsuming 19 subcategories, as follows:

> 128 administrators. These included 32 presidents/chancellors and 37 chief academic officers plus an additional 14 "deputy chief academic officers" (associate academic vice presidents or associate provosts or assistant deans of institutions). Among the 180 administrators were 77 deans; we limited these interviews to deans of graduate schools/divisions (15), liberal arts schools (26) and, where applicable, deans in two high-demand areas, business (21) and engineering (15).
>
> 127 department chairs. Aiming for disparate experiences within arts and sciences, we singled out chairs of biology, government, history, and physics.
>
> 225 faculty members. This group encompassed 34 faculty senate or council chairpersons and 11 chairs/presidents of faculty union locals. For rank-and-file faculty, we chose to concentrate on persons in five developmental stages of their academic careers: (1) junior faculty drawing close to a tenure decision; (2) faculty "nomads" who were teaching fulltime on a given campus, but who had not been able to secure, there or elsewhere, a tenure-track appointment; (3) mid-career faculty, which typically meant those recently promoted to the rank of associate professor with tenure; (4) senior professor/"highly productive," which meant a faculty member, almost always a full professor, who over a long career had continued to distinguish himself or herself by performing with excellence those tasks that the campus most prized; and (5) senior professor/"early-out," meaning a faculty member who had chosen to retire early but was not moving into another academic position.

1. This report is available for $4.00 from: Faculty in Education, Claremont Graduate School, Claremont CA 91711.

List and Characteristics of Sample Institutions for the Study of the Professoriate

Name of Institution	State	(A) Classification	(B) Control	(C) Coll. Barg.	(D) Admiss. Select. Index	(E) Enrol. Trend Index	(F) Compen. Change Index
1. Univ. of California, Berkeley	CA	RU I	Pub*	—	9	3	1
2. Univ. of Iowa	IO	RU I	Pub	—	4	7	1
3. Univ. of Michigan, Ann Arbor	MI	RU I	Pub*	—	7	3	4
4. New York Univ.	NY	RU I	Ind	a	6	4	5
5. Univ. of Pennsylvania	PA	RU I	Ind	—	10	5	4
6. Univ. of New Mexico	NM	RU I	Pub	—	4	6	1
7. Washington State Univ.	WA	RU II	Pub	—	4	3	4
8. American Univ.	DC	RU II	Ind(M)	—	5	2	5
9. Tulane Univ.	LA	RU II	Ind	—	6	5	7
10. Univ. of Louisville	KY	DG I	Pub	—	6	1	4
11. Fordham Univ.	NY	DG I	Ind	a	6	1	7
12. Southern Methodist Univ.	TX	DG I	Ind(M)	—	5	5	7
13. Univ. of Denver	CO	DG I	Ind	—	6	5	n.a.
14. Univ. of Nevada, Reno	NV	DG II	Pub	—	4	7	2
15. Univ. of the Pacific	CA	DG II	Ind	—	4	5	4
16. Calif. State Univ., Los Angeles	CA	CUC I	Pub*	b	4	5	1
17. Louisiana State Univ., Shreveport	LA	CUC I	Pub*	—	3	7	5
18. Northern Arizona Univ.	AZ	CUC I	Pub	—	4	4	3
19. Pratt Institute	NY	CUC I	Ind	c	4	2	3
20. Southern Connecticut State Univ.	CT	CUC I	Pub	d	4	1	7
21. Univ. of Alabama, Birmingham	AL	CUC I	Pub	—	3	3	2
22. Univ. of Hartford	CT	CUC I	Ind	—	4	1	4
23. Jersey City State Coll.	NJ	CUC II	Pub*	c	3	2	6
24. Southern Univ. in New Orleans	LA	CUC II	Pub*	—	3	4	4
25. Willamette Univ.	OR	CUC II	Ind	—	4	5	7
26. Colorado Coll.	CO	LA I	Ind	—	8	4	3
27. Grinnell Coll.	IO	LA I	Ind	—	8	2	3
28. Mills Coll.	CA	LA I	Ind	—	6	2	4
29. Swarthmore Coll.	PA	LA I	Ind	—	10	3	4
30. Alma Coll.	MI	LA II	Ind	—	5	1	1
31. Morehouse Coll.	GA	LA II	Ind	—	4	6	n.a.
32. Saint Mary's Coll.	IN	LA II	Ind(R)	—	5	6	6
33. Borough of Manhattan Community Coll.	NY	CC	Pub*	e	2	7	n.a.
34. City Coll. of San Francisco	CA	CC	Pub	c	1	7	n.a.
35. Cypress Coll.	CA	CC	Pub*	f	1	7	7
36. Henry Ford Community Coll.	MI	CC	Pub	c	1	1	2
37. Joliet Jr. Coll.	IL	CC	Pub	c	1	1	4
38. Montgomery Coll., Rockville Campus	MD	CC	Pub*	d	1	1	4

NOTES

(A) Classifications are from Carnegie Council's *A Classification of Institutions of Higher Education: Revised Edition* (1976).

   RU = Research Universities (I and II)

   DG = Doctorate-Granting Universities (I and II)

   CUC = Comprehensive Universities and Colleges (I and II)

   LA = Liberal Arts Colleges (I and II)

   CC = Two-Year Colleges and Institutes

(B) Control: Public (20), Independent (18). An asterisk (*) indicates a multicampus system or district, of which there are nine in our sample. Whether a campus is church-related is also shown in this column. Entries here apply only to current formal, legal church affiliations. Historical denominational ties, although possibly quite influential at present, are not included here, for instance: Alma Coll. (United Presbyterian), Fordham Univ. (Roman Catholic/Jesuit), Morehouse Coll. (American Baptist Convention), and Swarthmore Coll. (Society of Friends). In Column (B), (M) = United Methodist; (R) = Roman Catholic.

(C) Faculty Collective Bargaining Status[1]:
    a = independent agent for law school faculty only; no contract negotiated
    b = California Faculty Association, National Education Association (NEA), American Association of University Professors (AAUP)
    c = American Federation of Teachers (AFT)
    d = AAUP
    e = Professional Staff Congress, NEA, AAUP
    f = NEA

(D) Admissions Selectivity Index: 10-interval scale derived from *Barron's College Profiles*, 12th ed. (1980). These scores range from "Most Competitive" (10) to "Non-Competitive" (2) and "Not rated or open enrollment" (1).

(E) Enrollment Trend Index: Enrollment changes from Fall 1980 to Fall 1983, coded on a 7-interval scale, as follows:

Coded as	Percentage Change	Number of Campuses in Sample
7	10.0 or greater growth	8
6	9.9 to 5.0 growth	2
5	4.9 to 1.10 growth	9
4	0.9 growth to 0.9 decline	4
3	1.0 to 4.9 decline	4
2	5.0 to 9.9 decline	7
1	10.0 or greater decline	4

(F) Change in Compensation Index: Changes in Faculty Compensation from 1978-79 to 1983-83. 7-interval scale, ranging from greatest increase (7) to least increase (1) in compensation for all professorial ranks. The index was calculated from faculty compensation data published in *Academe*, Sept. 1979, pp. 337–67 and July–August 1984, pp. 20–63.

1. From Joel M. Douglas, *Directory of Faculty Contracts and Bargaining Agents in Institutions of Higher Education* (1983).

APPENDIX B

Data from Survey of Leading Graduate Departments

Change in quality and number of advanced graduate students enrolled and of tenure-track junior faculty, by discipline, 1983–84, compared with 1968–72, as reported by leading graduate departments.

(A)	(B)	Advanced Graduate Students				Junior Tenure-Track Faculty			
		(C) Qual-ity	(D) Rank	(E) Num-ber	(F) Rank	(G) Qual-ity	(H) Rank	(I) Num-ber	(J) Rank
92	**68 Biological Sciences**	**4.7**		**4.2**		**4.9**		**3.7**	
21	18 Biochemistry	4.9	(9)	3.9	(15)	4.4	(25)	3.5	(16)
12	10 Botany	4.4	(21)	4.5	(8)	5.0	(11)	4.3	(7)
13	9 Cellular/Molecular Biology	4.4	(23)	5.4	(2)	4.6	(24)	3.3	(20)
20	14 Microbiology	5.1	(8)	4.2	(11)	5.7	(2)	4.5	(3)
15	11 Physiology	4.7	(15)	3.8	(16)	4.7	(21)	3.3	(19)
11	6 Zoology	5.2	(6)	3.2	(24)	5.0	(11)	3.3	(18)
49	**35 Engineering**	**5.0**		**4.7**		**5.1**		**4.3**	
12	8 Chemical	5.6	(1)	5.4	(2)	4.9	(14)	4.4	(6)
11	9 Civil	5.3	(2)	4.7	(4)	4.1	(30)	3.9	(10)
14	9 Electrical	4.8	(14)	4.7	(4)	5.2	(6)	4.6	(2)
12	9 Mechanical	4.4	(20)	4.1	(12)	5.1	(9)	4.4	(4)
78	**60 Humanities**	**4.6**		**2.9**		**4.7**		**3.0**	
6	5 Art History	4.6	(19)	3.4	(22)	4.8	(18)	4.4	(5)
5	5 Classics	5.2	(3)	3.6	(19)	5.4	(4)	2.8	(27)
16	14 English	4.9	(10)	1.9	(31)	5.3	(5)	2.4	(30)
9	5 French	4.8	(13)	3.6	(19)	4.2	(28)	2.3	(31)
7	6 German	4.2	(26)	2.3	(28)	5.2	(7)	2.2	(32)
5	4 Linguistics	3.0	(32)	3.3	(23)	4.0	(31)	4.0	(8)
8	7 Music	3.7	(31)	4.4	(10)	4.8	(15)	3.5	(14)

(A)	(B)		Advanced Graduate Students				Junior Tenure-Track Faculty			
			(C) Qual-ity	(D) Rank	(E) Num-ber	(F) Rank	(G) Qual-ity	(H) Rank	(I) Num-ber	(J) Rank
12	10	Philosophy	4.9	(12)	2.8	(25)	3.3	(32)	3.2	(22)
10	4	Spanish	3.8	(30)	2.3	(30)	4.3	(26)	3.0	(24)
90	**71**	**Physical Sciences**	**4.6**		**4.2**		**5.0**		**3.6**	
22	20	Chemistry	4.7	(16)	4.7	(6)	5.1	(10)	3.6	(12)
9	6	Computer Science	5.2	(4)	6.7	(1)	4.7	(22)	6.0	(1)
14	13	Geological Science	4.9	(10)	4.5	(7)	5.4	(3)	3.8	(11)
17	12	Mathematics	4.2	(26)	2.3	(28)	5.0	(11)	2.4	(29)
18	14	Physics	4.6	(17)	4.0	(14)	4.8	(19)	3.2	(21)
10	6	Statistics	4.3	(24)	3.5	(21)	4.8	(15)	4.0	(8)
95	**82**	**Social/Behavioral Sciences**	**4.4**		**3.1**		**4.8**		**3.2**	
11	9	Anthropology	5.1	(7)	4.1	(12)	5.1	(8)	3.0	(24)
14	12	Economics	4.6	(17)	4.5	(9)	4.2	(29)	3.6	(13)
7	7	Geography	5.2	(4)	3.7	(17)	6.0	(1)	3.6	(14)
15	13	History	3.9	(28)	1.8	(32)	4.8	(17)	2.5	(28)
12	9	Political Science	3.9	(29)	3.7	(17)	4.2	(27)	3.5	(15)
22	21	Psychology	4.2	(25)	2.7	(26)	4.6	(23)	3.0	(26)
14	11	Sociology	4.4	(21)	2.6	(27)	4.7	(20)	3.4	(17)
404	316									

Column

(A) The number of academic departments ranking in the top 15 percent for each discipline. Based on unpublished standardized scores for departments' "program effectiveness," *Assessment of Quality-Related Characteristics of Research-Doctorate Programs in the U.S.* (National Research Council), 1982.

(B) The number of academic departments in Column A that provided usable responses to the Survey of Chairpersons of Leading Graduate Departments.

(C) Mean scores, by discipline (also aggregated for each of the 5 areas), of the change in quality of advanced graduate students. On a seven-point scale, 7 was "Much Better," 4 was "About the Same," and 1 was "Much Worse."

(D) The rank among the 32 disciplines as shown in Column C.

(E) Similar to Column C, but for change in number of advanced graduate students. Each discipline's rank is shown in Column F. On a seven-point scale, 7 was "Increased Sharply," 4 was "About the Same," and 1 was "Decreased Sharply."

(G) Similar to Column C, but for change in quality of junior tenure-track faculty. Each discipline's rank is shown in Column H. On a seven-point scale, 7 was "Much Better," 4 was "About the Same," and 1 was "Much Worse."

(I) Similar to Column C, but for change in number of junior tenure-track faculty. Each discipline's rank is shown in Column J. On a seven-point scale, 7 was "Increased Sharply," 4 was "About the Same," and 1 was "Decreased Sharply."

Illustrations: In both English and German, the leading academic departments report that the quality of advanced graduate students (Column C) and of junior tenure-track faculty (Col. G) has increased, while the corresponding number of advanced graduate students (Col. E) and of junior tenure-track faculty (Col. I) has declined sharply. In Chemical Engineering, the quality and number of persons in both categories has increased.

Note on response rate: Response rates for the 32 disciplines ranged from Classics and Geography (100% each) and Geological Sciences (92.9%) to French (55.5%), Zoology (54.4%), and Spanish (40%)—the only discipline in which fewer than half of the departments participated. The usable response rate for all disciplines was 78.2%.

SOURCE: Survey of Chairpersons of Leading Graduate Departments, 1984, by the authors.

APPENDIX C

Estimates of
Faculty Attrition

In this appendix, we assemble evidence on annual rates of faculty attrition, that is, on the percentage of faculty members departing from academe (not including those merely shifting from one college or university to another). Figures on attrition are essential in projecting the number of job openings available for faculty appointments. On the basis of the data we could muster, we concluded that two assumptions regarding attrition, a high and a low one, would be reasonable:

(1) a stable 4 percent of the faculty for the entire period 1985 to 2010 comprised of 1.3 percent for retirement and mortality and 2.7 percent for other types of attrition.

(2) A rising percentage from 4 to 6 percent over the years 1985 to 1994 and thereafter a stable 6 percent from 1995 to 2010; with attrition at 6 percent, 2.75 percent would represent retirement and mortality and 3.25 percent would represent other types of attrition.

In the following paragraphs we shall cite the evidence that brought us to these assumptions.

Freeman (1971, p. 178) found that the annual percentage of faculty persons leaving academe for reasons other than retirement and death were as follows:

1935–40	3.5%
1940–45 (war)	7.9
1945–50	7.1
1950–55	5.0
1955–60	3.5
1960–63	3.4

The fluctuations in these figures were influenced by World War II and its aftermath and by the depressed state of higher education in the early 1950s. The peacetime "norm" appeared to be about 3.5 percent. Note that these figures do not include separations by reason of retirement or death.

Cartter (1976, p. 119) reported that the flow of faculty from academe to take jobs outside was about 3 percent of total faculty. Klitgaard (1979, pp. 15–16) also adopted an estimate of 3 percent.

The National Education Association (1979, p. 9) asked faculty members in 1977–78 to indicate their plans for "next year." The responses indicated that 4 percent expected to leave academe for reasons other than retirement and death.

From the several sources, one may surmise that in the period of the late 1970s and early 1980s, about 3 to 4 percent of the faculty left academe each year for reasons other than retirement and death.

Data from the Carnegie Council on Policy Studies in Higher Education show that attrition due to retirement and death for science, engineering, and social science faculties averaged a bit above one percent in the years from 1976 to 1985 and was rising (Carnegie Council, 1980, p. 374).

Adding 3 percent for attrition not related to retirement and death, and one percent for that related, gives a total of about 4 percent. This figure may be compared with the results of a study of the National Science Foundation referring only to full-time academic scientists, engineers, and social scientists, which reported that about 3.6 percent of full-time faculty left academe in 1978–79 (National Science Foundation, 1981c, pp. 7–8). The corresponding figures for the several disciplines represented in the study varied between 3.2 in physical science and 4.3 percent in engineering.

Radner and Kuh (1979) estimated the attrition rate to be 3 percent. These estimates and others like them, however, probably understate the attrition in that the comings and goings of community college faculty and of part-time faculty are either not included or underrepresented. A study conducted by the National Research Council (Syverson and Forster, 1984) revealed the following rates of attrition over the period 1981–83, by disciplines, among Ph.D.s employed in academe:

Physical Science	5.4%
Mathematics & Computer Science	3.3
Life Sciences	4.3
Social Sciences	4.8
Engineering	2.2
History	2.9
Foreign Language & Literature	4.5
English & American Literature	5.3
Other Humanities	4.2
Music	4.7

The authors added that "at the present level of market activity . . . the current faculty will be renewed in 15 to 20 years."

Taking the evidence into account, including data on attrition for particular colleges and universities, we believe that 4 percent is a reasonable estimate of attrition as it was actually occurring in the early 1980s. This estimate includes all elements of the faculties—full-time and part-time, tenured, and non-tenured, and four-year and two-year institutions. We believe that this figure would provide a reasonable assumption regarding the amount of attrition over the years from 1985 to 2010 to be used in estimating the needed number of faculty replacements.

There are reasons, however, to believe that attrition will increase in the years ahead—especially after 1995—because the average age of the faculties will probably rise and the rate of retirements and deaths will probably increase. For example, one projection of combined retirement and death rates as a percentage of total faculty shows increases in these rates from 1985 to 2000 by selected discipline as follows (Carnegie Council, 1980 p. 374):

Mathematics	1.04% to 2.84%
Physical sciences	1.49 to 3.03
Engineering	1.29 to 2.50
Life sciences	1.44 to 2.55
Social sciences	1.73 to 3.12

It could turn out also that the working conditions and compensation of the professoriate will not keep pace with conditions and earnings outside and that the number of faculty members leaving academe will grow. Along this line, Toombs (1979, p. 12) has suggested that the shifting of faculty to non-academic positions will be facilitated by two important influences: (1) the gradual dispersion throughout the economy of knowledge-based activity such as research and development, and educational tasks connected with business firms and government agencies; and (2) the rise of two-income families which contributes to flexibility in search-out, training, and moving. In this connection, it is interesting to note that the National Center for Education Statistics (*Projections of Education Statistics*, 1982, pp. 78–79) adopted *for the next ten years*, 4.5 percent as their low assumption of faculty attrition and 6 percent as their high assumption. All of these considerations and these figures suggest that attrition rates might rise to 6 percent or even higher over the rest of the twentieth century. We think these figures may be a bit too high because they may assume an unrealistically large increase in the average age of faculty and may not give sufficient weight to the possible changes in retirement practices. On the other hand, if positions in higher education should become less attractive, relative to positions in the economy generally, the rate of attrition for reasons other than retirement and death could rise substantially. Taking all these factors into consideration, a high-side assumption of attrition rising from 4 percent in the early 1980s to 6 percent seems conservative.

APPENDIX D

Estimates of
Faculty Numbers, 1980–2010

In estimating the number of new faculty appointments during the twenty-five years (1985-2010) the first step was to make projections of the number of faculty members over this same period (see Chapter 10). Our underlying assumption was that the number of faculty numbers would be closely related to student enrollments. Our first task therefore was to estimate enrollments. We assumed that college attendance would vary with the number of persons in the population of ages 16 to 34, each age cohort weighted according to actual attendance in 1980. The basic data for deriving these estimates, presented in Table A–D1, show (a) estimated population trends by age groups, (b) the percentage of persons in each age group attending college in 1980, and (c) the weighted estimates of enrollments calculated by applying the 1980 percentage of attendance to each corresponding age group of the population. The index numbers of enrollments for 1980–81 through 2010–11 (the important figures for present purposes) are shown in the right-hand column of the lower portion of the table. These index numbers suggest that enrollments will decline by about 15.5 percent between 1980–81 and 1995–96, and will then rise by about 14.3 percent between 1995–96 and 2010–11. In our effort to explore the full range of possibilities, we noted also the effect of other assumptions concerning enrollments, including "steady state" and enrollments fluctuating with the populations of persons of ages 18 to 21 (as shown in Table A–D1). Though enrollment is the dominant variable affecting faculty numbers, there are several other variables at work as well, and we experimented with several of these.

One of these variables is change in the mix of part-time and full-time faculty which would affect the number of *full-time* career persons in the profession. Colleges and universities are of course faced with a dilemma in their use of part-timers. On the one hand, institutions have increased the proportion of part-timers to avoid heavy tenure commit-

Table A–D1. Enrollment Estimates by Age Groups, 1980 to 2010

(a) *Projected population of college age, by age groups* (000 omitted)[1]

	16–17	18–21	22–24	25–29	30–34	Sum: ages 16–34	Index 1980 = 100
1980–81	8,157	17,117	12,346	21,459	18,400	77,479	100.0
1985–86	7,220	15,658	13,056	21,830	19,951	77,715	100.3
1990–91	6,493	14,676	11,101	21,503	22,003	75,776	97.8
1995–96	6,756	13,059	10,625	18,807	21,682	70,929	91.6
2000–01	7,618	14,611	9,979	17,380	19,007	68,595	88.5
2005–06	7,682	16,555	11,317	17,540	17,739	70,833	91.4
2010–11	7,330	16,847	11,515	20,369	17,910	73,971	95.5

(b) *Percentage of persons in each age group attending college in 1980–81.*[2]

	16–17	18–21	22–24	25–29	30–34		
1980–81	3.0%	32.9%	15.8%	8.9%	6.2%	—	—

(c) *Enrollment estimates by age groups, applying 1980 attendance percentages to projected population of college age* (000 omitted)[3]

1980–81	245	5,631	1,951	1,910	1,141	10,878	100.0
1985–86	217	5,151	2,063	1,943	1,237	10,611	97.6
1990–91	195	4,828	1,754	1,914	1,364	10,055	92.4
1995–96	203	4,296	1,679	1,674	1,344	9,196	84.5
2000–01	228	4,807	1,577	1,547	1,178	9,337	85.8
2005–06	230	5,447	1,788	1,561	1,100	10,126	93.1
2010–11	220	5,543	1,819	1,813	1,110	10,505	96.6

[1] Bureau of the Census, July 1977, Oct. 1982; *Statistical Abstract of the United States, 1980*, p. 30. Some of the figures for 1980, 2005 and 2010 are the authors' estimates calculated by interpolation. All the figures are based on the middle of three separate projections made by the Bureau of the Census.
[2] Bureau of the Census, May 1981.
[3] Computed by applying percentages of persons in each age group attending college to the projected population in each age group. Total enrollment as computed for 1980–81 is slightly less than the actual enrollment (12,087) because students above 34 years of age are not included in these calculations.

ments. In so doing, they have planned in the event of retrenchment to drop part-timers rather than tenured full-timers. But part-time faculty are probably less costly than full-timers, and so in an era of financial stringency institutions are tempted to increase the proportion of part-time faculties. We believe that the higher educational community should, in the interests of academic excellence, try to reverse the increase in the proportion of part-timers that has occurred in recent years. Therefore, in exploring future possibilities for faculty, an interesting assumption is that the proportion of the faculty employed part-time would be reduced.

Another variable affecting the number of faculty is the ratio of students to faculty. This ratio has been fairly stable over many years in the range of 13 to 14 students per faculty member—both expressed in full-time-equivalents.[1] Recently, the ratio has been above 14 and there have

1. There are many different versions of the student-faculty ratio depending on the way the two elements are defined. The one we have chosen comes from H. R. Bowen, 1980b, p. 42.

been many proposals in the interest of economy to raise the ratio still further. Our judgment is that the long-standing ratio of 13 or 14 to 1 should not be raised but rather should be lowered—especially because the student population in decades ahead will probably contain an increasing proportion of persons needing remedial attention. Therefore, one of our assumptions is a reduced student-faculty ratio.

Finally, still another variable that might affect the number of faculty is the amount of participation by the higher educational community in research and public service. We have included an assumed increase in support for research and public service.

Given the several variables—enrollment, ratio of part-time to full-time faculty, ratio of students to faculty, and support for research and public service—we have adopted several sets of assumptions pertaining to faculty numbers over the twenty-five-year period from 1985 to 2010, and have estimated the number of faculty persons implicit in each set. The numbers are shown in Table A–D2.

Table A–D2. Projections of Number of Faculty, U.S., 1980–81 to 2010–11, with Varying Assumptions** (in thousands)

	Projection I (Steady State)				Projection II (Traditional Student Cohort)		
Year	*Total FTE*[1]	*Full Time*	*Part Time*	*Year*	*Total FTE*[1]	*Full Time*	*Part Time*
1980–81	537	466	212	1980–81	537	466	212
1985–86	537	466	212	1985–86	491	426	194
1990–91	537	466	212	1990–91	460	399	182
1995–96	537	466	212	1995–96	410*	356	162
2000–01	537	466	212	2000–01	459	398	181
2005–06	537	466	212	2005–06	519	451	205
2010–11	537	466	212	2010–11	528	459	209

	Projection III (Contemporary Student Cohort)				Projection IV (Reduced Part-Time Faculty)		
Year	*Total FTE*[1]	*Full Time*	*Part Time*	*Year*	*Total FTE*[1]	*Full Time*	*Part Time*
1980–81	537	466	212	1980–81	537	466	212
1985–86	524	455	207	1985–86	524	455	207
1990–91	496	431	196	1990–91	496	443	159
1995–96	454*	394	179	1995–96	454*	417	111
2000–01	461	400	182	2000–01	461	423	113
2005–06	500	434	197	2005–06	500	459	122
2010–11	518	450	205	2010–11	518	476	127

** See text (Chapter 10) for assumptions underlying each projection.
 * Low period
[1] Full-time-equivalent. Each part-time faculty member counted as one-third of a full-time member.

Table A–D2. continued

	Projection V (Reduced Student-Faculty Ratio)				Projection VI (Increase in Older Adult Learners)		
Year	Total FTE[1]	Full Time	Part Time	Year	Total FTE[1]	Full Time	Part Time
1980–81	537	466	212	1980–81	537	466	212
1985–86	524	455	207	1985–86	524	455	207
1990–91	521	452	206	1990–91	528	458	208
1995–96	499	433	197	1995–96	525*	455	207
2000–01	507	440	200	2000–01	542	470	214
2005–06	550	477	217	2005–06	577	501	228
2010–11	570	495	226	2010–11	601	521	237

Projection VII
(Increase in Research and Public Service)

Year	Total FTE[1]	Full Time	Part Time
1980–81	537	466	212
1985–86	524	455	207
1990–91	519	452	201
1995–96	500*	435	195
2000–01	530	461	207
2005–06	569	495	222
2010–11	587	511	228

From Table A–D2, changes in the number of faculty and the amount of faculty attrition over each five-year period 1980–81 to 2010–11 were estimated. Finally from these results, the number of new faculty appointments for each five-year period were computed as the algebraic sum of changes in the number of faculty plus faculty attrition. See Table 10–1.

Bibliography

Abel, Emily K. *Terminal Degrees: The Job Crisis in Higher Education.* New York: Praeger, 1984.

Albert, Louis P. 'Part-Time Faculty Policies, Practices, and Incentives in Maryland's Community Colleges.' Doctoral dissertation, University of Maryland, 1982.

Altbach, Philip G., and Robert O. Berdahl (eds.). *Higher Education in American Society.* Buffalo, N.Y.: Prometheus Books, 1981.

American Assembly of Collegiate Schools of Business. *Final Report of the Task Force on Doctoral Supply and Demand.* St. Louis: AACSB, 1982.

American Association of University Professors. 'Academic Freedom and Tenure, 1940 Statement of Principles and Interpretive Comments.' *AAUP Redbook.* Washington, 1977.

———. 'An Era of Continuing Decline: Annual Report on the Economic Status of the Profession.' *Academe,* Sept. 1979, pp. 319–67.

———. Subcommittee of Committee A on Academic Freedom and Tenure. 'The Status of Part-Time Faculty.' *Academe,* Feb.–March 1981, pp. 29–39.

———. '1982 Recommended Institutional Regulations on Academic Freedom and Tenure.' *Academe,* Jan.–Feb. 1983a, pp. 15a–20a.

———. 'Retirement Reflections.' *Academe,* Nov.–Dec. 1983b, pp. 3–10.

———. 'Bottoming Out? The Annual Report on the Economic Status of the Profession, 1983–84.' *Academe,* July–Aug. 1984, pp. 2–63.

———, 'The Annual Report on the Economic Status of the Profession.' *Academe,* July–Aug. issues, annual.

American Council on Education, Office of Minority Concerns. *Fact Book on Higher Education.* Washington: 1980, 1981–82, and 1984–85.

———. *Minorities in Higher Education: Second Annual Status Report.* Washington: American Council on Education, n.d.

Andersen, Charles J. *Student Quality in the Humanities: Opinions of Senior Academic Officials.* Washington: American Council on Education, 1984.

———, and Frank J. Atelsek. *Sabbatical and Research Leaves in Colleges and Universities.* Washington: American Council on Education, 1982.

Anderson, Raymond B., and Allen R. Sanderson. 'Financial Issues in Graduate Education and an Agenda for Research.' Cambridge, Mass.: Consortium on Financing Higher Education, 1982 (mimeo).

Anderson, Richard E. *Finance and Effectiveness: A Study of College Environments.* Princeton: Educational Testing Service, 1983.

Association of American Colleges. *Integrity in the College Curriculum: A Report to the Academic Community.* Washington: Association of American Colleges, 1985.

Association of American Colleges and American Association of University Professors. 'Statement of Principles on Academic Retirement and Insurance Plans.' *Academe,* 1980, pp. 321–23.

Astin, Alexander W. *Minorities in American Higher Education.* San Francisco: Jossey-Bass, 1982.

——— et al. *The American Freshman: National Norms.* Los Angeles: Higher Education Research Institute, annual editions, 1966–84.

Astin, Helen. Survey of Academic Personnel, 1973 and 1980. Unpublished tabulations supplied by courtesy of the author.

———, and M. B. Snyder. 'Affirmative Action 1972–1982—A Decade of Response.' *Change,* July–Aug. 1982, p. 26.

Atelsek, Frank J. *Student Quality in the Sciences and Engineering: Opinions of Senior Academic Officials.* Washington: American Council of Education, 1984.

———. *Tenure Practices at Four-Year Colleges and Universities.* Washington: American Council on Education, 1980.

———. *Selected Characteristics of Full-Time Humanities Faculty, Fall 1979.* Washington: American Council on Education, 1981a.

———. *An Analysis of Travel by Academic Scientists and Engineers to International Scientific Meetings in 1979–80.* Washington: American Council on Education, 1981b.

———, and Irene L. Gomberg. *New Full-Time Faculty 1976–77: Hiring Patterns by Field and Educational Attainment.* Washington: American Council on Education, 1978.

Austin, Ann E., and Zelda F. Gamson. *Academic Workplace: New Demands, Heightened Tensions.* Washington: Association for the Study of Higher Education, 1984.

Aydelotte, Frank. *The American Rhodes Scholarships: A Review of the First Forty Years.* Princeton: Princeton University Press, 1946.

Ayres, Q. Whitfield, and Ronald W. Bennett. 'University Characteristics and Student Achievement.' *Journal of Higher Education,* Sept.–Oct. 1983, pp. 516–32.

Bailey, Stephen K. 'Helping Professors (and Therefore Students) To Grow.' *Chronicle of Higher Education,* May 28, 1974a, p. 24.

———. 'People Planning in Post-Secondary Education: Human Resource Development in a World of Decremental Budgets.' In *More for Less: Academic Planning with Faculty Without New Dollars.* New York: Society for College and University Planning, 1974b.

Baldridge, J. Victor, David Curtis, George Ecker, and Gary Riley. *Policy Making and Effective Leadership.* San Francisco: Jossey-Bass, 1978.

Baldridge, J. Victor, and Frank R. Kemerer and Associates. *Assessing the Impact of Faculty Collective Bargaining.* Washington: American Association for Higher Education, 1981.

Bayer, Alan E. 'College and University Faculty: A Statistical Description.' *ACE Research Reports,* vol. 5, no. 5 June 1970.

———. *Teaching Faculty in Academe: 1972–73.* Washington: American Council on Education, 1973.

Beazley, Richard M. *Numbers and Characteristics of Employees in Institutions of Higher Education, Fall 1967.* Washington: U.S. Department of Health, Education and Welfare, Office of Education, National Center for Education Statistics, 1970.

Bennett, William J., and Study Group on the State of Learning in the Humanities in Higher Education. *To Reclaim a Legacy*. Washington: National Endowment for the Humanities, 1984.

Berelson, Bernard. *Graduate Education in the United States*. New York: McGraw-Hill, 1960.

Birnbaum, Robert. 'The Effects of a Neutral Third Party on Academic Bargaining Relationships and Campus Climate.' *Journal of Higher Education*, Nov.–Dec. 1984, pp. 719–34.

Blackburn, John O., and Susan Schiffman. *Faculty Retirement at the COFHE Institutions: An Analysis of Age 70 Mandatory Retirement and Options for Institutional Response*. Cambridge, Mass.: Consortium on Financing Higher Education, 1980.

Blatt, Burton. *In and Out of the University*. Baltimore: University Park Press, 1982.

Boberg, Alice L., and Robert I. Blackburn. 'Faculty Work Dissatisfactions and Their Concern for Quality.' Unpublished manuscript, ca. 1983.

Bok, Derek. *Beyond the Ivory Tower*. Cambridge, Mass.: Harvard University Press, 1982.

Bonner, Thomas N. 'The Distinctly Urban University: A Bad Idea?' *Chronicle of Higher Education*, Sept. 23, 1981, p. 48.

Bowen, Howard R. 'Faculty Salaries: Past and Future.' *Educational Record*, Winter 1968, pp. 9–21.

———. 'Manpower Management and Higher Education.' *Educational Record*, Winter 1973, pp. 5–14.

———. 'Higher Education: A Growth Industry?' *Educational Record*, Summer 1974, pp. 147–58.

———. *Investment in Learning*. San Francisco: Jossey-Bass, 1977.

———. *Academic Compensation*. New York: Teachers Insurance and Annuity Association and College Retirement Equities Fund, 1978.

———, *Adult Learning, Higher Education, and the Economics of Unused Capacity*. New York: College Entrance Examination Board, 1980a.

———. *The Costs of Higher Education*. San Francisco: Jossey-Bass, 1980b.

———. 'Deferred Maintenance of Human and Physical Assets.' An address presented to the National Association of State Universities and Land Grant Colleges, Atlanta, Nov. 16, 1980c.

———. *The State of the Nation and the Agenda for Higher Education*. San Francisco: Jossey-Bass, 1982.

———. 'The Art of Retrenchment.' *Academe*, Jan.–Feb. 1983, pp. 21–24.

Bowen, Howard R., and Gordon K. Douglass. *Efficiency in Liberal Education*. New York: McGraw-Hill, 1971.

———, and W. John Minter. *Private Higher Education: Annual Report on Financial and Educational Trends in the Private Sector of American Higher Education*. Washington: Association of American Colleges, 1975 and 1976.

Bowen, Howard R., and Jack H. Schuster. 'Outlook for the Academic Profession.' *Academe*, Sept.–Oct. 1985, pp. 9–15.

———. 'Whither the Gifted? The Changing Career Interests of the Nation's Intellectual Elite.' *The Key Reporter*, vol. 51, no. 1, Autumn 1985, pp. 1–4.

Bowen, William G. *Report of the President*. Princeton: Princeton University, 1981a (with separate Technical Appendix).

———. 'Graduate Education in the Arts and Sciences: Prospects for the Future.' *Change*, July/Aug. 1981b, pp. 40–44.

———. 'The Junior Faculty: A Time for Understanding and Support.' *Change*, July/Aug. 1983, pp. 22–23, 30–31.

Boyer, Carol, and Darrell R. Lewis. 'Faculty Consulting: Responsibility or Promiscuity?' *Journal of Higher Education*, Sept.–Oct. 1984, pp. 637–59.

Boyer, Ernest L., and Arthur Levine. *A Quest for Common Learning*. Washington: Carnegie Foundation for the Advancement of Teaching, 1981.

Braxton, John M., and William Toombs. 'Faculty Uses of Doctoral Training'. *Research in Higher Education*, vol. 16, no. 3, 1982, pp. 265–82.

Brookes, Michael C. F., and Katherine L. German. *Meeting the Challenges: Developing Faculty Careers*. Washington: ERIC Clearinghouse, 1983.

Brown, David G. *Market for College Teachers*. Chapel Hill: University of North Carolina Press, 1965.

————. *The Mobile Professors*. Washington: American Council on Education, 1967.

————. *Leadership Vitality: A Workbook for Academic Administrators*. Washington: American Council on Education, 1979.

Brown, Ralph S. 'Report on the Conference on Hard Times.' *Academe*, Jan.–Feb. 1983, pp. 4–9.

Bureau of the Census. *Projections of the Population of the United States, 1977 to 2050*. Current Population Reports, Series P25, No. 704, July 1977.

————. *School Enrollment—Social and Economic Characteristics of Students: October 1980*, Current Population Reports, Series P25, No. 922, May 1981.

————. *Projections of the Population of the United States, 1982–2050*. Current Population Reports, Series P25, No. 922, 1982a.

————. *Money Income of Households, Families, and Persons in the United States, 1982*. Current Publication Reports, Series P60, No. 142, 1982b.

————. *Statistical Abstract of the United States*. Washington: U.S. Government Printing Office, annual.

Burnett, Collins W. 'The Trojan Horse Phenomenon Reconsidered.' *Community College Review*, Summer 1982, pp. 18–23.

Business/Higher Education Forum, *America's Competition Challenge*. Washington: American Council on Education, 1983.

Butler-Nalin, Paul, Allen R. Sanderson, and David N. Redman. *Financing Graduate Education*. Cambridge, Mass.: Consortium on Financing Higher Education, 1983.

Cahn, Steven M. 'The Ethical Thicket of Academic Autonomy.' *Chronicle of Higher Education*, Feb. 2, 1983, p. 64.

California Community Colleges, Office of the Chancellor. Report on 'Number of Full-Time and Part-Time Faculty,' Sept. 9, 1980.

California Postsecondary Education Commission. *Capital Renewal and Replacement in California Higher Education*. Sacramento: California Postsecondary Education Commission, 1983a.

————. *Final Annual Report on Faculty and Administrative Salaries in California Public Higher Education, 1983–84*. Sacramento: California Postsecondary Education Commission, 1983b.

Calvin, Allen. 'Age Discrimination on Campus.' *AAHE Bulletin*, Nov. 1984, pp. 8–12.

Caplow, T., and R. J. McGee. *The Academic Marketplace*. New York: Basic Books, 1958.

Careers, Inc. *Career Brief: Teacher, College*. Largo, Fla.: 1984.

Carnegie Council on Policy Studies in Higher Education. *A Classification of Institutions of Higher Education* (Revised Edition). Berkeley: Carnegie Council on Policy Studies in Higher Education, 1976.

————. *Three Thousand Futures*. San Francisco: Jossey-Bass, 1980.

Carnegie Foundation for the Advancement of Teaching. *Missions of the College Curriculum*. San Francisco: Jossey-Bass, 1977.

————. *Common Learning*. Washington, 1981.

Cartter, Allan M. 'The Supply and Demand for College Teachers.' *Journal of Human Resources*, Summer 1966, pp. 22–37.

——. *Ph.D.'s and the Academic Labor Market.* New York: McGraw-Hill, 1976.

——. and R. Farrell. 'Higher Education in the Last Third of the Century.' *Educational Record,* Spring 1965, pp. 119–28.

Chait, Richard P., and Andrew T. Ford. *Beyond Traditional Tenure.* San Francisco: Jossey-Bass, 1982.

Chickering, Arthur W., and Associates. *The Modern American College.* San Francisco: Jossey-Bass, 1981.

Clark, Shirley M., Carol M. Boyer, and Mary Corcoran. 'Faculty and Institutional Vitality in Higher Education.' In Shirley M. Clark and Darrell R. Lewis (eds.), *Faculty Vitality and Institutional Vitality.* New York: Teachers College Press, 1985.

Clark, Shirley M., and Darrell R. Lewis (eds.). *Faculty Vitality and Institutional Productivity.* New York: Teachers College Press, 1985.

Cohen, Arthur M., and Florence B. Brawer. *The Two-Year College Instructor Today.* New York: Praeger, 1977.

——. *The American Community College.* San Francisco: Jossey-Bass, 1982.

Cole, Charles C., Jr. *Improving Instruction: Issues and Alternatives for Higher Education.* Washington: ERIC Clearinghouse, 1982.

Cole, Jonathan R. *Fair Science: Women in the Scientific Community.* New York: Free Press, 1979.

Commission on Academic Tenure. *Faculty Tenure. A Report and Recommendations.* San Francisco: Jossey-Bass, 1973.

Commission on the Higher Education of Minorities. *Final Report.* Los Angeles: Higher Education Research Institute, 1982.

Commission on the Humanities. *The Humanities in American Life.* Berkeley: University of California Press, 1980.

Commission on Strengthening Presidential Leadership. *Presidents Make a Difference.* Washington: Association of Governing Boards of Universities and Colleges, 1984.

Conference Board of Associated Research Councils, Committee on an Assessment of Quality-Related Characteristics of Research-Doctorate Programs in the United States. *An Assessment of Research-Doctorate Programs in the United States: Social and Behavioral Sciences.* Washington: National Academy Press, 1982.

Conrad, Clifton F., and Jean C. Wyer. 'Incest in Academe: The Case for Selective Inbreeding.' *Change,* Nov.–Dec. 1982, pp. 45–48.

Consortium on Financing Higher Education. *Faculty Retirement: Proceedings from the COFHE Retirement Conference.* Cambridge, Mass., 1981.

——. *Beyond the Baccalaureate.* Cambridge, Mass., 1983.

——. *Nine Disciplines, 1981–82 and 1982–83.* Cambridge, Mass., 1984.

Cook, Thomas J. *Public Retirement Systems: Summaries of Public Retirement Plans Covering Colleges and Universities—1983.* New York: Teachers Insurance and Annuity Association, 1983.

Cornford, F. M. *Microcosmographia Academica,* 4th ed. Cambridge, Eng.: Bowes and Bowes, 1949.

Corwin, T. M., and P. R. Knepper. *Finance and Employment Implications of Raising the Mandatory Retirement Age for Faculty.* Washington: American Council on Education, 1978.

Council of Graduate Schools and Graduate Record Examination Board. 'Report on Survey of Graduate Enrollment.' In *CGS Communicator,* Special Report, Vol. XV, No. 11, annual, Dec. 1982 and earlier years.

Council of Graduate Schools in the United States. *CGS Communicator,* Jan. 1983, p. 5.

——. *Major Issues Affecting Graduate Education and Research.* Washington: Council of Graduate Schools in the U.S., 1984.

Cox, Virginia B., and Bernard V. Khoury. *Report on the Council of Graduate Schools–Graduate Record Examination Board 1982–83 Survey of Graduate Enrollment.* Princeton: Graduate Record Examination Board, Dec. 1982.

Crosson, Patricia H. *Public Service in Higher Education: Practices and Priorities.* Washington: Association for the Study of Higher Education, 1983.

The Danforth Foundation. 'The Danforth Graduate Fellowship Program.' *Danforth News and Notes.* March 1979.

Dave, R. H. *Foundations of Lifelong Education.* Paris: UNESCO, 1976.

Deskins, Donald R. *Minority Recruitment Data: An Analysis of Baccalaureate Degree Production in the United States.* Totowa, N.J.: Rowman and Allenheld, 1983.

Douglas, Joel M. *Directory of Faculty Contracts and Bargaining Agents in Institutions of Higher Education.* New York: National Center for the Study of Collective Bargaining in Higher Education and the Professions, Baruch College, City University of New York. vol. 9 (Jan. 1983).

Dressel, Paul L. *College Teaching as a Profession: The Doctor of Arts Degree.* New York: Carnegie Corporation of New York, 1982.

Drew, David E. *Strengthening Academic Science.* New York: Praeger, 1985.

Drucker, D. C. 'Engineering Education.' *Science,* Oct. 28, 1983, p. 375.

Dunham, R. E., P. S. Wright, and M. O. Chandler. *Teaching Faculty in Universities and Four-Year Colleges.* Washington: U.S. Office of Education, 1966.

Edgerton, Russell. *Perspectives on Faculty.* An Address to the Kansas Conference on Postsecondary Education, Topeka, Nov. 18, 1980.

El-Khawas, Elaine H., and W. Todd Furniss. *Faculty Tenure and Contract Systems: 1972 and 1974.* Washington: American Council on Education, 1974.

Enarson, Harold L. 'The Last Word.' *AGB Reports,* Sept.–Oct. 1982, p. 48.

Eurich, Alvin C., and A. Kraetsch Gayla. 'A 50-Year Comparison of University of Minnesota Freshmen's Reading Performance.' *Journal of Educational Psychology,* vol. 74, no. 5 (October 1982), pp. 660–65.

Evangelauf, Jean. 'On Most Campuses, Faculty Work Conditions Are Said To Be Unchanged or Improving.' *Chronicle of Higher Education,* Sept. 26, 1984, pp. 25, 28.

Eymonerie, Maryse. *The Availability of Fringe Benefits in Colleges and Universities.* Washington: American Association of University Professors, 1980.

Farrell, Charles S. 'Administration Seeks To End Tax Exemptions for Tuition, Housing, and Health Benefits.' *Chronicle of Higher Education,* July 6, 1983, p. 7.

————. 'Minorities Seen Making No Gain in Campus Jobs.' *Chronicle of Higher Education,* June 13, 1984, pp. 1, 20.

Feldman, K. A., and T. M. Newcomb. *The Impact of College on Students* (2 vols.). San Francisco: Jossey-Bass, 1976.

Fellman, David. 'The Association's Evolving Policy on Faculty Exigency.' *Academe,* May–June 1984, pp. 14–22.

Fernandez, Luis. *U.S. Faculty After the Boom: Demographic Projections to 2000.* Berkeley: Carnegie Council on Policy Studies in Higher Education, 1978.

Fink, L. Dee. *First Year on the Faculty: A Study of 100 Beginning College Teachers.* Norman: University of Oklahoma, Office of Instructional Services, 1982.

Finkelstein, Martin J. *Understanding American Academics.* Buffalo: State University of New York at Buffalo, Department of Higher Education, 1980.

————. *The American Academic Profession: A Synthesis of Social Scientific Inquiry Since World War II.* Columbus: Ohio State University Press, 1984.

————. 'The Status of Academic Women: An Assessment of Five Competing Explanations.' *Review of Higher Education.* Forthcoming.

Flather, Paul. 'The Missing Generation.' *Times* [London] *Higher Education Supplement.* Sept. 24, 1982, pp. 8–9.

Folger, John K., Helen S. Astin, and Alan E. Bayer. *Human Resources and Higher Education.* New York: Russell Sage Foundation, 1970.

Frances, Carol. 'The Economic Outlook for Higher Education.' *AAHE Bulletin,* Dec. 1984, pp. 3–5.

Frazer, Catherine S. *Guidelines for Review of Tenured Faculty.* Grinnell, Iowa: Grinnell College, 1984.

Freeman, Richard B. *The Market for College-Trained Manpower: A Study in the Economics of Career Choices.* Cambridge, Mass.: Harvard University Press, 1971.

Friedlander, Jack. 'Instructional Practices of Part-Time Faculty.' In M. H. Parsons (ed.), *New Directions for Community Colleges: Using Part-Time Faculty Effectively.* San Francisco: Jossey-Bass, 1983, pp. 27–36.

Fuller, Jon W. (ed.). *Issues in Faculty Personnel Policies.* San Francisco: Jossey-Bass, 1983.

Furniss, W. Todd. 'Retrenchment, Layoff, and Termination.' *Educational Record,* Summer 1974, pp. 159–70.

Gaff, J. G., and R. C. Wilson. 'The Teaching Environment.' *AAUP Bulletin,* vol. 57, no. 4 (December 1971), pp. 475–93.

Galenson, Walter. 'Forced Retirement for Professors Only?' *New York Times,* Sept. 3, 1983, p. 3.

Gappa, Judith M. 'Employing Part-Time Faculty: Thoughtful Approaches to Continuing Problems.' *AAHE Bulletin,* Oct. 1984, pp. 3–7.

―――, and Barbara S. Uehling. *Women in Academe: Steps to Greater Equality.* Washington: American Association for Higher Education, 1979.

Garet, Michael S., and Paul Butler-Nalin. *Graduate and Professional Education: A Review of Recent Trends.* Cambridge, Mass.: Consortium on Financing Higher Education (Prepared for the National Commission on Student Financial Assistance), 1982.

Garland, James C. 'What Financial Exigency Means.' *Academe,* Jan.–Feb. 1983, pp. 24–26.

Gast, Linda K. *Criteria in the Job Selection Processes of Engineers.* Bethlehem, Pa.: CPC Foundation, 1983.

Gilford, Dorothy M., and Joan Snyder. *Women and Minority Ph.D.'s in the 1970's: A Data Book.* Washington: National Academy of Sciences, 1977.

Gleazer, Edmund J., Jr. *The Community College.* Washington: American Association of Community and Junior Colleges, 1980.

Glenny, Lyman A. 'Decision-Making in Panic Times.' *AGB Reports,* May–June 1982, pp. 20–24.

Goldberg, Frank, and Roy A. Koenigsknecht. *Highest Achievers.* Cambridge, Mass.: Consortium on Financing Higher Education, 1985.

Gooler, Dennis D. 'A Question of Vitality'. Unfinished manuscript.

Graham, Patricia Albjerg. 'Expansion and Exclusion: A History of Women in American Higher Education.' *Signs: Journal of Women in Culture and Society,* vol. 3 (Summer 1978), pp. 759–73.

Grant, Gerald, et al. *On Competence.* San Francisco: Jossey-Bass, 1979.

Grant, Gerald, and David Riesman. *The Perpetual Dream: Reform and Experiment in the American College.* Chicago: University of Chicago Press, 1978.

Greeley, Andrew W. 'Intellectuals as an Ethnic Group.' *New York Times Magazine,* July 21, 1970, pp. 22–32.

Green, Kenneth C. *Government Support for Minority Participation in Higher Education.* Washington: American Association for Higher Education, 1982.

―――. 'Entering Freshmen and the Migration of Talent Across Careers and Academic Disciplines, 1973–1984,' 1985, unpublished.

Guthrie, R. Claire. 'Can Tenure Be "Decoupled" from Retirement? An Analysis of the Law.' *Chronicle of Higher Education,* Sept. 29, 1982, p. 28.

Hansen, W. Lee. 'Mandatory Retirement Age Legislation for Tenured Faculty: The Policy Issues and Their Context.' *Faculty Retirement.* Cambridge, MA.: Consortium on Financing Higher Education, 1981, pp. 19–31.

————. 'A Blip on the Screen.' *Academe,* July–Aug. 1983, pp. 3–21.

Hartnett, Rodney T. *Trends in Student Quality in Doctoral and Professional Education.* New Brunswick, N.J.: Project on Trends in Academic Talent, Rutgers University, 1985.

Healy, Timothy. 'In Danger of Going Grey.' *The* [London] *Times Higher Education Supplement,* Feb. 25, 1983.

Heim, Peggy. 'The Economic Decline of the Professoriate in the 1970's.' In American Association for Higher Education, *Differing Perspectives on Declining Faculty Salaries.* Washington: American Association for Higher Education, 1980.

Hendrickson, Robert M. 'Five Reasons for Faculty Tension.' *AGB Reports,* May-June 1982, pp. 25–30.

Hendrickson, Robert M., and Barbara A. Lee. *Academic Employment and Retrenchment: Judicial Review and Administrative Action.* Washington: Association for the Study of Higher Education, 1983.

Henry, David D. *Challenges Past, Challenges Present.* San Francisco: Jossey-Bass, 1975.

Herman, Joyce, Ebba McArt, and Lawrence Belle. 'New Beginnings: A Study of Faculty Career Changes.' *Improving College and University Teaching,* Spring 1983, pp. 53–60.

Herrnstein, Richard J. *I.Q. in the Meritocracy.* Boston: Little, Brown, 1973.

Heveron, Eileen D. Part-Time Faculty: Quantitative and Qualitative Issues in Higher Education for the 1980's (Unpublished report). Claremont: Claremont Graduate School, 1983.

Hildebrand, Milton, Robert E. Wilson, and Evelyn R. Dienst. *Evaluating University Teaching.* Berkeley: Center for Research and Development in Higher Education, 1971.

Hofstadter, Richard, and Walter Metzger. *The Development of Academic Freedom in the United States.* New York: Columbia University Press, 1955.

Howard, Suzanne. *But We Will Persist: A Cooperative Research Report on the Status of Women in Academe.* Washington: American Association of University Women, 1978.

Hufstedler, Shirley. Address to the Association of American Colleges, Washington, Jan. 3, 1983.

Institute for Research in Social Behavior. *Retirement Plans and Related Factors Among Faculty at COFHE Institutions.* Cambridge, Mass.: Consortium on Financing Higher Education, 1980.

Jelinck, Mariann, Linda Smircich, and Paul Hirsch. 'A Code of Many Colors.' *Administrative Science Quarterly,* Sept. 1983, pp. 331–38.

Jencks, Christopher, and David Riesman. *The Academic Revolution.* Garden City, N.Y.: Doubleday, 1968.

Jenny, Hans H., Peggy Heim, and Geoffrey C. Hughes. *Another Challenge: Age 70 Retirement in Higher Education.* New York: Teachers Insurance and Annuity Association, 1979.

Jones, Lyle V., Gardner Lindzey, and Porter E. Coggeshall. *An Assessment of Quality-Related Characteristics of Research-Doctorate Programs in the United States.* Washington: National Academy Press, 1982.

Kaiser, Harvey H. *Crumbling Academe.* Washington: Association of Governing Boards, 1984.

Kanter, Rosabeth Moss. 'Changing the Shape of Work: Reform in Academe.' *Current Issues in Higher Education,* American Association for Higher Education, no. 1, 1979.

Kaplan, Wilfred. *Comments at Regents Meeting, June 17, 1982.* Ann Arbor: University of Michigan, 1982.

Kauffman, Joseph F. *Some Perspectives on Hard Times.* Presidential Address, Association for the Study of Higher Education, March 2, 1982.

Kearl, Bryant. 'Remarks.' *Academe*, Nov.–Dec. 1983, pp. 8a–10a.

Keast, William R., and John W. Macy, Jr. (eds.). *Faculty Tenure.* San Francisco: Jossey-Bass, 1973.

Kellman, Steven G. 'Rating, Rating, Rating.' *Academe*, Nov.–Dec. 1982, p. 29.

Kerchner, Charles T., and Jack H. Schuster. 'The Uses of Crisis: Taking the Tide at the Flood.' *Review of Higher Education*, Spring 1982, pp. 121–41.

Kieffer, Jarold A. 'Longer Life: An Opportunity for Higher Education.' *AAHE Bulletin*, Nov. 1983, pp. 3–6.

King, Francis P. 'Faculty Retirement: Early, Normal, and Late.' In Jon W. Fuller (ed.), *Issues in Faculty Personnel Policies.* San Francisco: Jossey-Bass, 1983, pp. 81–97.

Kirschling, Wayne R. (ed.). *Evaluating Faculty Performance and Vitality.* San Francisco: Jossey-Bass, 1978.

Klitgaard, Robert E. *The Decline of the Best?: An Analysis of the Relationship Between Declining Enrollments, Ph.D. Production, and Research.* Cambridge, Mass.: Office of the President, Harvard University, 1979.

Knoell, Dorothy M. *Through the Open Door. A Study of Patterns and Performance in California's Community Colleges*, Report 76–1. Sacramento: California Postsecondary Education Commission, Feb. 1976.

Kuhlmann, Roberta. 'What Price Tenure?' Doctoral dissertation, Claremont Graduate School, 1982.

Ladd, Everett C., and Seymour M. Lipset. *The Divided Academy.* New York: McGraw-Hill, 1975.

——————. 'The Ladd-Lipset Survey.' *Chronicle of Higher Education*, weekly issues, 1975–76 (from Sept. 15, 1975, to May 31, 1976).

Ladd, Everett Carll, Jr. 'The Work Experience of American College Professors: Some Data and an Argument.' *Current Issues in Higher Education.* American Association for Higher Education, no. 2, 1979.

Larsen, Charles M. 'Remarks.' *Academe*, Nov.–Dec. 1983, pp. 10a–11a.

Lazarsfeld, Paul F., and Wagner Thielens, Jr. *The Academic Mind.* Glencoe, Ill.: *Free Press, 1971 edition.*

Lazear, Edward P. 'Pensions as Severance Pay,' *The NBER Digest.* National Bureau of Economic Research, Jan. 1983, p. 1.

Leape, Martha P. *Report on the Class of 1982.* Harvard College (unpublished), 1982.

Leslie, David W. 'Part-Time Faculty: Legal and Collective Bargaining Issues.' *AAHE Bulletin*, Oct. 1984, pp. 8–12.

——————, Samuel E. Kellams, and G. Manny Gunne. *Part-Time Faculty in American Higher Education.* New York: Praeger, 1982.

Levine, Arthur. *Handbook on Undergraduate Curriculum.* San Francisco: Jossey-Bass, 1978.

Lewis, Darrell R., and William E. Becker, Jr. *Academic Rewards in Higher Education.* Cambridge, Mass.: Ballinger, 1979.

Light, Barbara Koolmees. The Female Professoriate. (Unpublished report). Claremont: Claremont Graduate School, 1983.

Light, Walter S. 'Business Leader Blames Tenure, Low Standards for "Piteous Shape" of Canadian Universities.' *Chronicle of Higher Education*, May 16, 1984, p. 30.

Linnell, Robert H. *Dollars and Scholars.* Los Angeles: University of Southern California Press, 1982.

Lipset, Seymour Martin. 'The Academic Mind at the Top: The Political Behavior and Values of Faculty Elites.' *Public Opinion Quarterly*, Summer 1982, pp. 143–68.

Lovett, Clara M. 'Vitality Without Mobility: The Faculty Opportunities Audit.' *Current Issues in Higher Education*, American Association for Higher Education, no. 4. 1983–84.

———. *Parting Ways with Academe: A Study in Career and Life Transitions*. Washington: Columbian College, George Washington University, unpublished manuscript, 1984.

Loyd, Sally. Faculty Working Conditions and Working Environment. (Unpublished report). Claremont: Claremont Graduate School, 1985.

Lynton, Ernest A. *Universities Today: A Crisis of Purpose*. (Unpublished address delivered at the University of Illinois on Oct. 6, 1982.) Boston: University of Massachusetts.

Mackay-Smith, Anne. 'Large Shortage of Black Professors in Higher Education Grows Worse.' *Wall Street Journal*, June 12, 1984, p. 37.

Marty, Myron A. 'Work, Jobs and the Language of the Humanities.' In *Strengthening the Humanities in Community College*. Washington: American Association of Community and Junior Colleges, 1980.

Marwer, James E., and Carl V. Patten. 'The Correlates of Consultation: American Academics in the "Real World".' *Higher Education*, Aug. 1976, pp. 319–35.

Maul, Ray C. *Teacher Supply and Demand in Universities, Colleges and Junior Colleges 1963–64 and 1964–65*. Washington: National Education Association, Research Report 1965–R4, April 1965.

May, Ernest R., and Dorothy G. Blaney. *Careers for Humanists*. New York: Academic Press, 1981.

Mayhew, Lewis B. *Surviving the Eighties*. San Francisco: Jossey-Bass, 1980.

Mayville, William V. 'Changing Perspectives on the Urban College and University.' *AAHE Bulletin*, April 1980, no page numbers.

McCain, Bruce E., Charles O'Reilly, and Jeffrey Pfeffer. 'The Effects of Departmental Demography on Turnover: The Case of a University.' *Academy of Management Journal*, 1983, vol. 26, no. 4, pp. 626–41.

McKeachie, Wilbert J. 'Perspectives from Psychology: Financial Incentives Are Ineffective for Faculty.' In D. R. Lewis and W. E. Becker, Jr. (eds.). *Academic Rewards for Higher Education*. Cambridge, Mass.: Ballinger, 1979a, pp. 3–20.

———. 'Student Ratings of Faculty: A Reprise.' *Academe*, Oct. 1979b, pp. 384–97.

———. 'The Role of Faculty Evaluation in Enhancing College Teaching.' *National Forum* (Phi Kappa Phi Journal), Spring 1983a, pp. 37–39.

———. 'Older Faculty Members: Facts and Prescriptions.' *AAHE Bulletin*, Nov. 1983b, pp. 8–10.

———. 'An Alternative to Forced Retirement.' *Academe*, Jan.–Feb. 1985.

McPherson, Michael S. *The State of Academic Labor Markets*. Washington: Brookings Institution, 1984 (unpublished).

———, and Gordon C. Winston. 'The Economics of Academic Tenure: A Relational Perspective.' *Journal of Economic Behavior and Organization*, vol. 4, nos. 2–3 (June-Sept. 1983), pp. 163–84.

Melchiori, Gelinda S. *Planning for Program Discontinuance: From Default to Design*. Washington: American Association for Higher Education, 1982.

Melendez, Winifred A., and Rafael M. de Guzman. *Burnout: The New Academic Disease*. Washington: Association for the Study of Higher Education, 1983.

Menges, Robert J., and William H. Exum. 'Barriers to the Progress of Women and Minority Faculty.' *Journal of Higher Education*, March–April 1983, pp. 124–44.

Minter, John, Associates. 'Changes in Size of Colleges' Work Force in the Past Year.' *Chronicle of Higher Education*, Aug. 3, 1983a, p. 20.

―――. 'Faculty Views of Trends in Their Departments,' *Chronicle of Higher Education*, Nov. 23, 1983b, p. 20.

―――. 'Faculty Pay Increases 6 Pct. in Year; Average Tops $28,000.' *Chronicle of Higher Education*, March 7, 1984.

Minter, W. John. *Faculty Salaries 1980–81 and Additional Earnings 1979–80*. Boulder: John Minter Associates, ca. 1981.

―――. 'Assessing Their Institutions.' *Chronicle of Higher Education*, May 26, 1982, pp. 8, 10.

Minter, W. John, and Howard R. Bowen. *Private Higher Education*. Washington: Association of American Colleges, 1977.

―――. *Independent Higher Education*. Washington: National Institute of Independent Colleges and Universities, 1978, 1979, 1980a.

―――. *Preserving America's Investment in Human Capital*. Washington: American Association of State Colleges and Universities, 1980b.

―――. 'The Minter-Bowen Report.' *Chronicle of Higher Education*. Part I, May 12, 1982, pp. 5–8; Part II, May 19, 1982, pp. 7–8; Part III, May 26, 1982, pp. 8–10; Part IV, June 2, 1982, pp. 9–10.

Mix, Marjorie C. *Tenure and Termination in Financial Exigency*. Washington: American Association for Higher Education and ERIC Clearinghouse on Higher Education, 1978.

Mommsen, Kent G. 'Black Ph.D.s in the Academic Market Place.' *Journal of Higher Education*, April 1974, pp. 253–67.

Moodie, Clara. 'The Overuse of Part-Time Faculty Members.' *Chronicle of Higher Education*, March 10, 1980, p. 72.

Moore, Kathryn M., Ann M. Salimbene, Joyce D. Marlier, and Stephen M. Bragg. 'The Structure of Presidents' and Deans' Careers.' *Journal of Higher Education*, Sept.–Oct. 1983, pp. 500–516.

Morgan, Anthony W. 'College and University Planning in an Era of Contraction.' *Higher Education*, vol. 11, no. 5 (September 1982).

Morris, William. 'Useful Work Versus Useless Toil.' In Asa Briggs (ed.). *William Morris: Selected Writings and Designs*. Harmondsworth: Penguin Books, 1962, pp. 117–36.

Mortimer, Kenneth P., and Michael L. Tierney. *The Three 'R's' of the Eighties: Reduction, Reallocation, and Retrenchment*. Washington: American Association for Higher Education, 1979.

Mulanaphy, James M. *Plans and Expectations for Retirement of TIAA-CREF Participants*. New York: Teachers Insurance and Annuity Association and College Retirement Equities Fund, 1981.

National Center for Education Statistics (U.S. Department of Education). *Digest of Education Statistics*. Washington: U.S. Government Printing Office, Annual.

―――. *Projections of Education Statistics*. Washington: U.S. Government Printing Office, annual.

―――. *The Condition of Education*. Washington: U.S. Government Printing Office, annual.

―――. *Salaries, Tenure, and Fringe Benefits of Full-Time Instructional Faculty in Institutions of Higher Education, 1975–76*. Washington: U.S. Government Printing Office, 1977.

―――. *Participation of Black Students in Higher Education: A Statistical Profile from 1970–71 to 1980–81*. Washington: U.S. Department of Education, Nov. 1983.

―――. 'Three Years of Change in College and University Libraries, 1978–79 to 1981–82.' *NCES Bulletin*, Feb. 1984.

National Commission on Excellence in Education. *A Nation at Risk.* Washington: U.S. Government Printing Office, 1983.

National Commission for Excellence in Teacher Education. *Report. Chronicle of Higher Education,* March 6, 1985, pp. 13–22.

National Commission on Higher Education Issues. *To Strengthen Quality in Higher Education.* Washington: American Council on Education, 1984.

National Commission on Student Financial Assistance. *Signs of Trouble and Erosion: A Report on Graduate Education in America.* Washington, 1983.

National Education Association. *Teacher Supply and Demand.* Washington: NEA, 1967.

———. *Extent of Faculty Dissatisfaction with the Provision of Benefits at Their Institutions.* Memo HE14, Dec. 1972.

———. *Higher Education Faculty: Characteristics and Opinions.* Washington: NEA, 1979.

National Endowment for the Humanities. 'To Reclaim a Legacy.' Report of Study Group on Humanities in Education. *Chronicle of Higher Education,* Nov. 28, 1984, pp. 16–21.

National Labor Relations Board vs. Yeshiva University 444 U.S. 672, 1980.

National Research Council. *Summary Report on Doctorate Recipients from United States Universities.* Washington: National Academy Press, annual, 1967 through 1982.

———. *Research Excellence Through the Year 2000: The Importance of Maintaining a Flow of New Faculty into Academic Research.* Washington: National Academy of Sciences, 1979.

———. *Employment of Humanities Ph.D.s: A Departure from Traditional Jobs.* Washington: National Academy of Sciences, 1980.

———. *Employment of Minority Ph.D.s: Changes over Time.* Washington: National Academy Press, 1981.

———. *Science, Engineering, and Humanities Doctorates in the United States: 1981 Profile.* Washington: National Academy Press, 1982.

National Science Foundation. *Employment Patterns of Academic Scientists and Engineers, 1973–78.* Washington: National Science Foundation, 1980.

———. *Academic Science 1972–81.* Washington: U.S. Government Printing Office, 1981a.

———. *Activities of Science and Engineering Faculty in Universities and 4-Year Colleges: 1978/79.* Washington: NSF, 1981b.

———. *Young and Senior Science and Engineering Faculty, 1980.* Washington: NSF, 1981c.

Nelsen, William C. *Renewal of the Teacher Scholar.* Washington: Association of American Colleges, 1981.

Newby, James E. *Teaching Faculty in Black Colleges and Universities.* Washington: University Press of America, 1982.

Newman, John Henry. *The Scope and Nature of University Education.* New York: Dutton, 1958 (originally published 1859).

Oldham, Greg R., and Carol T. Kulik. 'Motivation Enhancement Through Work Redesign,' *Review of Higher Education,* Summer 1983, pp. 323–42.

Olswang, Steven G., and Barbara A. Lee. *Faculty Freedoms and Institutional Accountability.* Washington: Association for the Study of Higher Education, 1984.

O'Toole, James. *Work, Learning, and the American Future.* San Francisco: Jossey-Bass, 1977.

Parsons, Talcott, and Gerald M. Platt. *The American University.* Cambridge, Mass.: Harvard University Press, 1975.

Patton, Carl V. *Academia in Transition: Mid-Career Change or Early Retirement.* Cambridge, Mass.: Abt Books, 1979.

————. 'Consulting by Faculty Members.' *Academe*, May 1980, pp. 181–85.

————. 'Voluntary Alternatives to Forced Termination.' *Academe*, Jan.–Feb. 1983, pp. 1a–8a.

————, and Kell, Zellon, and Palmer. In Margot Sanders Eddy, *Faculty Response to Retrenchment.* AAHE ERIC, Higher Education Currents. Washington: AAHE, 1982.

————, and James D. Marver. 'Paid Consulting by American Academics.' *Educational Record*, Spring 1979, pp. 175–84.

Pellins, Glenn R., Alice L. Boberg, and Colman O'Connell. *Planning and Evaluating Professional Growth Programs for Faculty.* Ann Arbor: University of Michigan, Center for the Study of Higher Education, 1981.

Peterson, Marvin W. 'Faculty and Academic Responsiveness in a Period of Decline: An Organizational Perspective.' *Journal of the College and University Personnel Association*, Spring 1980, p. 95.

Peterson, Richard, et al. Institutional Functioning Inventory: Preliminary Technical Manual. Princeton: Educational Testing Service, 1970.

Phillips, E. Lakin. *Stress, Health and Psychological Problems in the Major Professions.* Washington: University Press of America, 1982.

Price, J. L. *The Study of Turnover.* Ames: Iowa State University Press, 1977.

Prince, H. J. *Faculty Retirement Profile Report.* Lansing: American Association of University Professors (Michigan Conference), 1984.

Project on Engineering College Faculty Shortage. *Catalog of Industrial Programs To Aid Graduate Engineering Education.* Washington: Project on Engineering College Faculty Shortage, 1983.

Project on Redefining the Meaning and Purpose of Baccalaureate Degrees, *Integrity in the College Curriculum.* Washington: Association of American Colleges, 1985.

Radner, Roy, and Charlotte V. Kuh. *Market Conditions and Tenure for Ph.D.s in U.S. Higher Education.* Berkeley: Carnegie Council on Policy Studies in Higher Education, 1977.

————. *Reconcilable Differences? An Examination of Alternative Projections of Academic Demand for Recent Science and Engineering Ph.D.s in the 1980s.* Washington: National Research Council, May 1, 1979.

————. *Market Conditions and Tenure in U.S. Higher Education, 1955–1973.* Berkeley: Carnegie Council on Policy Studies in Higher Education, 1977.

————. *Preserving a Lost Generation: Policies to Assure a Steady Flow of Young Scholars Until the Year 2000.* Carnegie Council on Policy Studies in Higher Education, October 1978.

Radner, Roy, and Leonard S. Miller. *Demand and Supply in U.S. Higher Education.* New York: McGraw-Hill, 1975.

Rand Corporation. 'Simple Justice: Pittsburgh's Model Arbitration Program.' *Rand Research Review*, Spring 1984, pp. 3, 6.

Reagan, Gerald M. 'Comtemporary Constraints on Academic Freedom.' *Educational Forum*, Summer 1982, pp. 391–402.

Reskin, Barbara F. 'Review of the Literature on the Relationship Between Age and Scientific Productivity.' In Commission on Human Resources, National Research Council. *Research Excellence through the Year 2000.* Washington: National Academy of Sciences, 1979, pp. 187–208.

Riday, George E. *Job Satisfaction: A Comparative Study of Community College Faculty to Secondary School and Four-Year College Faculty.* Doctoral dissertation, Brigham Young University, 1981.

Riesman, David. *On Higher Education.* San Francisco: Jossey-Bass, 1980.

————. 'Some Personal Thoughts on the Academic Ethic.' *Minerva: A Review of Science, Learning, and Policy*, Summer–Autumn 1983, p. 265.

————. Notes on Academic Ethics. Unpublished manuscript, 1984.

Rothkirch, Christoph von. *Field Disaggregated Analysis and Projections of Graduate Enrollments and Higher Degree Production.* Berkeley: Carnegie Council on Policy Studies in Higher Education, 1978.

Rudolph, Frederick. *The American College and University.* New York: Knopf, 1962.

Ruml, Beardsley, and Donald H. Morrison. *Memo to a College Trustee: A Report on the Financial and Structural Problems of the Liberal Arts College.* New York: McGraw-Hill, 1959.

Ruskin, John. *Unto This Last.* London: George Allen and Unwin, 1960, pp. 105–74.

Schuster, Jack H. 'Liberal Learning, Vocationalism, and Institutional Coping Strategies.' In David E. Drew (ed.). *Competency, Careers, and College.* San Francisco: Jossey-Bass, 1978, pp. 47–55.

———. 'Faculty Vitality: Observations from the Field.' In Roger W. Baldwin (ed.), *Incentives for Faculty Vitality.* San Francisco: Jossey-Bass, 1985a, pp. 21–32.

———. 'Studying the Professoriate: Notes on Methods Used for Campus Visits.' Claremont, CA: Faculty in Education, Claremont Graduate School, 1985b.

———, and Howard R. Bowen. 'The Faculty at Risk.' *Change,* Sept.–Oct. 1985, pp. 13–21.

Schwebel, M. *Who Can Be Educated?* New York: Grove Press, 1968.

Scully, Malcolm G. 'Possible Faculty Shortage in 1990's Worries Today's Academic Leaders.' *Chronicle of Higher Education,* Oct. 6, 1982, p. 10.

———. 'Graduate Schools Report Increases in Applications.' *Chronicle of Higher Education,* March 30, 1983a, p. 1.

———. '4000 Faculty Members Laid off in 5 Years by 4-Year Institutions, Survey Shows.' *Chronicle of Higher Education,* Oct. 26, 1983b, p. 21.

Seidman, Earl, Patrick J. Sullivan, and Mary B. Schatzkamer. *The Work of Community College Faculty: A Study Through In-Depth Interviews* (an unpublished report). Amherst: University of Massachusetts, School of Education, 1983.

Shapiro, Harold T. 'The Privilege and the Responsibility: Some Reflections on the Nature, Function, and Future of Academic Tenure.' *Academe,* Nov. Dec. 1983, pp. 3a–9a.

Sharp, Laure M. *The Employment Situation of Humanists, 1979–1981.* Washington: Bureau of Social Science Research, 1984.

Shulman, Carol Herrnstadt. *Old Expectations, New Realities: The Academic Profession Revisited.* Washington: American Association for Higher Education, 1979.

———. 'Fifteen Years Down, Twenty-five To Go: A Look at Faculty Careers.' *AAHE Bulletin,* Nov. 1983, p. 11.

Smelser, Neil, and Robin Content. *The Changing Academic Market.* Berkeley: University of California Press, 1980.

Smith, Bardwell L., and Associates. *The Tenure Debate.* San Francisco: Jossey-Bass, 1973.

Smith, Bruce L. R., and Joseph J. Karesky. *The State of Academic Science.* Change Magazine Press, ca. 1978, 2 vol. paperback.

Smith, Hoke L. 'The Care and Nurturing of Faculty.' *AGB Reports,* March-April 1984, pp. 15–19.

Solmon, Lewis C. 'Ph.D.s in Nonacademic Careers: Are There Good Jobs?' In American Association for Higher Education, *Current Issues in Higher Education.* Washington: American Association for Higher Education, no. 7, 1979.

———. *U.S. Science Manpower and R and D Capacity: New Problems on the Horizon.* Los Angeles: Higher Education Research Institute, no date (ca. 1980).

————, and William Zumeta, Laura Kent, Nancy L. Ochsner, and Margo-Lea Hurwicz. *Underemployed Ph.D.s.* Lexington, Mass.: D. C. Heath, 1981.

Southworth, J. Russell, and Ronald A. Jagmin. *Potential Financial and Employment Impact of Age 70 Mandatory Retirement Legislation on COFHE Institutions.* Cambridge, Mass.: Consortium on Financing Higher Education, 1979.

Spofford, Tim. 'The Field Hands of Academe,' *Change*, Nov.–Dec. 1979, pp. 14–16.

Stadtman, Verne A. *Academic Adaptations.* San Francisco: Jossey-Bass, 1980.

Stathis-Ochoa, Roberta. 'Universities as Work Places: Changing Conditions.' Paper delivered at a joint meeting of the Association for the Study of Higher Education and the American Educational Research Association, San Francisco, Oct. 20, 1983.

Stecklein, J. E., and R. Willie. 'Minnesota Community College Faculty Activities and Attitudes, 1956–1980.' *Community/Junior College Quarterly of Research and Practice*, April–June 1982, pp. 217–37.

Story, Robert C., and Ann Guewa. *Survey of Salaries and Occupational Attitudes of Faculty Personnel of Higher Education, 1947–48.* Circular No. 254. Washington: Federal Security Agency, Office of Education, March 15, 1949.

Strohm, Paul. 'Faculty Roles Today and Tomorrow.' *Academe*, Jan.–Feb. 1983, pp. 10–15.

Stroup, S. W., N. Van Gieson, and P. A. Zirkel. *Deficits, Declines, and Dismissals: Faculty Tenure and Fiscal Exigency.* Washington: ERIC Clearinghouse, SP019495, Feb. 1982.

Study Group on the Conditions of Excellence in American Higher Education. *Involvement in Learning: Realizing the Potential of American Higher Education.* Washington: U.S. Department of Education, Natonal Institute of Education, 1984.

Syverson, Peter D., and Lorna E. Forster. *New Ph.D.s and the Academic Labor Market.* Washington: National Research Council, 1984.

Tickton, Sidney G. *1982 Idea Handbook: Attracting and Retaining Highly Qualified Young Faculty Members at Colleges and Universities.* New York: Academy for Educational Development, 1982.

Tillinghast, Neson and Warren, Inc. *Potential Financial and Employment Impact of Age 70 Mandatory Retirement Legislation on COFHE Institutions.* Cambridge, Mass.: Consortium on Financing Higher Education, 1979.

'Too Many Administrators.' *Chronicle of Higher Education*, Oct. 5, 1983, p. 21.

Toombs, William E. *Faculty Career Change: A Pilot Study of Individual Decisions.* Address Presented to American Association for Higher Education, Washington, 1979.

————, and Joyce Marlier. *Career Change Among Academics: Dimensions of Decision.* Center for the Study of Higher Education, Pennsylvania State University, 1981.

Trow, Martin (ed.). *Teachers and Students.* New York: McGraw-Hill, 1975.

Tuckman, Howard P. *Publication, Teaching and the Academic Reward Structure.* Lexington, Mass.: Lexington Books, 1976.

————, Jaime Caldwell, and James Gapinski. 'The Wage Rates of Part-Timers in Higher Education: A Preliminary Inquiry,' *Proceedings of the American Statistical Association*, 1978.

————, and Cyril F. Chang. *Own-Price and Cross Elasticities of Demand for College Faculty.* Address Presented to the American Economic Association, San Francisco, Dec. 30, 1983.

————, and William D. Vogler. 'The "Part" in Part-Time Wages.' *AAUP Bulletin*, May 1978, pp. 70–77.

———, William D. Vogler, and Jaime Caldwell. *Part-Time Faculty Series.* Washington: American Association of University Professors, ca. 1978.

U.S. Department of Commerce, *Earned Degrees Conferred: An Examination of Recent Trends.* Washington: National Technical Information Service, no date (ca. 1982).

University of California at Los Angeles, Office of Public Communications. *UCLA News.* Los Angeles: Jan. 23, 1983.

Veblen, Thorstein. *The Instinct of Workmanship.* New York: Viking Press, 1914.

———. *Theory of the Leisure Class.* New York: Random House, 1931.

Volkwein, J. Fredericks. 'Responding to Financial Retrenchment.' *Journal of Higher Education,* May–June 1984, pp. 389–401.

Waggaman, John S. *Faculty Recruitment, Retention, and Fair Employment: Obligations and Opportunities.* Washington: Association for the Study of Higher Education, 1983.

Warren, Jonathan R. 'The Faculty Role in Educational Excellence.' Unpublished paper, Educational Testing Service, June 1982.

Weiler, William C. *Faculty Turnover in Higher Education Institutions.* Minneapolis: University of Minnesota, Office of Management Planning and Information Services, 1983.

Weinschrott, David J. *Demand for Higher Education in the United States: A Critical Review of the Empirical Literature.* Santa Monica: Rand Corporation, 1977.

Whitman, Neal, and Elaine Weiss. *Faculty Evaluation: The Use of Explicit Criteria for Promotion, Retention, and Tenure.* Washington: ERIC Clearinghouse, 1982.

Wildavsky, Aaron. 'The Debate over Faculty Consulting.' *Change,* June–July 1978, pp. 13–14.

Williams, Edgar Trevor (ed.). *A Register of Rhodes Scholars, 1903–1981.* Oxford: Alden Press, 1981.

Williams, G., T. Blackstone, and D. Metcalf. *The Academic Labor Market.* New York: Elsevier, 1974.

Willie, R., and J. E. Stecklein. 'A Three-Decade Comparison of College Faculty Characteristics, Satisfactions, Activities, and Attributes.' *Research in Higher Education,* 1982, no. 1, pp. 81–93.

Wilson, Logan. *American Academics Then and Now.* New York: Oxford University Press, 1979.

Wilson, R. C., J. G. Gaff, E. R. Dienst, L. Wood, and J. L. Bavry. *College Professors and Their Impact on Students.* New York: Wiley, 1975.

Wolfle, Dael, and Charles V. Kidd. 'The Future Market for Ph.D.s,' *Science,* Aug. 1971, pp. 784–93.

Zey-Ferrell, Mary. 'Predictors of Faculty Intent To Exit the Organization: Potential Turnover in a Large University.' *Human Relations,* May 1982, pp. 349–72.

Zumeta, William. *Market Developments at the Ph.D. and Postdoctoral Levels and Their Implications for Universities.* Unpublished address before the American Council on Education, Washington, 1981.

———. *Anatomy of the 'Boom' in Postdoctoral Appointments During the 1970s.* Los Angeles: Higher Education Research Institute, 1982a.

———. 'Doctoral Programs and the Labor Market, or How Should We Respond to the "Ph.D. Glut"?' *Higher Education,* vol. 11, no. 3 (July 1982b), pp. 321–43.

———. *Extending the Educational Ladder: The Changing Quality and Value of Post-Doctoral Study.* Lexington, Mass.: Lexington Books, 1984.

Zur-Muehlen, Max von. *A Profile of Full-Time Teachers at Canadian Universities: A Statistical Review for the Eighties.* Unpublished paper, 1982.

Index